Evolutionary Social Psychology

Evolutionary Social Psychology

Edited by

Jeffry A. Simpson
Texas A & M University

Douglas T. Kenrick
Arizona State University

Psychology Press
Taylor & Francis Group

New York London

First published by
Lawrence Erlbaum Associates, Inc., Publishers
10 Industrial Avenue
Mahwah, NJ 07430

Transferred to Digital Printing 2009 by Psychology Press
270 Madison Ave, New York NY 10016
27 Church Road, Hove, East Sussex, BN3 2FA

Cover design by Kathryn Houghtaling

Library of Congress Cataloging-in-Publication Data

Evolutionary social psychology / edited by Jeffry A. Simpson
and Douglas T. Kenrick.
     p. cm.
    Includes bibliographical references.
    ISBN 0-8058-1905-3 (cloth : alk. paper). — ISBN
0-8058-2420-0 (paper : alk. paper)
    1. Social psychology. 2. Genetic psychology. I.
Simpson, Jeffry A. II. Kenrick, Douglas T.
HM251.E846   1996
302—dc20   96-22748

Contents

Preface

Social psychology is a remarkably diverse and important field in the social and behavioral sciences. Students and laypersons alike find its incredibly fascinating. Researchers in the field seek answers to questions about how we persuade others toward our viewpoints, how we form impressions of strangers as well as our closest friends, how we fall in love, how we rise and fall in social status, and how we become angry enough to hurt or even kill one another. Over the years, the harvest of empirical answers to these intriguing and important questions has yielded what often seems to be rather isolated, disconnected pieces of trivia. This realization has led many social psychologists to experience periods of crisis and self-doubt about how all of the often fragmented empirical data fits together and where the field is—and should be—headed.

In 1978, Paul Meehl[1] reflected on why the "softer" areas of psychology (including clinical, counseling, social, and personality) tend to make slower scientific progress than, and do not have the same cumulative character as, the "harder" areas (such as physiology and neuroscience). Meehl articulated 20 reasons why the softer areas of psychology are more difficult to "scientize." They included problems associated with classifying and sampling environments and situations, the complexity of accounting for idiographic individual differences, divergent causality, the operation of multiple feedback loops, random events experienced during development, the large number of variables that can impinge on any single behavior, open theoretical concepts, and ethical constraints inherent in human research. Although Meehl acknowledged that his list was not complete, he did not mention perhaps the gravest impediment to scientific progress in any field—the lack of a grand theory capable of integrating and explaining diverse—and what may appear to be discrepant—empirical results.

We believe that the intermittent "crises of confidence" in social psychology as well as its slower "cumulative character" can be traced to the same fundamental source: the failure to develop or adopt a powerful theoretical perspective, one capable of *integrating* existing findings across different domains and *guiding* researchers toward new empirical hypotheses. What is ironic is that the general framework of such a grand, integrative theory—Darwin's theory of evolution by natural selection—has been around for more than 130 years, yet, until recently, it has been largely ignored or overlooked by most social psycholo-

---

[10]Meehl, P. E. (1978). Theoretical risks and tabular asterisks: Sir Karl, Sir Ronald, and the slow progress of soft psychology. *Journal of Consulting and Clinical Psychology, 46*, 806–834.

gists. Recent refinements and extensions of Darwin's original ideas have allowed evolutionary principles to be more precisely stated and tested by social and behavioral scientists (see the introductory chapter by Kenrick and Simpson, this volume). As a result, there has been a tremendous proliferation of research guided by various middle-level evolutionary theories during the past decade. Some of this research is showcased in this volume.

This volume is intended to provide readers with a glimpse of how various topics in social psychology can be understood and integrated within an evolutionary framework. The volume is divided into five thematic sections: social perception, interpersonal attraction, pair bonding and mating strategies, kinship and social relations, and groups and group selection. These topics, of course, do not canvass all of the domains in social psychology to which an evolutionary perspective can be profitably applied. However, they do reflect most of the major content areas in which specific evolutionary hypotheses have been tested or new evolutionary models are being proposed.

Besides developing a deeper appreciation for how different evolutionary principles bear on many central topics in social psychology, we also hope that readers will gain a clearer and more accurate understanding of what evolutionary psychology is and what it is not. Many people who are unfamiliar with the basic ideas and principles of evolutionary psychology still harbor strong misconceptions about what an evolutionary perspective entails and how it can clarify and further our understanding of human social behavior. These misconceptions must be identified and cast aside if evolutionary psychology is going to make a rapid and lasting contribution to the social and behavioral sciences. We hope that this book begins to accomplish these goals.

ACKNOWLEDGMENTS

The editors would like to thank the following people for the considerable time and effort they devoted to reviewing preliminary abstracts of chapter proposals. These people include: Judith Anderson, Diane Berry, Thomas Bouchard, Marilynn Brewer, Eugene Burnstein, David Buss, Nancy Cantor, Avshalom Caspi, Jonathon Cheek, Charles Crawford, Michael Cunningham, Martin Daly, Patricia Draper, David Funder, Steven Gangestad, William Graziano, Robert Hogan, Richard Keefe, Dennis Krebs, Robert McCrae, Steve Neuberg, Robert Plomin, Felicia Pratto, W. Steven Rholes, David Rowe, Philippe Rushton, Auke Tellegen, Melanie Trost, David Sloan Wilson, and Leslie Zebrowitz.

—*Jeffry A. Simpson*
—*Douglas T. Kenrick*

I

Introduction

1

Why Social Psychology and Evolutionary Psychology Need One Another

Douglas T. Kenrick
Arizona State University

Jeffry A. Simpson
Texas A&M University

What a pity it would have been if biologists had refused to accept Darwin's theory of natural selection. This theory has been essential in helping biologists understand a wide range of phenomena, including the resplendent display of the peacock, the melodious song of the meadowlark, the bloody competitions of elephant seals, and the cooperative architectural skills of termites, who build airconditioned mounds standing 15 feet above the ground! These days, to study any animal species while refusing to consider the evolved adaptive significance of their behavior would be considered pure folly. That is, of course, unless the species in question is *homo sapiens*. Graduate students training to study this particular primate species may never take a single course in evolutionary theory, although they often take two undergraduate and four graduate courses in statistics. These methodologically sophisticated students then embark on a career studying human aggression, cooperation, mating behavior, family relationships, or altruism with little or no understanding of the general evolutionary forces and principles that shaped the behaviors they are investigating. When a handful of brave social psychologists tried to introduce evolutionary thinking into the field several years ago (see e.g., Cunningham, 1981; Rajecki, 1977), the reaction of most social psychologists either was to ignore or attack the suggestion that our field might profit from applying perhaps the most powerful and integrative set of ideas in the life sciences.

For the field of social psychology, this state of affairs has been more than just selfhandicapping; it has impeded us from making our rightful contribution to the revolutionary developments that now promise to connect the life sciences with the cognitive sciences.

Instead of holding up the rear guard in the struggle to understand how the human mind is designed and structured, we should be leading the charge. We say this because even the most cursory exposure to modern evolutionary principles suggests that the human mind is primarily designed for social life. Rather than trying to piece together the nature of the mind by studying the ability of people to recognize the letter *a* presented at 250 ms, cognitive scientists should be studying people's ability to send and interpret signals of friendship, cooperation, aggressiveness, and love as they transpire at the speed of natural human conversation. Rather than handicapping ourselves with assumptions that cognition, learning, and normative influences on behavior are somehow "outside" the realm of biology, social psychologists should be trying to understand how social cognition and social learning blend together to produce evolved psychological mechanisms. In short, it is time for a fully evolutionary social psychology, one that stops jeering from the periphery and finally assumes its important position at the center of the new evolutionary paradigm.

Because of deficient or nonexistent training in evolution, most social psychologists are only dimly aware of recent advances in evolutionary models, especially the numerous advances that have occurred in biology since the mid-1960s (see Trivers, 1985, for a review). In the last three decades, Darwin's original ideas about natural selection have been refined and expanded into a modern, neo-Darwinian perspective. Built on the foundation of Inclusive Fitness Theory (Hamilton, 1964), the modern perspective contains a rich network of interrelated middle-level theories of evolution such as reciprocal altruism (Trivers, 1971), parental investment and sexual selection (Trivers, 1972), parent–offspring conflict (Trivers, 1974), and parasite–host coevolution (Hamilton & Zuk, 1982) to name a few. The lack of exposure to these important developments in evolutionary theory is often magnified by a set of strongly held misconceptions, some of which are reviewed in the following discussion.

REASONS FOR IGNORING EVOLUTIONARY
SOCIAL PSYCHOLOGY

According to some psychologists, there are several reasons why evolutionary principles should not or cannot be incorporated into the field of social psychology, yet none of them are cogent ones. Here are some of the misguided reasons that have been advanced:

- Genes cannot influence complex human behavior.

- Evolutionary explanations imply genetic determinism and ignore the role of the environment.
- Culture is more important than (and independent of) evolved psychological mechanisms.
- Social psychologists can ignore evolutionary explanations because the latter are at a different level of analysis.
- Evolutionary principles are relevant only to a narrow range of phenomena in social psychology.
- Evolutionary models provide explanations that are merely common sense, post hoc, or untestable.

Let us consider each misconception in greater detail.

Misconception 1: Genes Cannot Influence Complex Human Behavior

The least sophisticated arguments relevant to this first misconception presume that strings of genetic proteins could not possibly influence something as complicated as human behavior. Such arguments were laid to rest long ago by behavioral genetic evidence regarding genetically based individual differences. Disorders such as Down's syndrome, Huntington's disease, and schizophrenia all show substantial genetic inheritance. As for normal behavior, cross-cultural studies have shown that several behaviors are seen in virtually all cultures and, therefore, may reflect species-specific patterns of behavior in humans (see Eibl-Eibesfeldt, 1989, for a review). These behaviors include basic emotional expressions (Ekman & Friesen, 1971; Ekman et al., 1987) as well as a host of nonverbal gestures (Eibl-Eibesfeldt, 1975; Simpson, Gangestad, & Biek, 1993). They also include more complex actions, such as the tendency for males to commit considerably more violent and homicidal acts than females (Daly & Wilson, 1988a), the tendency for females to select mates based on characteristics associated with their social status and resources (Buss, 1989), the tendency for humans to fall in love (Jankowiak & Fisher, 1992), and the tendency for men, but not for women, to prefer progressively younger mates across the lifespan (Kenrick, Gabrielidis, Keefe, & Cornelius, in press; Kenrick & Keefe, 1992).

Misconception 2: Evolutionary Explanations Imply Genetic Determinism and Ignore the Role of the Environment

To claim that variance underlying certain behaviors may have a genetic basis is not to adopt a genetic determinist doctrine. From the moment of conception, genes have their impact on behavior in the context of other genes as well as environmental influences. In fact, evolutionary theory is as much a theory about the environment as it is about genetic predispositions (Crawford & Anderson, 1989). This allows evolutionary psychology to explain cross-cultural variation in certain kinds of behaviors. Cross-cultural differences in behavior tend to arise because different cultures have had to solve slightly different adaptive problems posed by their unique social or physical environments. Consider this example. Physical attractiveness should be a good index of a prospective mate's current health and perhaps his or her resistance to diseases. Accordingly, people who live in environments in which parasites are plentiful (e.g., heavily populated areas with climates that are warm, wet, or at low elevation) should place more importance on a mate's physical attractiveness than should people who live in low pathogenic environments (e.g., sparely populated areas with climates that are cold, dry, or at high elevation). Examining 37 different cultures located in different geographical areas around the world, Gangestad and Buss (1993) tested and confirmed this prediction. Thus, even cultural differences in certain social behaviors can be predictable and interpretable from an evolutionary perspective.

In this volume, Caporael and Baron discuss some problems associated with genetic determinism, which, although controversial, have gained popularity among some biologists. According to their position, natural selection does not necessarily operate at the level of genes; rather, it operates at a higher level in complex dynamic systems. Although this does not imply that heritable traits are irrelevant in the process of natural selection, it does highlight an important point embedded in most evolutionary models—namely, that genetic traits, considered individually, may not fully account for natural selection.

Misconception 3: Culture is More Important Than (and Independent of) Evolved Psychological Mechanisms

If individual genetic traits do not provide the complete picture, can we conclude that the effects of genes on behavior are less important than the effects of culture? The answer is, not necessarily. Human cultures are created by organisms with a particular set of inherited cognitive and affective mechanisms (Barkow, Cosmides, & Tooby, 1992; Lumsden & Wilson, 1981). This explains, at least in part, why there are a number of behavioral regularities across vastly different cultures. Hence, rather than working against biological influences, culture often fosters and promotes the expression of evolved psychological mechanisms. Indeed, when cultural rules are created to counter specific, genetically selfish tendencies, these rules should tell us a great deal about the nature of these pan-cultural tendencies (Campbell, 1975).

Cultural rules also may be developed to monitor and to control selfish actions on the part of other group members, such as rules to punish cheaters and those who engage in excessive social loafing. Wilson (chapter 13, this volume) addresses some of the complex relationships between genetic selfishness and group interests. In doing so, he overviews the controversial topic of group selection, connecting it with some well-known findings in social psychology.

Misconception 4: Social Psychologists Can Ignore Evolutionary Explanations Because they Represent a Different Level of Analysis

Another argument advanced against applying evolutionary principles to human social behavior is that social psychologists and evolutionary biologists operate at different levels of analysis. This myopic reasoning implies that we can fully understand events at a cognitive level without ever considering what the human mind was designed to do and that we can understand the operation of norms without considering the thought processes of people who adopt or reject these normative influences.

It is a mistake to be so insular. If we wish to understand learning, we must ask why some things are easier to learn than others (e.g., phobias or food aversions versus a reliable place to put one's keys), and why some species can easily learn things that others cannot (e.g., dogs learn to fetch more easily than cats; cats learn to bury their feces better than dogs). If we want to understand cognitive processes, we need to ask why people devote so much cognitive attention to some topics (e.g., the traits of their colleagues, sex, possible losses of status or resources) compared to others (e.g., the devastating consequences of destroying the oceans and the rain forests, the crucial distinctions between Petty and Cacioppo's [1986] and Chaiken, Liberman, & Eagly's [1989] models of persuasion). In examining the social cognition literature, for instance, one comes across several findings indicating that "personal relevance" often leads to more effortful, controlled, and systematic information processing (Petty & Cacioppo, 1986). Unfortunately, personal relevance is all too often left undefined, or it is defined in either ad hoc or purely operational ways. An evolutionary perspective would generate a number of hypotheses about which kinds of situations and events ought to enhance personal relevance in different kinds of people. Moreover, an evolutionary perspective would suggest that topics pertinent to differential survival and major reproductive goals should be processed differently (Kenrick, Sadalla, & Keefe, in press). Shackelford's treatment (chapter 4, this volume) of the different links between behavior and perceptions of betrayal in romantic partners versus friends demonstrates this novel emphasis.

Misconception 5: Evolutionary Principles Are Relevant to Only a Narrow Range of Phenomena in Social Psychology

When presented with the counterarguments discussed so far, some social psychologists still might argue that evolutionary models may be applicable to more basic social behaviors such as aggression and sexuality, but not to others such as group behavior, social cognition, or family relationships. The chapters in this volume suggest that, to the contrary, an evolutionary perspective can offer fresh approaches to all of these phenomena. Krebs and Denton consider the evolutionary significance of in-group and out-group biases and how such biases tie into self-deception. Wilson (in chapter 13), Haslam (in chapter 11), and Caporael and Baron (in chapter 12) discuss implications for research on intergroup and intragroup relationships. Viewed together, these chapters suggest ways in which most research areas in social psychology could profit from an infusion of evolutionary theory.

Misconception 6: Evolutionary Models Provide Explanations That are Only Common Sense, Post Hoc, and/or Untestable

Another set of rationalizations for ignoring evolutionary theory is the claim that evolutionary models provide post hoc explanations that simply fit with common sense, but are not heuristically useful in the same ways that, say, cognitive dissonance theory, balance theory, or social identity theory presumably are. There are several responses to this misconception. Darwin's (1859) original theory was an integrative theory that made sense out of many existing observations, explaining in a parsimonious fashion diverse findings from the fossil record, botany, zoology, and geology. Such synthesis is a central goal of a good scientific theory, and Darwin's original theory was in some sense post hoc. But since its inception, the theory has been remarkably successful at explaining many new and unexpected findings. It also has generated many important and insightful hypotheses that are still being tested and expanded by contemporary geneticists, zoologists, biologists, botanists, and psychologists.

There now exists enough evidence in support of Inclusive Fitness Theory to warrant its status as the major integrative theory in the life sciences. However, contrary to what some critics have claimed, the basic assumptions of the theory could have been—and can still be—falsified. For example, doubt would be cast on some central tenets of the theory if evidence were to reveal that complex life forms were created too quickly in geological time (e.g., in seven days instead of over millions of years). Furthermore, Inclusive Fitness Theory would be in trouble if evolution produced adaptations in which organisms routinely benefited intrasexual competitors or other species more than themselves (Buss, 1995). In terms of derivations from middle-level evolutionary theories, Trivers' theory of sexual selection would be called into question if, across different contexts, the sex who invested more in offspring (usually, but not always, females in most species) was less discriminating in mate selection than the sex who invested less (usually males).

In our own research, we have usually confirmed—but occasionally have disconfirmed—hypotheses derived from various middle-level evolutionary theories. For example, evolutionary predictions about the occurrence of contrast effects in perceptions of one's current relationship were confirmed in one study, yet disconfirmed in another (Kenrick, Neuberg, Zierk, & Krones, 1994; Zuckschwerdt, 1993, described in Kenrick, 1994). Although the confirmation increased our confidence in the viability of the first hypothesis and decreased our confidence in the latter one, the results did not undermine our belief in the higher-order theory of inclusive fitness. Thus, as is true of theories that focus on ontogenetic and proximate causation, evolutionary theories are testable and refutable.

Buss (1995) noted that there are many possible hypotheses about social behavior that can be derived from different middle-level evolutionary theories. Some of these predictions could not have been derived from existing social science models. For example, contrary to norm-based models of mate preference, Kenrick and Keefe (1992) found systematic gender differences in age preferences. Previous models had posited straightforward gender differences in age preferences, whereby women should prefer older men, and men ought to prefer younger women. These predicted differences were explained in terms of the sex-typed norms of American culture, which specify that men often have more power in relationships than women do. From an evolutionary perspective, women should prefer higher-status mates because such men could better assist them in providing for offspring. In contrast, men should prefer females who are somewhat younger and less socially powerful than themselves but high in fertility. Consistent with this evolutionary hypothesis, men in all cultures examined to date tend to seek and find mates near their own age when they are young, yet seek and find progressively younger women as they age (e.g., Broude, 1992; Harpending, 1992; Kenrick & Keefe, 1992). Also consistent with the evolutionary hypothesis, but not with previous sociocultural–power explanations, teenaged males are attracted to substantially older women (Kenrick, Gabriedilis, et al., in press).

If these predictions merely fit with common-sense notions, they do so only if common sense includes 20/20 hindsight. Previous theories did not—and in many cases cannot—make these predictions. Furthermore, evolutionary-based hypotheses, just like those derived from nonevolutionary models, frequently vie with one another. In this volume, we see many examples of this. Gangestad and Thornhill (chapter 7) pit several models of physical attractiveness against one another. Although all of their models are interpretable within an evolutionary framework, only one seems to fit the data well. Showing that evolution-inspired hypotheses are not all of the same stripe, Miller and Fishkin (chapter 8) argue against certain assumptions made by Buss and other evolutionary theorists, proposing that both males and females are designed primarily for monogamous relationships. Although the amount and nature of evidence in support of this hypothesis is meager at present, it is entirely legitimate to advance competing evolutionary models as long as they can be tested empirically. Along the same lines, Graziano and his colleagues (chapter 6) take issue with previous evolutionary hypotheses advanced by Kenrick and his associates (Sadalla, Kenrick, & Vershure, 1987). Because these hypotheses are testable, Graziano et al. marshal support for their position, both from new findings they present and from prior research. Springer and Berry (chapter 3) clearly reveal how some evolutionary assumptions underlying ecological models of social perception may need serious reconsideration.

REASONS WHY SOCIAL PSYCHOLOGISTS SHOULD CONSIDER EVOLUTIONARY PSYCHOLOGY

There are several good reasons why all social psychologists should pay some attention to the evolutionary context of the behaviors or phenomena they study. To begin with, adopting an evolutionary perspective would help us to avoid saying things that are either wrong or incomplete. It also would allow us to understand and explain many phenomena that, at present, are inexplicable from the standpoint of traditional theories or existing models of social behavior. Furthermore, social psychologists are in a unique position to make important and meaningful contributions toward answering one major set of questions currently facing evolutionary psychology: What sorts of psychological mechanisms underlie and govern social thoughts, emotions, and behavior? Finally, adopting an evolutionary perspective would allow us to understand and explain how each subarea of our seemingly scattered field fits with the others, with other areas of psychology, and with the other social and behavioral sciences. We consider each of these reasons below.

To Avoid Saying Things That Are Wrong

Social psychologists make some mistakes they could avoid if they were aware of the interdisciplinary literature on evolutionary psychology. One of the most common mistakes is to attribute a phenomenon to an arbitrary norm such as *our culture* or *modern media images* when the same phenomenon is observed across cultures and sometimes across different species. Daly and Wilson (1988a, 1988b) noted this tendency with regard to the male–female difference in homicide. In the United States, males commit more than 80% of homicides. Social scientists routinely attribute this phenomenon to aspects of North American culture transmitted through the media. Yet when Daly and Wilson examined homicide data from other cultures and from other historical time periods, they found that, despite wide cultural variations in the frequency of homicides, the sex difference remained remarkably constant. Other cross-cultural patterns of behavior alluded to earlier, such as males' preference for relatively younger mates as they grow older and the tendency for both men and women to experience romantic love, have also been attributed to arbitrary cultural norms, norms that could not possibly explain the universality of these behaviors (Jankowiak & Fischer, 1992). In chapter 10, Daly, Salmon, and Wilson consider kin relationships in the same light. Even though the patterning of kin relationships has frequently been assumed to be under the control of arbitrary local norms, Daly, Salmon, and Wilson present a number of cross-culturally robust behavioral patterns that seem to be better explained by Inclusive Fitness Theory.

A more subtle error that could be averted by increased exposure to interdisciplinary thinking is the presumption that if something varies from one culture to the next, biological influences can be ruled out. A recent and otherwise very thoughtful treatment of cross-cultural variations in social behavior makes this error in a number of places (Moghaddam, Taylor, & Wright, 1993). This error is analogous to saying that because domestic dogs vary in their tendency to act aggressively from one situation to the next, canine aggressiveness does not have a biological basis. As we noted earlier, genetic proclivities are designed for adaptation to changing environments. In the case of cross-cultural variation, just as in the case of cross-situational variation in behavior, different local environments should activate different kinds of evolved mechanisms with differential probabilities.

Not all human cultures, for example, are exclusively monogamous. This does not disprove the hypothesis that humans have evolved inclinations toward pair-bonding (see the chapters by Miller and Fishkin and by Zeifman and Hazan, this volume). Indeed, all human cultures, even polygamous ones, have some form of marriage (Daly & Wilson, 1983). Cultural variations may be associated with systematic differences in ecological characteristics that shift the relative benefits of monogamy versus polygyny (which is common) and monogamy versus polyandry (which is rare, but found in a few cultures). Even bird species that commonly form monogamous pair bonds will form polygamous mateships depending on factors such as food distribution and the variation in resources across territories (Daly & Wilson, 1983). In humans, men with more than one wife in the household tend to be high in social status. If a woman has more than one husband in the household, her husbands are likely to be brothers.

Following these regularities, one may hypothesize that the reason why there is late marriage and little premarital sex on the Irish island of Inis Beag as compared to a great deal of premarital sex and early marriage on the Polynesian island of Mangaia could stem from arbitrary variation in local cultural norms. Yet these differences might also reflect evolutionary principles that link mating strategies to variations in climate, food availability, or the distribution of status and resources within each society. Furthermore, there may be genetic differences between the two cultures. Although many evolutionary psychologists assume the existence of a single, common human nature (Tooby & Cosmides, 1990a, 1990b), there are clear variations in physical features between people from different geographical regions, and there are probably variations in chronic hormonal levels. Hormonal levels could vary across groups owing to genetic drift or natural selection in particular habitats. These are topics about which we know relatively little. Nevertheless, we cannot ignore gene–environment interactions if we want to fully understand person–environment fits.

Several other erroneous assumptions hamper our field. These include the assumption that an understanding of the general processes underlying nonsocial cognition (such as those involved in lexical priming and other reading tasks) is sufficient to elucidate social cognition (Glass & Holyoak, 1986; Markus & Zajonc, 1985). Evolutionary psychologists have treated the problems with this assumption in considerable depth (Kenrick, Sadalla,et al., in press; Tooby & Cosmides, 1992). Another incorrect assumption is that all gender differences in behavior can be explained solely in terms of power differences between men and women, whereby males make the rules and females are helpless pawns. Recent evolutionary work suggests that social scientists have overemphasized the importance of male power and female helplessness, and that an under-standing of female choice may be more important than an understanding of male choice in tracing the origins of many gender differences (Kenrick, Trost, & Sheets, in press; Small, 1992; Smuts, 1985).

To Avoid Saying Things That Are Incomplete

Although the failure to consider the evolutionary literature sometimes has led to explanations that are wrong, more often it has led to explanations that are incomplete. For instance, Moghaddam et al. (1993) raised several very inter-esting points about how social psychologists, because of their emphasis on short-term, voluntary relationships in Western culture, have largely overlooked the long-term and involuntary relationships that most people in the world have with their relatives. These involuntary, kin-based relationships are experienced, not only by people in non-Western cultures, but also by many women in North America, the majority of whom list relatives as their most intimate confidants. Moghaddam et al. also note that when people need money, they most often turn to relatives. The lack of attention to kin relationships would never have occurred if social psychologists were more familiar with evolutionary theory, as Daly, Salmon, and Wilson point out in chapter 10. Ironically, Moghaddam, et al. also do not consider the evolutionary significance of the worldwide tendency to have strong, lifelong relationships with kin. An evolutionary perspective not only would have helped these authors to organize and integrate this literature; it also would have helped them to make sense of some specific cultural variations, such as why people in urban cultures are more violent toward one another. Perhaps the experience of living among strangers in large cities fails to activate kin-related cooperative psychological mechanisms or triggers out-group exploitation mechanisms of the sort discussed by Krebs and Denton in chapter 2.

As another example, consider a recent book that reviews studies suggesting that aggressive behavior is frequently triggered by threats to one's self-esteem. In this book, Baumeister (1993) noted that males seem more susceptible than females to esteem-induced aggression. After reviewing this literature, however, he left one large part of the story untold. Evolutionary theorists previously documented the same relationships (Daly & Wilson, 1988a, 1988b; Wilson & Daly, 1985), but they connected them to a much wider literature spanning different cultures and even different species. The fact that males are more concerned about their esteem, and the unfortunate consequences of this hypersensitivity in terms of frequent intramale violence, can be better understood in the light of recent cross-cultural and cross-species findings indicating a strong, positive relationship between male status and female mate choice. Mammalian males who fail to secure, defend, and maintain an adequate position in their social hierarchy are less successful at attracting and retaining mates.

Other examples can be found in the literature on sex differences. Abbey (1982) showed that males viewing a heterosexual interaction are more likely to attribute sexual intent to the actors than females are. This result is more easily understood if it is interpreted within the voluminous body of evolutionary-based research on sex differences across cultures and species. It is still common to find social psychologists discussing sex differences in behavior without even considering powerful evolutionary models that could place some sex differences into a clearer perspective. Even when such models are mentioned, the level of understanding is often inadequate or riddled with the sorts of misconceptions discussed earlier.

Social psychologists are often satisfied with offering only proximate explanations of social behavior, such as saying that someone did something "because it was reinforcing," "for sexual gratification," or "to verify their self-concept." To claim merely that people do things because their actions are reinforcing begs the more basic question of why some things are more reinforcing than others in the first place. Likewise, to claim that social situations that are "personally relevant" generate more extensive and in-depth cognitive processing than situations that are not relevant fails to consider the reasons why certain things are more relevant than others. It also fails to make important distinctions between different types of "relevance" (Kenrick, 1994; Kenrick, Sadalla, & Keefe, in press). In chapter 2, Krebs and Denton consider in detail the problems with incomplete explanations.

To Help Explain Things That Are Otherwise Inexplicable

Many patterns of social behavior are very difficult to explain using traditional social models, but they make more sense when viewed from an evolutionary perspective. We have already discussed a number of cross-cultural regularities in social behavior that are difficult to explain solely in terms of the impact of "American culture" or idiosyncratic societal rules, expectations, or norms.

Consider some violations of the similarity–attraction principle that are difficult to explain in terms of traditional principles of reinforcement or balance notions, which are commonly used to explain attraction phenomena. Take the case of incest avoidance. Although there are laws against incest in our society, the incidence of biological incest (intercourse between first-degree blood relatives, as opposed to legally defined incest between steprelatives) is quite low (Thornhill, 1990; Van den Berghe, 1983). Indeed, some societies have no laws prohibiting incest (Thornhill, 1990). When unrelated children are raised as siblings in the same home environment, they tend to become good friends but rarely become romantically involved with one another (Van den Berghe, 1983). In a classic study involving almost 3,000 kibbutz marriages, Shepher (1971) found that children raised from birth in the same kibbutz pods never married each other, despite the fact that they did marry neighbors quite frequently and that intrapod attractions were not discouraged by the norms of kibbutz society. This striking violation of otherwise powerful principles of interpersonal attraction makes sense, however, from an evolutionary perspective. Because of the deleterious genetic consequences associated with inbreeding, Shepher argued that humans, like other species in which young are raised around one another, have an evolved psychological mechanism that inhibits strong incestuous attraction.

Another violation of the similarity–attraction principle is the finding that people often choose mates who differ from them in age. This violation, however, tends to be in one direction, with females showing relatively more attraction to older men, and males being more attracted to younger women. Traditional models that initially attributed these results to gender differences in power within Western societies have a difficult time explaining why this pattern appears to be universal. Even more problematic for traditional models is the finding that young males show little preference for younger females and no aversion to older females. Indeed, teenage males are actually more interested in older females than in younger ones, even though older females do not generally reciprocate their interests (Kenrick, Gabrielidis, et al., in press). These findings make sense, however, in terms of differential parental investment in offspring. Women should be attracted to older men because male age is positively correlated with status and resources that men have traditionally contributed to the pair-bond; teenage men should be attracted to older women, and men older than 30 years of age should be increasingly attracted to relatively younger women, because female fertility is highest when women are in their 20s, and decreases considerably after age of 40 (Simpson, 1993).

An evolutionary perspective can even help us to understand several apparently paradoxical findings in homosexual mate choice. Similar to heterosexual men, homosexual men place greater emphasis on youth and physical attractiveness than on status and resources in choosing sexual partners. Similar to heterosexual women, lesbians stress and display greater fidelity and monogamy in their relationships, despite the lack of concern about becoming pregnant. These findings make sense from the vantage point of emerging evolutionary models, which posit that humans do not possess a single, monolithic psychological mechanism designed to maximize their general fitness (which would be very difficult to program and highly inefficient to use). Instead, humans likely have several specific psychological mechanisms, each of which evolved to solve specific adaptive problems (Bailey, Gaulin, Agyei, & Gladue, 1994; Buss, 1995).

Because Social Psychologists Have Special Skills Essential for the Emerging Paradigm

We are not advocating that all social psychologists should dig up Australopithecine bones in the Olduvai gorge, take up residence with bands of wild chimpanzees on the Gombe Stream Reserve, or drop everything to join the gene mapping project. The remnants of our evolutionary heritage are etched in the human brain. Therefore, our ancestral past may be best revealed not by looking for bones or strings of genes but by studying the nature of human psychological mechanisms (Buss, 1995; Buss & Kenrick, in press; Tooby & Cosmides, 1992). The most important scientific mission for the next decade will be to identify the major psychological mechanisms, to understand how they are differentially activated or terminated by specific situational cues, and to explore how they develop in different kinds of environments across time. Furthermore, because many of the strongest pressures on human evolution probably originated from problems associated with group living (Hogan, 1982; Lancaster, 1976), some of the most powerful and pervasive psychological mechanisms should have evolved to deal with complex social relationships. Who has the methodological skills required to identify and investigate these mechanisms? Who is best able to delve below the surface of self-deception and social desirability to demarcate different levels of social cognition and tease apart cognitive, motivational, and affective components underlying human social behavior? The answer is not geneticists, field anthropologists, or ethologists, but social psychologists. Evolutionary psychology, therefore, needs social psychology.

To See How It All Fits Together

Social psychology is a remarkably diverse field. It covers topics that range from altruistic generosity to homicidal aggression, from interpersonal attraction to intergroup conflict, from obedience and conformity to group pressure to minority dissent, from the way people persuade others to the way they avoid being persuaded, and from how people perceive and think about others to what they actually do in myriad social situations. Given this diversity, one might expect that most research in social psychology would be organized around a few comprehensive metatheories. Indeed, students in our social psychology courses occasionally ask some very insightful questions: Why are so many minitheories (e.g., reactance theory, gain–loss theory, dissonance theory) used to explain isolated empirical findings that often seem confined to specific areas of research (e.g., resistance to persuasion, interpersonal attraction, attitude change)? Why is there so little discussion in social psychology textbooks about how different mintheories relate to one another? Why are the minitheories in one chapter of a social psychology text often discontinuous with minitheories presented in other chapters? Why does such an important field in the social and behavioral sciences—one that studies so many significant topics that are vital to understanding human behavior—not have a metatheory, one capable of tying different research areas and disparate findings together?

In a recent review of research on interpersonal relationships, Kenrick and Trost (in press) showed how the principles of evolutionary psychology are tightly interconnected and how they can be used to comprehend phenomena ranging from altruism to aggression, romantic love, and deception in relationships. These interconnected principles serve as organizing themes in the chapters of this volume. Moreover, these chapters demonstrate how several major areas of research in social psychology can be organized and understood from a single—although multifaceted—theoretical perspective: Inclusive Fitness Theory (Hamilton, 1964), which is derived from Darwin's (1859) theory of evolution by natural selection. This volume showcases only a few of the research areas—social perception, interpersonal attraction, mating, close relationships, and group behavior—that can be more profitably studied and more clearly understood when viewed from an evolutionary perspective. Nonetheless, it offers a representative and promising glimpse of what evolutionary models can do to unify theory and research in social psychology.

To argue that an evolutionary perspective can integrate diverse research areas in social psychology does not imply that minitheories are useless. However, we must make more concerted efforts to understand how various minitheories relate to each other. To do so, we must be more cognizant of the different levels of analysis at which research can be conducted. The question "Why does organism A engage in behavior X?" can be asked at the following three different levels of analysis (Sherman, 1988; Tinbergen, 1963).

Questions of *proximate causation*, which social psychologists typically address, focus on how factors in an individual's immediate environment activate, maintain, and regulate a given behavior. For example, one may ask what specific features of men and women are associated with judgments that they are physically attractive (as Cunningham, Druen, & Barbee do in chapter 5), or how a man's likeability interacts with his social dominance to influence his attractiveness to women (addressed by Graziano, Jensen-Campbell, Todd, & Finch in chapter 6)?

Questions of *ontogeny* center on how a behavior develops and changes as an individual grows and matures across the life span. For example, how do children learn which characteristics signify status in their particular society? What developmental processes trigger the aversion to strong romantic attraction between siblings? Zeifman and Hazan (chapter 9) and Miller and Fishkin (chapter 8) address ontogenetic questions of how different childhood attachment experiences might elicit different orientations to close relationships in adulthood.

Questions of *ultimate causation* deal with the evolutionary origins and history of a specific behavior. Why, for example, did ancestral humans benefit from mating with partners with whom they developed strong emotional bonds? Questions of this nature are addressed by Zeifman and Hazan (chapter 9) and by Miller and Fishkin (chapter 8). Why does kinship moderate several social exchange relationships and social group biases? Questions pertinent to these issues are discussed by Daly, Salmon, and Wilson (chapter 10), Haslam (chapter 11), Krebs and Denton (chapter 2), and Wilson (chapter 13).

The failure to distinguish adequately between these different levels of analysis has led to several misunderstandings and frivolous debates over the years. Perhaps the most famous example is the fallacious nature–nurture debate. In the 1950s, Konrad Lorenz and Daniel Lehrman argued over whether certain behaviors displayed by newborn chicks (e.g., orienting behaviors toward their parents in the first days of life) are innate or acquired through experience with the environment (Lehrman, 1953). The debate ended when it became clear that the dispute hinged more on semantic and conceptual differences than on factual discrepancies. Lorenz was asking and trying to answer ultimate causation and phylogenetic questions about the behavior of newborn chicks (i.e., *why* questions), whereas Lehrman was asking and attempting to answer questions concerning ontogeny and proximate causation (i.e., *how* questions). The perspectives advocated by both Lorenz and Lehrman were legitimate and probably correct from the standpoint of their respective levels of analysis.

As mentioned previously, social psychologists historically have focused on proximate questions. Because of this strong focus, social psychologists receive rigorous and highly specialized training in experimental and quasi-experimental laboratory research methods and techniques. This training has allowed us to ask and answer questions of proximal causation with considerable skill and precision. Yet research questions must eventually go beyond a proximal level of analysis if we want to thoroughly understand human social behavior from a comprehensive scientific perspective. One theme of this volume is that if social psychology is going to make important contributions to the wider scientific community, social psychologists must begin think about their programs of research from all three levels of analysis, not exclusively from the confines of proximate causation.

It is important to emphasize that competition between alternative theoretical models typically occurs within a given level of analysis, not between different levels (Mayr, 1982; Sherman, 1988). Let us return to the case of incest avoidance. In terms of ontogenetic development, different theories make novel predictions about how incest avoidance is instilled in children. Kitcher (1985) proposed that children learn to avoid incestuous relationships because their parents teach them to do so. Eibl-Eibesfeldt (1989), in contrast, suggested that incest aversion develops from early and prolonged exposure to siblings (and close playmates) during critical periods of development. It is possible that neither of these theories adequately explains how incest avoidance develops, both may do so, or one theory might receive more empirical support than the other. The point is that the theories compete with each other within a given level of analysis. If, instead, these theories sought to answer questions at different levels (e.g., pitting questions about the functional significance of incest aversion against questions about its ontogenetic development), they could not be compared directly.

Evolutionary principles also can elucidate the behavior of higher-order dynamic systems, explaining phenomena that range from interactions between genes, to predator–prey relationships, to stock market crashes (Lewin, 1993; Waldrop, 1992). Recently, dynamic systems analyses have begun to be applied to social psychological processes (Eiser, 1994; Latane & Nowak, 1994; Vallacher & Nowak, 1994). These analyses, which are discussed by Caporael and Baron (chapter 12), do not provide an alternative to evolutionary models, but they do call attention to the importance of higher levels of analysis. As we noted earlier, asking insular questions can produce not only limited answers, but incorrect ones as well. One cannot postulate processes at one level of analysis that simply do not fit with those at higher levels. For example, one cannot seriously consider cognitive processes that regularly act against individuals' self-interest and kin-interests unless one also has higher-level models that can explain how and why these non-egocentric proclivities should exist. As discussed by Wilson (chapter 13), some self-sacrificial cooperation could have been selected for, but only under limited conditions.

CONCLUSION

Richard Nisbett recently co-organized a colloquium series on evolution and social cognition at the University of Michigan. Nisbett introduced the series by saying that he once thought every psychology department would need to hire an evolutionary psychologist, but he had changed his mind. Instead, Nisbett predicted that evolutionary theory will come to play the same role in psychology as it currently assumes in biology: "Not every psychologist will be an evolutionary psychologist, but every psychologist will be aware of the perspective and will have to address its explanations and constraints in his or her own work" (Nisbett, 1995, personal communication).

The theoretical and empirical work presented in this volume demonstrates why Nisbett is likely to be right. The time has come for our field to move beyond the old misconceptions and realize the full potential of a truly evolutionary social psychology.

ACKNOWLEDGMENT

We thank William Graziano for the helpful comments he provided on an earlier draft of this chapter.

REFERENCES

Abbey, A. (1982). Sex differences in attributions for friendly behavior: Do males misperceive females' friendliness? *Journal of Personality & Social Psychology, 42,* 830–838.

Bailey, J. M., Gaulin, S., Agyei, Y., & Gladue, B. A. (1994). Effects of gender and sexual orientation on evolutionarily relevant aspects of human mating psychology. *Journal of Personality and Social Psychology, 66,* 1074–1080.

Barkow, J. H., Cosmides, L., & Tooby, J. (1992). *The adapted mind: Evolutionary psychology and the generation of culture.* New York: Oxford University Press.

Baumeister, R. (1993). *Self-esteem: The puzzle of low self-regard.* New York: Plenum.

Broude, G. J. (1992). The May–September algorithm meets the 20th century actuarial table. *Behavioral and Brain Sciences, 15,* 94–95.

Buss, D. M. (1989). Sex differences in human mate preferences: Evolutionary hypotheses tested in 37 cultures. *Behavioral and Brain Sciences, 12,* 1–49.

Buss, D. M. (1995). Evolutionary psychology: A new paradigm for psychological science. *Psychological Inquiry, 6,* 1–30.

Buss, D. M., & Kenrick, D. T. (in press). Evolutionary social psychology. In D. Gilbert, S. Fiske & G. Lindzey (Eds.), *Handbook of social psychology* (4th ed.). New York: McGraw-Hill.

Campbell, D. T. (1975). On the conflicts between biological and social evolution and between psychology and moral tradition. *American Psychologist, 30,* 1103–1126.

Chaiken, S., Liberman, A., & Eagly, A. H. (1989). Heuristic and systematic information processing within and beyond the persuasion context. In J. S. Uleman & J. A. Bargh (Eds.), *Unintended thought* (pp. 212–252). New York: Guilford.

Crawford, C. B., & Anderson, J. L. (1989). Sociobiology: An environmentalist discipline. *American Psychologist, 44,* 1449–1459.

Cunningham, M. R. (1981). Sociobiology as a supplementary paradigm for social psychological research. In L. Wheeler (Ed.), *Review of personality and social psychology* (Vol. 2, pp. 69–106). Beverly Hills, CA: Sage.

Daly, M., & Wilson, M. (1983). *Sex, evolution, and behavior* (2nd ed.). Belmont, CA: Wadsworth.

Daly, M., & Wilson, M. (1988a). *Homicide.* New York: Aldine de Gruyter.

Daly, M., & Wilson, M. (1988b). Evolutionary social psychology and family homicide. *Science, 242* (October), 519–524.

Darwin, C. (1859). *The origin of species.* London: Murray.

Eibl-Eibesfeldt, I. (1975). *Ethology: The biology of behavior* (2nd ed.). New York: Holt, Rinehart, & Winston.

Eibl-Eibesfeldt, I. (1989). *Human ethology.* New York: Aldine de Gruyter.

Eiser, J. R. (1994). *Attitudes, chaos, and the connectionist mind.* Oxford, UK: Blackwell.

Ekman, P. & Friesen, W. V. (1971). Constants across cultures in the face and emotion. *Journal of Personality and Social Psychology, 17,* 124–129.

Ekman, P., Friesen, W. V., O'sullivan, M., Chan, A., Diacoyanni-Tarlatzis, I., Heider, K., Krause, R., LeCompte, W. A., Pitcairn, T., Ricci-Bitti, P. E., Scherer, K., Tomita, M., & Tzavaras, A. (1987). Universals and cultural differences in the judgments of facial expressions of emotion. *Journal of Personality and Social Psychology, 53,* 712–717.

Gangestad, S. W., & Buss, D. M. (1993). Pathogen prevalence and human mate preferences. *Ethology and Sociobiology, 14,* 89–96.

Glass, A. L., & Holyoak, K. J. (1986). *Cognition* (2nd ed.). New York: Random House.

Hamilton, W. D. (1964). The evolution of social behavior. *Journal of Theoretical Biology, 7,* 1–52.

Hamilton, W. D., & Zuk, M. (1982). Heritable true fitness and bright birds: A role for parasites? *Science, 218,* 384–387.

Harpending, H. (1992). Age differences between mates in southern African pastoralists. *Behavioral and Brain Sciences, 15,* 102–103.

Hogan, R. (1982). A socioanalytic theory of personality. In M. Page (Ed.), *Nebraska symposium on motivation* (pp. 55–89). Lincoln, NE: University of Nebraska Press.

Jankowiak, W. R., & Fischer, E. F. (1992). A cross-cultural perspective on romantic love. *Ethnology, 31,* 149–155.

Kenrick, D. T., (1994). Evolutionary social psychology: From sexual selection to social cognition. In M. P. Zanna (Ed.), *Advances in experimental social psychology, vol. 26.* (pp. 75–122). San Diego, CA: Academic Press.

Kenrick, D. T., Gabrielidis, C., Keefe, R. C., & Cornelius, J. S. (in press). Adolescents' age preferences for dating partners: Support for an evolutionary model of life-history strategies. *Child Developmen,.*

Kenrick, D. T., & Keefe, R. C. (1992). Age preferences in mates reflect sex differences in reproductive strategies. *Behavioral and Brain Sciences, 15,* 75–133.

Kenrick, D. T., Neuberg, S. L., Zierk, K. L., & Krones, J. M. (1994). Evolution and social cognition: Contrast effects as a function of sex, dominance, and physical attractiveness. *Personality and Social Psychology Bulletin, 20,* 210–217.

Kenrick, D. T., Sadalla, E. K., & Keefe, R. C. (in press). Evolutionary cognitive psychology. In C. Crawford & D. Krebs (Eds.), *Evolution and human behavior.* Hillsdale, NJ: Lawrence Erlbaum Associates.

Kenrick, D. T., & Trost, M. R. (in press). Evolutionary approaches to relationships. In S. Duck (Ed.) *Handbook of Personal Relationships* (2nd ed.). London: Wiley.

Kenrick, D. T., Trost, M. R., & Sheets, V. (in press). Power, harassment, and trophy mates: The feminist advantages of an evolutionary perspective. In D. M. Buss & N.

Malamuth (Eds.) *Sex, power, conflict: feminist and evolutionary perspectives.* Oxford University Press.

Kitcher, P. (1985). *Vaulting ambition.* Cambridge, MA: MIT Press.

Lancaster, J. B. (1976). *Primate behavior and the emergence of human culture.* New York: Holt, Rinehart, & Winston.

Latane, B., & Nowak, A. (1994). The stream of social judgment. In R. R. Vallacher & A. Nowak (Eds.), *Dynamical systems in social psychology* (pp. 219–249). San Diego: Academic Press.

Lehrman, D. S. (1953). A critique of Konrad Lorenz's theory of instinctive behavior. *Quarterly Review of Biology, 28,* 337–363.

Lewin, R. (1993). *Complexity: Life at the edge of chaos.* London: J. M. Dent.

Lumsden, C. J., & Wilson, E. O. (1981). *Genes, mind, and culture: The coevolutionary process.* Cambridge, MA: Harvard University Press.

Markus, H., & Zajonc, R. B. (1985). The cognitive perspective in social psychology. InLindzey & E. Aronson (Eds.), *Handbook of social psychology,* (vol. 1, pp.137–230). New York: Random House.

Mayr, E. (1982). *The growth of biological thought.* Cambridge, MA: Harvard University Press.

Moghaddam, F. M., Taylor, D. M., & Wright, S. C. (1993). *Social psychology in cross-cultural perspective.* New York: Freeman.

Petty, R. E., & Cacioppo, J. T. (1986). The elaboration likelihood model of persuasion. In L. Berkowitz (Ed.), *Advances in experimental social psychology, vol. 19* (pp. 123–205). New York: Academic Press.

Rajecki, D. W. (1977). Ethological elements in social psychology. In C. Hendrick (Ed.), *Perspectives on social psychology* (pp. 223–303). Hillsdale, NJ: Lawrence Erlbaum Associates.

Sadalla, E. K., Kenrick, D. T., & Vershure, B. (1987). Dominance and heterosexual attraction. *Journal of Personality and Social Psychology, 52,* 730–738.

Shepher, J. (1971). Mate selection among second generation kibbutz adolescents and adults: Incest avoidance and negative imprinting. *Archives of Sexual Behavior, 1,* 293–307.

Sherman, P. W. (1988). The levels of analysis. *Animal Behavior, 36,* 616–619.

Simpson, J. A. (1993). Male reproductive success as a function of social status: Some unanswered evolutionary questions. *Behavioral and Brain Sciences, 16,* 307–307.

Simpson, J. A., Gangestad, S. W., & Biek, M. (1993). Personality and nonverbal behavior: An ethological perspective of relationship initiation. *Journal of Experimental Social Psychology, 29,* 434–461.

Small, M. F. (1992). Female choice in mating. *American Scientist, 80,* 142–151.

Smuts, B. B. (1985). *Sex and friendship in baboons.* New York: Aldine de Gruyter.

Thornhill, N. (1990). The evolutionary significance of incest rules. *Ethology and sociobiology, 11,* 113–129.

Tinbergen, N. (1963). On the aims and methods of ethology. *Z. Tierpsychol., 20,* 410–433.

Tooby, J., & Cosmides, L. (1990a). The past explains the present: Emotional adaptations and the structure of ancestral environments. *Ethology and Sociobiology, 11,* 375–424.

Tooby, J., & Cosmides, L. (1990b). On the universality of human nature and the uniqueness of the individual: The role of genetics and adaptation. *Journal of Personality, 58,* 17–68.

Tooby, J., & Cosmides, L. (1992). The psychological foundations of culture. In J. H. Barkow, L. Cosmides, & J. Tooby (Eds.), *The adapted mind: Evolutionary psychology and the generation of culture* (pp. 19–136). New York: Oxford University Press.

Trivers, R. (1971). The evolution of reciprocal altruism. *Quarterly Review of Biology, 46,* 35–57.

Trivers, R. (1972). Parental investment and sexual selection. In B. Campbell (Ed.), *Sexual selection and the descent of man: 1871–1971* (pp. 136–179). Chicago: Aldine.

Trivers, R. (1974). Parent-offspring conflict. *American Zoologist, 14,* 249–264.

Trivers, R. (1985). *Social evolution.* Menlo Park, CA: Benjamin/Cummings.

Vallacher, R. R., & Nowak, A. (1994). *Dynamical systems in social psychology.* San Diego: Academic Press.

Van den Berghe, P. L. (1983). Human inbreeding avoidance: Culture in nature. *Behavioral and Brain Sciences, 6,* 91–123.

Waldrop, M. M. (1992). *Complexity: The emerging science at the edge of order and chaos.* New York: Touchstone Books.

Wilson, M., & Daly, M. (1985). Competitiveness, risk taking, and violence: The young male syndrome. *Ethology and Sociobiology, 6,* 59–73.

Zuckschwerdt, M. (1993). *Contrast effects, attractiveness, and pregnancy: Tests of an evolutionary hypothesis.* Unpublished honors thesis, Arizona State University, Tempe.

II

Social Perception

2

Social Illusions and Self-Deception: The Evolution of Biases in Person Perception

Dennis L. Krebs
Kathy Denton
Simon Fraser University

When I look at my friend who is your enemy, I see a person quite different from the person you see. My image of you, also, is quite different from your image of yourself. Wherein lies the truth, or is it all socially constructed illusion? The purpose of this chapter is to introduce an account of how cognitive structures designed to process information about self and others evolved in the human species. We begin by reviewing research showing that people categorize members of their in-groups differently from members of their out-groups. We demonstrate that such categorization influences social judgments: People view in-group members, including themselves, in more positive ways than they view out-group members. We consider psychological and evolutionary explanations of biases in social cognition, then sketch a model of mutual social support for self-serving illusions.

SOCIAL CATEGORIZATION

Research in social psychology has revealed that when we encounter other people, we immediately and automatically classify them as in-group or out-group (e.g., Bruner, 1957; Devine, 1989), and this categorization structures our subsequent perceptions of them (Sherif, Harvey, White, Hood, & Sherif, 1961; Tajfel, 1982; Vine, 1992). As we might expect, enduring characteristics such as race or gender may form the basis for group categorization, but so also may more trivial and arbitrary attributes. For example, Tajfel and his colleagues (Tajfel,

1982; Tajfel & Billing, 1974; Tajfel & Turner, 1979) randomly divided participants into two groups on the basis of a coin toss and found that their subjects
immediately identified with the group to which they were assigned, even though
they knew that the group assignment was random.

After we categorize others as members of out-groups, we view them in terms
of our stereotypes of the group to which we assign them (Allen & Wilder, 1979).
This process occurs automatically unless we make a conscious effort to refute
or disregard our preconceptions (Devine, 1989), and it prevails in the presence
of disconfirming information (Secord, 1959). We view out-group members as
"all alike" (Anthony, Copper, & Mullen, 1992; Chance & Goldstein, 1981) and
different from us (Langer & Imber, 1980).

We are more variable in our perceptions of in-group members. Although we
perceive in-group members as more similar to us than we perceive out-group
members to be, we do not perceive them as all alike. Members of in-groups are
viewed as individuals who possess unique and complex sets of traits and abilities
(Judd & Park, 1988).

Negative Biases in the Perception of Out-group Members

The "differences" we attribute to out-group members are generally negative in
nature (Judd, Ryan, & Park, 1991; Linville, Fischer, & Salovey, 1989). We tend
to view out-group members as less deserving than members of in-groups
(Gudykunst, 1989; Wilder, 1981). We tend to make dispositional attributions
for their failures (Ross & Fletcher, 1985), and we tend to characterize their
undesirable behaviors in abstract terms (e.g., hostile, irresponsible; Maass,
Milesi, Zabbini, & Stahlberg, 1995; Maass, Salvi, Arcuri, & Semin, 1989). We
tend to hold them responsible when they are victimized by crime (Wagstaff,
1982), disease (Gruman & Sloan, 1983), and other misfortunes (e.g., Furnham
& Gunter, 1984), and we tend to minimize their credit for positive outcomes
by attributing the outcomes to situational factors or to luck (Hewstone, Bond,
& Wan, 1983; Hewstone & Jaspars, 1982). Pettigrew (1979) labeled negatively
biased causal attributions for out-group members the *ultimate attribution error*.

Positive Biases in the Perception of In-group Members

Members of in-groups also are viewed in biased ways, but the bias tends to be
positive. A spate of research in social psychology (see Berscheid, 1985, for a
review) demonstrates that similarity breeds liking, which channels social perception in favorable ways. For example, Kandel (1978) reported that people
tend to be more attracted to those who are similar to them than to those who
are different from them in race and religion. When victims of crime or disease
are highly similar to us, we tend to make "defensive attributions" (Burger, 1981)

for their victimizations by attributing the victimizations to external factors such as bad luck. In contrast to the negative attributions we make about the achievements of out-group members, we tend to credit in-group members for their successes and externalize their failures. This group-serving or ethnocentric bias has been demonstrated in studies of close friends, relationship partners (Burger, 1981; Lau & Russell, 1980; Schlenker & Miller, 1977; Winkler & Taylor, 1979), and teammates (see Mullen & Riordan, 1988, for a review).

Positive Biases in Self-Perception

Evidence demonstrating that positive biases toward others are fostered by a perception of similarity implies that people are positively biased toward themselves, and, in general, they are. Positive biases in self-perception tend to be more extreme than positive biases in the perception of others. Psychologists have identified at least eight biases in self-perception (see Table 2.1). Compared to others, we tend to view ourselves as more successful, more efficacious, more insightful, more consistent, more appropriate, more capable, more productive,

TABLE 2.1

Biases in Self-Perception

Bias	Description and Prototypical Example
Self-serving bias	The tendency for people to take credit for their successes by attributing them to internal dispositions (e.g., intelligence) and to deny responsibility for their failures by attributing them to external circumstances (e.g., unfair test conditions). "I am a capable person, able to succeed." "I failed because the situation was unfair."
Self-centered bias	The tendency to take more than one's share of the credit for outcomes (both successful and unsuccessful) that involved a joint or group effort. "I have contributed more than others."
Egocentricity bias	The tendency to recall one's role in past events as positive and causally significant. "I was important; I had an impact on others."
False consensus effect	The tendency for people to see their own attitudes, values, and behavioral choices as relatively common and appropriate. "Most people would agree with me; most people would have done as I did."
False uniqueness effect/ Assumption of uniqueness	The tendency for people to view their identity-defining traits and abilities as relatively rare/distinctive/unique. "I have rare qualities; I am special."
Illusion of control	The tendency for people to believe they can influence events beyond their control to produce desireable outcomes (e.g., winning a lottery) or to avoid undesirable outcomes (e.g., becoming a victim of crime). "I am in control."
Hindsight bias	The tendency to find outcomes inevitable in retrospect. "I knew this would happen."
Self-righteous bias	The tendency to view oneself as possessing more moral integrity than others. "I am more likely to abide by moral principles than others."

more invulnerable, more moral, and more uniquely talented (see Taylor, 1989, for a review).

PSYCHOLOGICAL EXPLANATIONS
FOR BIASES IN SOCIAL PERCEPTION

Social psychologists have offered several explanations for biases in social cognition, which they have divided into those that involve "hot" and "cold" mechanisms. According to hot explanations, people's needs and motives interfere with and slant the ways in which they process information. For example, people process information in ways that enhance and protect self-esteem or increase their sense of control. We demean out-group members and view ourselves and our friends through rose-colored glasses to allay our anxieties and to make ourselves feel good.

In contrast, cold explanations attribute biases in social cognition to inherent limitations of human information-processing structures. For example, self-serving attributional biases have been attributed to the fact that people are more aware of their expectations for success than they are of others' expectations for success (Miller & Ross, 1975), and stereotypes have been viewed as overgeneralizations similar to those that we make when we categorize other things.

EVOLUTIONARY EXPLANATIONS FOR BIASES
IN SOCIAL PERCEPTION

Evolutionary Epistemology

Evolutionary psychology offers a conceptual framework for situating psychological explanations of biases in social cognition. Consistent with cold explanations, evolutionary psychologists attend to the design of the cognitive mechanisms that process social information. Like computer programs, such mechanisms sometimes "cooly" mediate information processing according to hardwired instructions in (genetic) software. In other cases, however, the mechanisms may run hot, generating emotional reactions and motivational states that induce individuals to engage in behaviors that enhanced the fitness of our hominid ancestors in their Pleistocene environments.

The main difference between psychological and evolutionary explanations is that evolutionary theorists attend to the original sources and ultimate functions of cognitive processes rather than to their immediate causes. From the perspective of evolution, conclusions such as "people enhance their self-images because it makes them feel good" beg the overriding question of why enhancing self-images make people feel good. This question is equivalent to asking why sugar is sweet (Barash, 1982). Evolutionary theory leads us to ask why the human brain evolved in a way that induces individuals to process information about themselves in one way and to process information about others in a different way. What adaptive functions did our inherited brain mechanisms, mental structures, and cognitive processes serve for our ancestors? How did the mechanisms help the individuals who passed them on to propagate their genes?

It is, of course, difficult to determine how cognitive processes were selected in our ancestors millions of years ago. Whereas paleontologists attempt to deduce the social behavior of our ancestors from fossil evidence, evolutionary psychologists attempt to deduce the adaptive functions current mechanisms served in Paleolithic environments from their functional designs. It is important in evolutionary analyses to compare the environments in which characteristics were selected to contemporary environments and ask whether the selected characteristics continue to enhance the fitness of individuals who possess them today.

Evolution works through selection. Early Darwinians focused on the natural selection of characteristics through the differential survival of individuals possessing differentially adaptive characteristics. But survival is significant in evolution mainly as a necessary condition for propagating genes—as a means to an end. The main way in which individuals propagate their genes is through sexual reproduction. For this reason a form of selection labeled *sexual selection*, which involves the selection of desirable characteristics in mates by members of the opposite sex, may play an important role in the evolution of many attributes. In addition, individuals also may propagate their genes by enhancing the reproductive success of those who possess copies of their genes—their relatives. This process has come to be called *kin selection*. The ultimate unit of analysis in evolution is *inclusive fitness*—the total representation of copies of an individual's genes that he or she has contributed to a population.

Obviously, individuals cannot propagate their genes through mental insemination, by sitting around thinking various thoughts. Thoughts (and feelings) are significant only in terms of their effects on fitness-enhancing behaviors. Although behaviors such as mating and caring for offspring are obvious means to the ultimate end of genetic propagation, a host of other behaviors are necessary for mating and parenting. To survive until sexual maturity, our ancestors had to protect themselves from predators, obtain food, find shelter, form social bonds, and so on. To reproduce sexually, our ancestors had to secure mates. Evolutionary analyses are difficult because long, complex sequences of events are necessary for sexual reproduction and kin selection, and ways of thinking and feeling may enhance an individual's fitness at any link in the complex chain.

Concerning the cognitive structures that have evolved to process information, we tend to assume it as normal and natural for people to process information as logically and as accurately as possible; we view inaccurate and illogical conclusions as mistakes caused by limitations of information-processing systems. It is, of course, possible for mistakes to occur in the process of evolution, but when all people systematically err in the same directions, evolutionary theorists suspect that the mechanisms producing the "mistakes" were selected to serve some adaptive function. From the perspective of evolution, the task of explaining cognitive mechanisms that systematically produce illogical inferences is identical to the task of explaining cognitive mechanisms that systematically produce logical inferences. In both cases evolutionary theorists ask what adaptive functions such forms of thought served in the past and whether they continue to serve such functions today. As Michael T. Ghiselin (1974) said:

> We are anything but a mechanism set up to perceive the truth for its own sake. Rather, we have evolved a nervous system that acts in the interest of our gonads, and one attuned to the demands of reproductive competition. If fools are more prolific than wise men, then to that degree folly will be favored by selection. And if ignorance aids in obtaining a mate, then men and women will tend to be ignorant. (p. 126)

When we observe a phenomenon from an evolutionary perspective—in the present case, biases in social cognition—we should entertain four general explanations: the phenomenon is caused by culture, the phenomenon is pathological—a maladaptive variation, the phenomenon is a byproduct of some adaptive process, and the phenomenon is an adaptation that evolved in our ancestral environment. As Buss (1995) explained, "'culture', 'learning,' and 'socialization' do not constitute explanations, let alone alternative explanations to those anchored in evolutionary psychology. Instead, they represent human phenomena that require explanation. The required explanation must have a description of the underlying evolved psychological mechanism at its core" (p. 14). To qualify as a pathology, a characteristic must be shown to be maladaptive. To qualify as a neutral by-product of an adaptive process, the adaptive process must be identified.

We argue that the biases in social cognition we discuss are adaptations that were selected in our ancestral environment. That is our theory. We adduce evidence to show that these adaptations are reflected in the design of the cognitive mechanisms we invoke to process social information, and that they serve essentially the same adaptive functions today as they served in our ancestral past in contexts that parallel those of our ancestral social environments. In essence, we describe a fit between the form of social information-processing structures and the adaptive functions we believe they served. At this early stage of theory construction, we seek mainly to sketch a plausible picture. Our implicit challenge is for critics to recommend more plausible alternatives.

Most contemporary evolutionary psychologists argue that the mechanisms of evolution tend to select domain-specific organs: "Organisms' brains are composed of some number of relatively independent modular mechanisms designed to deal specifically with particular cognitive problems confronting specific species" (Kenrick, Sadalla, & Keefe, in press). One theme of this chapter is that the cognitive mechanisms selected to process information about the self and one's friends are significantly different from those selected to process information about strangers and enemies.

The Adaptive Functions of Social Categorization

In evolutionary theory, other people are resources: They are significant in terms of the potential to enhance our inclusive fitness. The "survival of the fittest" versions of Social Darwinism popular in the 19th century portrayed individuals as unrelentingly bent on defeating competitors, usurping their resources, and in Tennyson's words, "red in tooth and claw." To prevail in the struggle for existence, hominids had to protect themselves from the ravages of others, secure resources, and defeat enemies: To the victor went the spoils. There are some parallels in this type of social relations in contemporary environments. Juvenile gangs fight for territory; people assault, steal from, and rape one another. The media are sickeningly replete with evidence of human capacity to exploit others in the most horrible ways during war. The depth of cognitive structures mediating competitive relationships is reflected in our enchantment with sports and games.

It does not, however, take a lot of insight to see that individuals who competed with everyone else would not fare very well in the struggle for existence against people who joined forces against them. Many evolutionary theorists have therefore argued dispositions fostering cooperation are among the most adaptive dispositions inherited by members of the human species (Alexander, 1987; Krebs, 1987; Krebs, in preparation; Leaky & Lewin, 1977). As game theorists such as Axelrod (1984) showed, cooperative strategies may fare better than noncooperative strategies in evolutionary competitions. The relative value of cooperation and competition would be expected to vary across contexts.

Paleontological evidence suggests that our ancestors lived in relatively small cooperative groups that competed against other relatively small cooperative groups (Lewin, 1993; Tooby & Devore, 1987). In such an environment, discriminating between friend and foe would have been valuable, as it is today. Making a mistake about whether another person was a friend or an enemy could have been lethal, so it is plausible to assume that it would have been adaptive to categorize others quickly and decisively and to be wary of strangers. As demonstrated in research on social perception, our initial categorization of others channels the way in which we orient ourselves toward them (Fiske & Newberg, 1990). It is as though the act of classifying others as in-group or out-group members activates two quite different brain circuits or decision-making programs.

The Adaptive Functions of Negative Illusions

As demonstrated earlier, people tend to perceive out-group members through dark-colored lenses. "They" are all alike; "they" are bad; and "they" are bent on exploiting or destroying us. In projecting hostile intentions onto our enemies, and in characterizing their negative behaviors in abstract terms that imply high stability (e.g., evil; see Maass et al., 1989, 1995), we feel justified in exploiting them. In viewing our enemies as dissimilar to us, we move them outside our domain of empathy. Indeed, when out-group members are perceived as a threat, it is not unusual to dehumanize them completely, viewing them as rats, pigs, weasels, snakes, and dogs. The mechanisms that structure our perceptions of out-group members generate affective reactions such as suspicion, anger, and hate, which give rise to discriminatory, prejudicial, aggressive, and exploitative behaviors.

Negative biases toward out-group members tend to be self-perpetuating. If I am wary of you because I do not know you, and if you are wary of me because you do not know me, then we will tend to avoid one another, thus perpetuating our wariness. If I believe that you are motivated to exploit me, then I will be more likely to exploit you, thus inducing you to retaliate. The result may be an endlessly iterated feud or arms race. As Kelley and Stahelski (1970) showed, it takes only one competitive response to transform cooperative relations into an enduring competitive standoff.

It would not, however, have been in our ancestors' adaptive interest to be too rigid in categorizing others because circumstances change. It follows that the cognitive programs we have inherited to categorize others, should be quite flexible. We may base our in-group–out-group categorizations on trivial cues (Tajfel, 1982), and, as demonstrated so often in the history of international relations, we may switch allegiances when it is in our vested interest to do so.

There is an inevitable tension in cooperative relations: It is in our interest to form cooperative alliances to compete against out-groups for the acquisition of resources, but it also is in our interest to compete against those with whom we have formed cooperative alliances on an individual basis when it comes to dividing the spoils. In the final analysis, there is only one stable in-group—our complement of genes and their copies in others. When our interests conflict with those of in-group members, we quickly recategorize them (Fiske & Neuberg, 1990), processing information about them through cognitive mechanisms designed to deal with out-groups. The spouse who was viewed as flawless during the honeymoon is the same person genetically who is vilified during the divorce.

It benefits us to regulate the severity of our negative judgments about others in terms of the extent to which it is in our interest to behave antagonistically toward them. In large part, this depends on the scarcity of resources. If the environment is replete with resources, the risks and expenditure of energy supporting the realistic conflict theory of prejudice and discrimination (Levine & Campbell, 1972) demonstrates that stereotypes of out-group members vary in accordance with the scarcity of resources and ensuing necessity of competition. For example, a study showed that negative stereotypes of Chinese in America varied historically in accordance with the extent to which the Chinese were involved in the competition for resources. And then, there is the negative correlation between lynching in the southern United States and economic indexes (e.g., value of cotton) reported by Hovland and Sears (1940; see also Hepworth & West, 1988).

Principles of social categorization are well exemplified in two classic studies. In the first, Hastorf and Cantril (1954) showed that Dartmouth fans, in essence, saw a different football game from Princeton fans—each seeing twice as many infractions against their team as against the opposing team. In the second study—the robber's cave experiment (Sherif et al., 1961)—12-year-old boys in a summer camp were divided arbitrarily into two groups: the Rattlers and the Eagles. The boys immediately identified with their groups. When resources were scarce—when there was a conflict of interest—Rattlers and Eagles vilified each other and competed for the resources until their antagonism got out of control. However, when there was a confluence of interest—both groups were needed to achieve a common goal—they moderated their negative perceptions of one another and engaged in mutually beneficial cooperative tasks.

The Adaptive Functions of Positive Illusions

Taylor (1989) labeled positive biases in self-perception *positive illusions*. Research on the salutary effects of positive illusions was well-reviewed by Taylor (1989) in her engaging book, *Positive Illusions: Creative Self-Deception and the Healthy Mind*, and by Goleman, (1985) in an equally engaging book, *Vital Lies, Simple Truths: The Psychology of Self-Deception*. Both writers concluded that positive illusions foster physical and mental health. From an evolutionary perspective, the positive effects documented by Taylor and Goleman are intermediate steps in the chains of consequences that fostered the fitness of our ancestors. In some cases, the association between these effects and the enhancement of fitness is obvious; in other cases not. In the remainder of this chapter we review research documenting the effects of positive illusions on physical and mental health (psychological explanations) then attempt to explain how these effects enhanced our ancestors' fitness (evolutionary explanations).

Physical Fitness.   Individuals can be seen as vehicles driven by their genes to replicate themselves. Like other vehicles, individuals do not function very well when they are in disrepair. Positive illusions help people keep themselves mentally and physically fit. Many studies found that positive illusions, especially illusions of control, are associated with physical health (see Snyder & Higgins, 1988; Taylor & Brown, 1988, for reviews). Although good health would be expected to engender positive illusions, studies have found that the causal relationship goes in the reverse direction as well: Positive illusions enhance health. For example, Peterson, Seligman, and Vaillant (1988) found that the immune systems of optimists are stronger than those of pessimists. Goleman (1987) reported that "Pollyannaish" patients who denied the risks of impending surgery "had fewer medical complications and were discharged sooner than were the more vigilant patients" (p. 26). Trotter (1987) found that more than twice as many women who denied the severity of their breast cancer with a "fighting spirit" prevailed over the disease than women who "helplessly accepted it" (p. 27). Kobassa (1979) found that "staying healthy under stress is critically dependent on a strong sense of commitment to self" (p. 4) buttressed by illusions of control. The apparently magical effects of placebos have a long history in medicine. Although scientists do not fully understand how positive illusions affect physical health, studies found that positive thoughts induce the release of various chemicals, especially neurotransmitters, that mediate the salutary effects.

In the past, psychologists have argued that mental health involves perceiving the world—including self and others—accurately. Drastically inaccurate perceptions of oneself, as in paranoid delusions, are clearly maladaptive. But recent research suggests that moderately unrealistic, overly positive perceptions of oneself may well be necessary for optimal mental health (Snyder & Higgins, 1988). Taylor (1989) derived four main criteria of mental health from the literature—happiness, the capacity for creative and productive work, the capacity for personal growth, and the capacity to care for and about other people—and adduced evidence to show that positive illusions enhance each of them. Building on Taylor's review, we attempt to explain how the four criteria of mental health may enhance fitness.

Happiness.   The sad truth is that life can be very cruel. People get raped, injured, betrayed, and murdered every day. From the perspective of evolution, life is not, in fact, as fair or as meaningful as most people assume. As stated by Dawkins (1976), individuals are "survival machines" programmed to propagate their genes. The purpose of chickens is to produce eggs, and the purpose of eggs is to produce chickens. Philosophers such as Becker and Camus discussed the dilemma of humans faced with the meaninglessness and absurdity of their existence and the inevitability of death.

Facing up to these truths—perceiving our prospects, purposes, and place in the order of things realistically—may induce anxiety, depression, despair, withdrawal, and resignation. Several studies have found that depressed people make more unfavorable, but more valid, attributions than those who are not depressed—a phenomenon labeled "depressive realism" or the "sadder but wiser effect" (Alloy & Abramson, 1979; but see also Sweeney, Anderson, & Bailey, 1986). Depression is physically, socially, and psycho-

"sadder but wiser effect" (Alloy & Abramson, 1979; but see also Sweeney, Anderson, & Bailey, 1986). Depression is physically, socially, and psychologically maladaptive. In the extreme, depression gives rise to the evolutionarily most maladaptive of all behaviors: loss of sexual appetite, parental negligence, and suicide. It follows that brain structures that induced our ancestors to perceive accurately and fully how insignificant they were and how absurd and unfair life was may well have been maladaptive. Those who inherited such forms of thought may have put less effort toward surviving and reproducing, causing the genes that mediated such accurate perceptions to become extinct. Conversely, those who inherited brain mechanisms that induced them to perceive themselves and the world in more optimistic terms may well have been those who prevailed.

Optimistic illusions allay depression by making life seem more hopeful and purposeful than it actually is. Karl Marx characterized religion as "the opiate of the people." Virtually all religions demand faith, which entails believing things will turn out well in the face of evidence to the contrary. Religious beliefs accepted on faith are antidotes to evidence of the meaninglessness and injustice in life. Such beliefs foster a false sense of security, a sense that someone "up there" will take care of us in the same way that our parents did when we were children. It is fascinating to see the extent to which people believe they are special in the eyes of their gods. Prizefighters pray before going into the ring; football players fall on their knees to thank God for a touchdown. Of course, most religions do not vouchsafe divine grace gratuitously; the Lord helps those who help themselves. Such a Protestant ethic supports optimistic illusions by inducing people to sustain goal-oriented behaviors. Even if your efforts do not pay off immediately, you should keep on trying; God will ensure that you are rewarded in the end.

From an evolutionary perspective, happiness is epiphenomenal. Like orgasm and the sense of sweetness we experience when tasting certain foods, happiness is a sign that our systems and circumstances are in a favorable state; our fitness-enhancing prospects are good, relative, at least, to our ancestral environments. Depression warns us that we are not adapting to our environment in ways that enhanced the fitness of our ancestors.

The Capacity for Productive Work.   In her review of the research on the effect of positive illusions on mental health, Taylor (1989) devoted considerable attention to the salutary effects of persistence:

> Some of the most dramatic effects of self-enhancing perceptions, a belief in personal control, and optimism concern motivation and persistence. The confidence that one can successfully accomplish a task, the belief that one has the means to do it, and the optimism that success will eventually ensue leads people to attempt tasks that they otherwise might avoid. (p. 59)

In support of these conclusions, Taylor (1989) cited research showing that people with high self-esteem are more optimistic about the probability of success than are people with low self-esteem (Bandura, 1977; Felson, 1984). Optimism has been found to mediate constructive thinking (Epstein, 1990); delay of gratification (Diener & Dweck, 1980); and complex, proactive coping strategies (Scheier & Carver, 1985). People prone to making positive attributions for their achievements have been found to be more persistent than those prone to making negative attributions (Dweck & Licht, 1980). Illusions of control have been found to foster success (Carsrud & Olm, 1986).

There are, of course, limits on the extent to which overestimations of the probability of success and overestimations of one's abilities are adaptive (Baumeister, 1989). In some situations, it is more adaptive to cut one's losses than to persevere. Extreme illusions about one's worth become delusions of grandeur. The illusions found by psychologists to mediate success are actually rather modest: It pays off to believe you are a little, but not too much, better than you actually are.

The relationship between perseverance and fitness is obvious and does not need belaboring. In current and ancestral environments, individuals who persevered would be more likely than those who gave up to have achieved their fitness-enhancing goals, whether these goals involved the acquisition of resources, status, high-quality mates, or fecund offspring. To explain fully the evolution of optimistic illusions, we need a model that outlines the adaptive limits of optimism. The general principle is obvious: Persist at tasks within your ability to accomplish. Judgments of attainability will certainly be based on past experience, as demonstrated by research on learned helplessness.

The Capacity for Personal Growth.   Ideas are important in evolutionary theory in terms of their effects on fitness. Among the most important positive illusions harbored by people are those illusions about what they could become. Researchers from many theoretical traditions have discussed the importance of aspirations about the self, whether they are called ego-ideal (Freud, 1933), ideal self (Van den Daele, 1968), or, more recently, possible selves (Markus & Nurius, 1986). Such ideas may have considerable motivational power, pulling people proactively toward their illusions of accomplishment, sexual success, fame, and fortune. As such, they may serve as instruments of self-actualization.

One of the main ways in which optimistic and self-enhancing beliefs foster inclusive fitness is by engendering self-fulfilling prophesies. Snyder (1984) reviewed evidence from a vast array of studies showing that "beliefs create reality." If you do not believe you can catch a fish, you will not throw your line into the water. Conversely, if you believe you will succeed, you will try harder. Believing in a result activates systems of thought, ways of feeling, motivational states, and behaviors that affect the probability of the desired effect. Frank (1961) and Aronoff and Lesse (1983) reviewed evidence demonstrating that self-fulfilling prophesies play an important role in the success of psychotherapy. Patients who have faith in their therapists and in the effects of psychotherapy, especially "true believers," tend to improve more than patients who do not.

The Capacity to Care for and About Other People. The final criterion of mental health cited by Taylor (1989) is the capacity to care for others. As acknowledged by Taylor, the idea that self-enhancing illusions induce people to care for others is paradoxical because such illusions would seem to induce a narcissistic focus on the self. This paradox is similar to the paradox of altruism in Darwinian theory: How can behaviors evolve that enhance the fitness of others at costs to one's own fitness? In an attempt to resolve this paradox, Taylor (1989) suggested:

> People with positive self-regard seem to achieve better social relationships. They have higher regard for others generally. For example, people with high self-esteem are generally better liked by others than people with low self-esteem. (p. 54)

Thus, in Taylor's view, harboring positive illusions about yourself induces you to care for yourself and for others, which pays off because caring for others induces them to care for you.

Recognizing the ultimate selfishness inherent in this process, Taylor searched for a purer example of caring and came up with parental nurturance. As stated by Lional Tiger (1979, p. 96) in *Optimism: The Biology of Hope:*

> When all is said and done, the act of being a parent involves a set of radically unselfish and often incomprehensibly inconvenient activities. Two adults who could otherwise employ their time and resources in pleasurable activities of various kinds elect to [indulge] completely dependent organisms.

But how do positive illusions about the *self* foster caring for one's offspring? Taylor (1989) suggested that positive illusions may mediate parental caring and sacrifice by fostering the belief that one is an effective parent and by fostering optimism about the future—the belief one's children will succeed. These illusions, however, relate more to the enhancement of *others* (i.e., one's children), than to self-enhancement. How do positive illusions about others enhance the fitness of those who harbor the illusions?

How Benevolence Toward Others Fosters Fitness: The Adaptive Functions of Positive Illusions About In-group Members

Evolutionary theory suggests four main mechanisms through which positive illusions about others enhance individuals' fitness: kin selection, sexual selection, group identity, and reciprocity.

Kin Selection.   The sacrifices that parents make for their children are easy
to explain in evolutionary terms. Children are the main vehicles for transporting
their parents' genes into future generations. Explaining why parents care for
their children is no different from explaining why people care for themselves.
Indeed, genetically speaking, parents are caring for themselves when they care
for their offspring.

Similarly, people can care for parts of themselves (i.e., copies of their genes)
by caring for other relatives. As Hamilton (1972) showed, the higher the
degree of relatedness between benefactor and recipient, the greater the genetic
payoff for self-sacrificial altruism. It follows that the favorableness of our biases
toward in-group members should, in general, vary with their degree of relatedness, qualified by perceived reproductive potential. In general this seems to
be the case. Indeed, the ability of parents to deny the faults of their children
sometimes seems to border on delusion. Witness, for example, the adamant
refusal of the mothers of mass murderers to acknowledge the guilt of their sons
in the face of extremely incriminating evidence.

The cognitive mechanisms that give rise to positive illusions about
in-group members are probably variations of the cognitive mechanisms that
engender positive illusions about the self. These mechanisms are probably
regulated by a gauge sensitive to the inclusive fitness-enhancing value of
favoring the recipient. In general, we should be most prone to harbor
favorable illusions about ourselves and our offspring, and, among our relatives, we should be prone to harbor the most positive illusions about those
to whom we are most closely related and about those we believe are most
likely to propagate our genes. Following this line of thought, we would expect
positive illusions about offspring to be stronger than positive illusions about
parents, even though both share 50% of our genes, and we might expect
positive illusions about ourselves and our relatives to decline with decreasing
reproductive potential.

One of the fortunate ironies of evolution is that genes cannot recognize
copies of themselves, a phenomenon that Dawkins (1976) called "the green
beard effect" (see also Krebs, 1989). Individuals must rely on cues to genetic
relatedness such as physical similarity or propinquity. This is fortunate
because it mediates the expansion of nepotism to those who possess cues to
relatedness (Krebs, 1987). A large body of research supports the idea that
we favor others who, in general, share characteristics our kin probably
possessed in our ancestral past, characteristics such as physical similarity,
proximity, and familiarity (Krebs, 1987). Inasmuch as these cues reliably
induce us to favor others who are more similar to us genetically than average,
they should mediate the selection of associated positive biases in person
perception. It is no coincidence that similarity, proximity, and familiarity
have been found to be the best predictors of interpersonal attraction (Berscheid, 1985).

Sexual Selection.   The only people we exalt more than our parents and our children are those we perceive as potential mates. Nowhere is the glorification of others more extreme than in people in love:

> She walks in beauty, like the night
>> Of cloudless climes and starry skies;
> And all that's best of dark and bright
>> Meet in her aspect and her eyes. (Byron, 1815/1994)

Viewed from an evolutionary perspective, we would expect such exaltation would vary with the perceived reproductive value of the mate, and that there would be sex differences in the characteristics that are valued. (As suggested in Byron's poem, males should value physical beauty more than females do; Buss, 1995). Males and females need one another to propagate their genes, so they have a vested interest in one another's welfare, at least until their offspring are born. Inasmuch as caring for potential mates increased the probability of our ancestors' reproductive success, adaptations mediating such dispositions would have been selected. Because of the helplessness of human infants and their need for parental care, relatively strong dispositions fostering pair-bonding appear to have been selected in the human species (Lovejoy, 1981). Illusions about the value of one's mate would be expected to support pair-bonding.

Group Identity.   The general principle underlying kin and sexual selection is that it benefits us to invest in those who carry our genes. It also may pay off to invest in those with whom we have formed what Heider (1958) called "unit relations." Inasmuch as other people perceive us in terms of our social identities and treat us as members of the groups to which we belong, it is in our interest to enhance the status of our groups. This does not involve group selection because we are, in effect, using the groups with which we are associated to foster our own fitness. When we enhance the value of our in-group, we, as a member of the group, may gain more than we have sacrificed on behalf of the group. In a similar vein, social psychologists have documented the tendency of people to associate with those of higher status and to "bask in their reflected glory" (Cialdini et al., 1976).

Reciprocity.   Finally, caring for others pays off because it induces others to repay the favor. Others may reward us in many ways for caring for them—by giving us gifts and other material resources, by mating with us, and, perhaps most importantly, by supporting our illusions about ourselves. Evolutionary psychologists join sociologists such as Gouldner (1960), social psychologists such as

Walster, Walster, and Berscheid (1978), and developmental psychologists such as Piaget (1932) in emphasizing the significance of reciprocity in social relationships. Cooperation pays off: Individuals who cooperate fare better than those who do not, and cooperative groups tend to fare better in intergroup competition than noncooperative groups (Alexander, 1987; Axelrod, 1984). Positive illusions that induce us to care for in-group members are adaptive because they encourage cooperation. The greater the potential for a member of our in-group to reward us, the more positive our illusions about him or her likely would be.

To summarize, we have evolved to view our relatives, our mates, and members of our in-groups in unrealistically glowing terms because such illusions give rise to behaviors that enhance our inclusive fitness by enhancing the fitness of those who carry copies of our genes, by inducing those we help to care for us, and by fostering fitness-enhancing cooperative relations.

SELF-DECEPTION THROUGH SOCIAL INTERACTION

The portrait of social relationships painted by this evolutionary brush is not a particularly pretty one. People are disposed to categorize others as in-group or out-group, and this categorization channels subsequent information processing. The cognitive processes activated when others are categorized as out-group members are relatively simple. They give rise to global, abstract, negative perceptions and characterizations (Maass et al., 1989; Maass et al., 1995) that tend to mediate guarded, competitive, exploitative, and aggressive behaviors. In contrast, the cognitive processes activated when others are categorized as in-group members are more complex and conditional. We discriminate among in-group members in terms of the mechanisms we have inherited that enabled our ancestors to enhance their fitness. In general, the greater the perceived potential of in-group members to enhance our fitness (in ways similar to those in ancestral environments), the more favorable our biases toward them.

Social relationships tend to be reciprocal; everything works also in reverse. Members of our in-groups invest in us when they believe it will benefit them. They evaluate us in much the same way they evaluate other resources such as stocks and bonds, in terms of our potential to pay back their investments with returns. As in the stock market, people's investment decisions are determined, not by the actual value of a resource but by the perceived value. It follows that it is in our interest to induce others to overestimate our value. In the currency of natural selection, what counts is how others treat you, not why they treat you that way. If manipulation and deceit enhanced our ancestors' fitness better than honesty, then adaptations fostering manipulation and deceit would have been selected, and some evolutionary theorists believe they actually were. For example, according to Alexander (1987), human society is a "network of lies and deception," although other theorists believe that the costs of excessive cheating would not have been worth the gains (Simon, 1990). We believe humans have inherited proclivities to anticipate the costs and benefits of cheating and to respond accordingly.

Impression Management: The Deception of Others

The sociologist Erving Goffman (1959) has, perhaps, captured the structure of human social relations more compellingly than any other theorist. In his book, *Presentation of Self in Everyday Life*, Goffman cast the world as a stage and the people in it as actors. Like actors, people play roles and manage the impressions they convey. They use makeup and costumes as props to disguise their true identities. In particular, as implied by Buss (1995), women attempt to represent themselves as more youthful and physically attractive than they are, and men attempt to present themselves as more resourceful. Face-lifts, tummy-tucks, fancy clothes, wigs, BMWs, Rolex watches, and other status symbols are the stock in the trade of self-presentation.

The evolutionary perspective directs us to ask what kinds of adaptations would have enhanced our ancestors' ability to deceive others about their worth. Interestingly, some evolutionary theorists have focused on what at first might seem to be an unlikely adaptation: self-deception. As Trivers (1985) explained:

> With powers to deceive and to spot deception being improved by natural selection, a new kind of deception may be favored: self-deception. Self-deception renders the deception being practiced unconscious to the practitioner, thereby hiding from other individuals the subtle signs of self-knowledge that may give away the deception being practiced. (p. 395)

Illusions about one's worth are adaptive because they help people deceive others about their worth: 'Sincere hypocrisy, with repressed awareness of guilt, will optimize individuals' inclusive fitness (Alexander, 1987, p. 123). Therefore, in addition to the adaptive consequences of positive biases in self-perception considered so far, positive illusions are adaptive because they help us fool others about our investment potential.

Self-Deception.   As pointed out by several scholars (Lockard & Paulhus, 1988), the idea that people deceive themselves is paradoxical. If the self-as-deceiver knows the truth, how can the self-as-deceived not know the truth? Self-deception is possible only if one assumes that the self can be split into self-as-deceiver and self-as-deceived, with the latter unaware of the former.

Krebs, Denton, and Higgins (1988) and Lockard and Mateer (1988) reviewed evidence demonstrating that the human brain is composed of mechanisms that can enable the kind of mental splitting necessary for self-deception. Lockard and Mateer (1988) reviewed neurological evidence showing that gating mechanisms in the brain may regulate the flow of information from unconscious to conscious domains. In addition, Lockard and Mateer (1988) described neurological disorders that appear to involve a splitting of awareness, with some parts of the brain knowing things that other parts do not know. These reviewers concluded that "different aspects of information are directed to different centers or systems and may or may not have access to portions of the brain important for awareness of that information..." (p. 36).

Research on self-perception (Bem, 1972) and social inference (Nisbett & Ross, 1980) indicates that people do not necessarily know why they do things; they are not necessarily aware of the causes of their behavior. As LeDoux, Wilson, and Gazzaniga (1979) stated; "It is as if the verbal self looks out and sees what the person is doing, and from that knowledge it interprets a reality" (p. 549). Mechanisms in one part of the brain (e.g., the left hemisphere) interpret behaviors produced by mechanisms in another part of the brain (e.g., the right hemisphere). LeDoux et al. (1979) suggested that the "verbal self" is the dominant level of consciousness (somewhat resembling the "tip of the iceberg"). These theorists asserted that people possess "multiple mental systems" capable of regulating behavior, and that these systems "are not necessarily conversant internally" (p. 550).

Whether one part of our brain deceives another part or not, believing we are better than we are affects the value that others place on us. Faced with a decision about how much to invest in us, other people read us for information about our value. The value we attach to ourselves—our level of confidence and self-esteem, our sense of deservingness—are important sources of information about our worth. If we do not believe in ourselves, who will? Positive illusions induce others to overvalue us. As Swann (1985) said, the self is an "architect of social reality." Individuals manipulate their social environments into verifying their conceptions of themselves by displaying "signs and symbols of who they are," seeking out "individuals whose appraisals confirm their self-views," and adopting "interaction strategies that elicit self-confirmatory reactions" (p. 105).

The Grand Conspiracy

Deception involves two parties: the sender and the receiver—the deceiver and the deceived. It is easy to understand why adaptations that enhance deception would foster the fitness of the deceiver, but should adaptations not also be selected that enhance the detection of deception? Should people not inherit deception-detecting mechanisms that prevent them from being beguiled? The evolutionary answer is this: only if it was in the genetic interest of our ancestors to detect the deception.

We believe humans inherit deception-detectors, which vary in sensitivity in accordance with the costs and benefits of detecting deception. When our interests conflict with those of others, we process information about them through cognitive channels designed for out-groups. The deception-detectors that evolved for this purpose seem very sensitive—indeed, overdetermined. We are prone to assume that those whose interests conflict with ours are up to no good. However, in general, it does not pay off to be as vigilant in detecting deception in those with whom we share interests; the costs outweigh the gains. Indeed, inasmuch as others' gains increase our own gains, it is in our interest to support their illusions. Goffman (1959) suggested that audiences actively collaborate in ensuring the success of the performances of the actors they observe, and recent social psychological research supports this idea. Investigators such as Denton and Zarbatany (in press), Swann (1987), and Snyder and Higgins (1988) found that people actively support the false images presented by their friends, exalting their successes and excusing their failures. Deception and self-deception work together hand in glove in in-group relationships. Self-deception enhances the deception of others, as Alexander (1987) and Trivers (1985) emphasized, and deception of others, in turn, enhances self-deception.

As argued earlier, supporting the positive illusions of in-group members may pay off in several ways, but none is more important than the benefit of reciprocity. People make implicit deals with their friends and allies: "You believe my false images, and I'll believe yours." "You're great; I'm great." Of course, greatness, like most social attributes, is a rather intangible quality, and the intangibility of valued social qualities is the fertile ground in which implicit conspiratorial social illusions thrive. I have an investment in the notion that I am great, so I present myself as greater than I actually am. Inasmuch as you share interests with me, you too have an investment in this misconception of my identity, so you support the inflated images I project. In the end, we deceive one another about my identity, and I end up believing my own lies. In return, I do the same for you. The product of this process is a benign folie a deux—a subtle and largely unconscious conspiracy between two (or more) people to support their misconceptions of themselves. Indeed, as suggested by Alexander (1987), as members of the human species, we all collaborate in positive illusions about human nature.

QUALIFICATIONS AND CONDITIONS

The thesis of this chapter is that humans are evolved to process social information in ways that deceive us about our worth and the worth of others, and it is in the interest of friends and relatives to collaborate with one another in supporting such deceptions. The brain mechanisms that give rise to biases in social cognition were selected because they enhanced the fitness of our ancestors in the social environments of their time. It is not unreasonable to assume that the social environments of today are similar to those of our ancestors in many, if not most, essential respects (e.g., in-groups and out-groups, males and females, young and old; relatives and nonrelatives), and that the biases in social cognition we inherit continue to serve similar adaptive functions. Although we undoubtedly are exposed to many more people than our ancestors were (e.g., through the media and modern modes of transportation), we still treat strangers with suspicion, categorize people as *us* or *them*, attend to status, compete for mates, favor our relatives, and form cooperative in-groups (e.g., unions, clubs, teams, and armies).

We do not attempt to present an empirically testable theory of social illusions in this chapter. Our goal has been to indicate how the types of biases in social cognition studied by social psychologists might look when observed through an evolutionary lens. To offer a testable theory, we would have to supply principles that explain the conditions in which social illusions fostered fitness in the social environments of our ancestors, and how such conditions selected the forms of information processing we invoke today. Clearly, it is sometimes adaptive to perceive oneself and others accurately, and it is sometimes more adaptive not to. Similarly, it is sometimes adaptive to be honest with others, and sometimes more adaptive to deceive them; it is sometimes adaptive to cooperate, sometimes to compete. If we examine the conditions that regulate the accuracy of perception in our current environment, we may be able to deduce the conditions in which the mental structures that mediate social perception evolved. The value of evolutionary theory is in keeping us mindful that the ultimate adjudicator of how we process social information is not accuracy or truth; it is the effects on the genes propagated by our ancestors in the environments in which they lived.

Our cognitive programs for categorizing others are flexible and sensitive to social conditions; they are not like fixed-action potentials evoked full blown by particular stimuli. For example, features such as proximity, similarity, and familiarity are not immutable or inextricably bound to particular individuals, and they are easily overridden. The principle that ultimately regulates our categorization of others is based in an implicit judgment about whether our interests will be advanced more by competing with them or by cooperating with them. Allies may become enemies and rivals may become friends. Indeed, the ways in which we view others may change from day to day, even minute to minute. The interesting thing about all of this is how radically our judgments about others change when they change. There is an inevitable tension in cooperative relations. When joining forces with others for mutual gain, people view them through their in-group lens. However, when the interests of in-group members conflict, such as when it is time to divide up the resources gained through cooperation, cooperators become competitors; in-groups become out-groups, and allies become enemies.

We have assumed relatively egalitarian social relationships throughout most of this chapter, but social relationships are often hierarchical. One of the main ways in which individuals foster illusions of security is to form myths about the power of their leaders, sometimes elevating them to the status of gods. The cognitive mechanisms underlying hero worship were probably derived from preadaptations mediating respect for parents, which probably became projected to conceptions of God.

Finally, it is important to repeat that the biases in social cognition that have evolved in the human species mediate only moderate distortions of reality. It would have been, and continues to be, adaptive to perceive reality accurately for many, if not for most, purposes. Furthermore, reality constrains biases in social perception. Extreme distortions put undo pressure on those called on to confirm them. It is too much to expect of your friends to view you as six feet tall when you are five foot one, but you can count on them to sustain the illusion that you are, at least, not terribly short.

The images we project are like promissory notes. We must live up to them, and this exerts a conservative influence on positive illusions. It is tragic to observe the decline of people who have been exalted far beyond what they are: movie stars who have passed their prime; athletes whose careers have ended through injury or scandal. Their "friends" abandon them in droves, and they lose confidence in themselves. Such disillusionment also occurs on a minor scale among friends and lovers. People project overly positive images of themselves to one another when they first meet, but as time goes by, they come to see one another more and more for what they really are. Reality imposes and interests diverge. Take marriage, for example. The illusions that helped people reach the altar have largely served their purpose after the children are born.

Social psychologists have identified two potentially conflicting goals of social cognition; self-enhancement, which is featured in this chapter, and self-verification, which is not. When people's self-concepts are positive, or when positive aspects of people's self-concepts are in question, self-enhancement and self-verification work in concert. However, according to Swann and his colleagues (Swann, 1983; Swann & Read, 1981), people with negative self-concepts sometimes process information in ways that verify their negative self-concepts rather than cultivate positive illusions. The question posed by evolutionary theory is how the cultivation of a negative self-concept or self-image could have enhanced the inclusive fitness of our ancestors.

One possibility is that believing one is better than one really is entails a responsibility to live up to one's ideal, especially in front of one's friends. Aronson (1992) and others found that people are more likely to engage in self-verification in the presence of those they expect to see again than in the presence of those they do not expect to see. Alternatively, self-handicapping, the disavowal of one's abilities (Arkin & Baumgardner, 1985), may lower expectations and standards, thereby increasing credit for success.

Does this all sound too cynical? Consistent with the theme of this chapter, perhaps we should collaborate on a more positive conclusion. As mentioned earlier, the process of self-deception may foster self-actualization. As with optimistic illusions, positive biases in self-conception may activate mechanisms that enable the biases to validate themselves. It is an interesting irony that positive illusions may be the most powerful of all inducements to socially desirable behaviors. Many, if not most, of the good things people do may, unconsciously, be directed toward the enhancement of their social images and the perpetuation of their myths about themselves, their relatives, and their friends. Moreover, friends may play an active role in this process. I present myself in the best possible light—not really what I am, but at least the way I might be, at my best. You validate my identity, which encourages me to work more toward achieving my ideal. We have collaborated in making me a better person, for the moment at least.

REFERENCES

Alexander, R. D. (1979). *Darwinism and human affairs.* Seattle, WA: University of Washington Press.

Alexander, R. D. (1987). *The biology of moral systems.* New York: Aldine de Gruyter.

Allen, V. L., & Wilder, D. A. (1979). Group categorization and attribution of belief similarity. *Small Group Behavior, 10,* 73–80.

Alloy, L. B., & Abramson, L. Y. (1979). Judgment of contingency in depressed and nondepressed students: Sadder but wiser? *Journal of Experimental Psychology: General, 108,* 441–485.

Anthony, T., Cooper, C., & Mullen, B. (1992). Cross-racial facial identification: A social cognitive integration. *Personality and Social Psychology Bulletin, 18,* 296–301.

Arkin, R. M., & Baumgardner, A. H. (1985). Self-handicapping. In J. H. Harvey & G. Weary (Eds.), Attribution: Basic issues and applications (pp. 169–202). New York: Academic Press.

Aronoff, M. S., & Lesse, S. (1983). Principles of psychotherapy. In B. Wolman (Ed.), *The therapist's handbook: Treatment methods of mental disorders* (pp. 50–64). New York: Van Nostrand Reinhold.

Aronson, E. (1992). The return of the repressed: Dissonance theory makes a comeback. *Psychological Inquiry, 3,* 303–311.

Axelrod, R. (1984). *The evolution of cooperation.* New York: Basic Books.

Bandura, A. (1977). *Social learning theory.* Englewood Cliffs, NJ: Prentice-Hall.

Barash, D. P. (1982). *Sociobiology and behavior* (2nd ed.). New York: Elsevier.

Baumeister, R. F. (1989). The optimal margin of illusion. *Journal of Social and Clinical Psychology, 8,* 176–189.

Bem, D. J. (1972). Self-perception theory. In L. Berkowitz (Ed.), *Advances in experimental social psychology* (Vol. 6, pp. 1–62). New York: Academic Press.

Berscheid, E. (1985). Interpersonal attraction. In G. Lindzey & A. Aronson (Eds.), *The handbook of social psychology* (3rd ed., Vol. 2, pp. 413–484). Reading MA: Addison-Wesley.

Bruner, J. S. (1957). On perceptual readiness. *Psychological Review, 64,* 123–152.

Burger, J. M. (1981). Motivational biases in the attribution of responsibility for an accident: A meta-analysis of the defensive-attribution hypothesis. *Psychological Bulletin, 90,* 496–512.

Buss, D. (1995). Evolutionary psychology: A new paradigm for psychological science. *Psychological Inquiry, 6*, 1–30.

Byron, G. G. (1994). In J. J. McGann (Ed.). *The new Oxford book of Romantic period verse.* Oxford: Oxford University Press. (Original work published 1815)

Carsrud, A. L., & Olm, K. W. (1986). The success of male and female entrepreneurs: A comparative analysis of the effects of multidimensional achievement motivation and personality traits. In R. Smilor & R. L. Kuhn (Eds.), *Managing take-off in fast growth companies* (pp. 147–161). New York: Praeger.

Chance, J. E., & Goldstein, A. G. (1981). Depth of processing in response to own- and other-race faces. *Personality and Social Psychology Bulletin, 7*, 475–480.

Cialdini, R. B., Borden, R. J., Thorne, A., Walker, M. R., Freeman, S., & Sloan, L. R. (1976). Basking in reflected glory: Three (football) field studies. *Journal of Personality and Social Psychology, 34*, 366–375.

Dawkins, R. (1976). *The selfish gene.* London: Oxford University Press.

Denton, K., & Zarbatany, L. (in press). Age differences in support processes in conversations between friends. *Child Development.*

Devine, P. G. (1989). Stereotypes and prejudice: Their automatic and controlled components. *Journal of Personality and Social Psychology, 56*, 5–18.

Diener, C. I., & Dweck, C. S. (1980). An analysis of learned helplessness: 2. The processing of success. *Journal of Personality and Social Psychology, 36*, 451–462.

Dweck, C. S., & Licht, B. G. (1980). Learned helplessness and intellectual achievement. In M. E. P. Seligman & J. Garber (Eds.), *Human helplessness: Theory and applications* (pp. 197–222). New York: Academic Press.

Epstein, S. (1990). Cognitive experiential self-theory. In L. A. Pervin (Ed.), *Handbook of personality: Theory and research* (pp. 165–192). New York: Guilford.

Felson, R. B. (1984). The effect of self-appraisals of ability on academic performance. *Journal of Personality and Social Psychology, 47*, 944–952.

Fiske, S. T., & Neuberg, S. L. (1990). A continuum of impression formation, from category-based to individuating processes: Influences of information and motivation on attention and interpretation. In M. Zanna (Ed.), *Advances in experimental social psychology* (Vol. 23, pp. 1–74). New York: Academic Press.

Frank, J. D. (1961). *Persuasion and healing: A comparative study of psychotherapy.* Baltimore, MD: Johns Hopkins University Press.

Freud, S. (1933). *New introductory lectures on psycho-analysis.* New York: Norton.

Furnham, A., & Gunter, B. (1984). Just world beliefs and attitudes towards the poor. *British Journal of Social Psychology, 23*, 265–269.

Ghiselin, M. T. (1974). *The economy of nature and the evolution of sex.* Berkeley, CA: University of California Press.

Goffman, E. (1959). *The presentation of self in everyday life.* New York: Anchor Books.

Goleman, D. (1985). *Vital lies, simple truths: The psychology of self-deception.* New York: Simon & Schuster.

Goleman, D. (1987). Who are you kidding? *Psychology Today, 21*(3), 24–30.

Gouldner, A. W. (1960). The norm of reciprocity: A preliminary statement. *American Sociological Review, 25*, 161–178.

Gruman, J. C., & Sloan, R. P. (1983). Disease as justice: Perceptions of the victims of physical illness. *Basic and Applied Social Psychology, 4*, 39–46.

Gudykunst, W. B. (1989). Culture and intergroup processes. In M. H. Bond (Ed.), *The cross-cultural challenge to social psychology* (pp. 165–181). Newbury Park, CA: Sage.

Hamilton, W. D. (1972). Altruism and related phenomena, mainly in the social insects. *Annual Review of Ecological Systems, 3,* 193–232.

Hastorf, A., & Cantril, H. (1954). They saw a game: A case study. *Journal of Abnormal and Social Psychology, 49,* 129–134.

Heider, F. (1958). *The psychology of interpersonal relations.* New York: Wiley.

Hepworth, J. T., & West, S. G. (1988). Lynchings and the economy: A time-series reanalysis of Hovland and Sears (1940). *Journal of Personality and Social Psychology, 55,* 239–247.

Hewstone, M., Bond, M. H., & Wan, K. (1983). Social facts and social attributions: The explanation of intergroup differences in Hong Kong. *Social Cognition, 2,* 142–157.

Hewstone, M., & Jaspars, J. M. F. (1982). Intergroup relations and attribution processes. In H. Tajfel (Ed.), *Social identify and intergroup relations* (pp. 99–133). Cambridge, England: Cambridge University Press.

Hovland, D. J., & Sears, R. R. (1940). Minor studies in aggression: VI. Correlation of lynchings with economic indices. *Journal of Psychology, 9,* 301–310.

Judd, C. M., & Park, B. (1988). Out-group homogeneity: Judgments of variability at the individual and group levels. *Journal of Personality and Social Psychology, 54,* 778–788.

Judd, C. M., Ryan, C. S., & Park, B. (1991). Accuracy in the judgment of in-group and out-group variability. *Journal of Personality and Social Psychology, 61,* 366–379.

Kandel, D. B. (1978). Similarity in real-life adolescent friendship pairs. *Journal of Personality and Social Psychology, 36,* 306–312.

Kelley, H. H., & Stahelski, A. J. (1970). The social interaction basis of cooperators' and competitors' beliefs about others. *Journal of Personality and Social Psychology, 16,* 66–91.

Kenrick, D. T., Sadalla, E. K., & Keefe, R. C. (in press). Evolutionary cognitive psychology: The missing heart of modern cognitive science. In C. Crawford & D. Krebs (Eds.), *Evolution and human behavior: Ideas; issues, and applications.* Mahwah, NJ: Lawrence Erlbaum Associates.

Kobassa, S. C. (1979). Stressful life events, personality, and health: An inquiry into hardiness. *Journal of Personality and Social Psychology, 37,* 1–11.

Krebs, D. L. (1987). The challenge of altruism in biology and psychology. In C. Crawford, M. Smith, & D. Krebs (Eds.), *Sociobiology and psychology: Ideas, issues and applications* (pp. 81–118). Hillsdale, NJ: Lawrence Erlbaum Associates.

Krebs, D. L. (1989). Detecting genetic similarity without detecting genetic similarity. *Behavioral and Brain Sciences, 12,* 533–534.

Krebs, D. L. (in preparation). The evolution of moral behaviors. In C. Crawford & D. Krebs (Eds.), *Evolution and human behavior: Ideas, issues, and applications.* Mahwah, NJ: Lawrence Erlbaum Associates.

Krebs, D. L., Denton, K., & Higgins, N. C. (1988). On the evolution of self-knowledge and self-deception. In K. B. MacDonald (Ed.), *Sociobiological perspectives on human development* (pp. 103–139). New York: Springer-Verlag.

Langer, E. J., & Imber, L. (1980). The role of mindlessness in the perception of deviance. *Journal of Personality and Social Psychology, 11,* 155–165.

Lau, R. R., & Russell, D. (1980). Attributions in the sports pages: A field test of some current hypotheses about attribution research. *Journal of Personality and Social Psychology, 39,* 29–38.

Leaky, R. E., & Lewin, R. (1977). *Origins.* New York: Dutton.

LeDoux, J. E., Wilson, D. H., & Gazzaniga, M. S. (1979). Beyond commissurotomy: Clues to consciousness. In M. S. Gazzaniga (Ed.), *Handbook of behavioral neurobiology* (Vol. 2, pp. 543–554). New York: Plenum.

Levine, R. A., & Campbell, D. T. (1972). *Ethnocentrism: Theories of conflict, ethnic attitudes, and group behavior.* New York: Wiley.

Lewin, R. (1993). *Human evolution: An illustrated introduction.* Cambridge, MA: Blackwell.

Linville, P. W., Fischer, G. W., & Salovey, P. (1989). Perceived distributions of the characteristics of in-group and out-group members: Empirical evidence and a computer simulation. *Journal of Personality and Social Psychology, 57,* 165–188.

Lockard, J. S., & Mateer, C. A. (1988). Neural bases of self-deception. In J. S. Lockard & D. L. Paulhus (Eds.), *Self-deception: An adaptive mechanism?* (pp. 23–39). Englewood Cliffs, NJ: Prentice-Hall.

Lockard, J. S., & Paulhus, D. L. (Eds.). (1988). *Self-deception: An adaptive mechanism?* Englewood Cliffs, NJ: Prentice-Hall.

Lovejoy, C. O. (1981). The origin of man. *Science, 211,* 341–350.

Maass, A., Milesi, A., Zabbini, S., & Stahlberg, D. (1995). Linguistic intergroup bias: Differential expectancies or in-group protection? *Journal of Personality and Social Psychology, 68,* 116–126.

Maass, A., Salvi, D., Arcuri, L., & Semin, G. (1989). Language use in intergroup contexts: The linguistic intergroup bias. *Journal of Personality and Social Psychology, 57,* 981–993.

Markus, H., & Nurius, P. (1986). Possible selves. *American Psychologist, 41,* 954–969.

Miller, D. T., & Ross, M. (1975). Self-serving biases in the attribution of causality : Fact or fiction? *Psychological Bulletin, 82,* 213–225.

Mullen, B., & Riordan, C. A. (1988). Self-serving attributions for performance in naturalistic settings: A meta-analytic review. *Journal of Applied Social Psychology, 18,* 3–22.

Nisbett, R. E., & Ross, L. (1980). *Human inference: Strategies and shortcomings of human judgment.* Englewood Cliffs, NJ: Prentice Hall.

Peterson, C., Seligman, M. E. P., & Vaillant, G. E. (1988). Pessimistic explanatory style is a risk factor for physical illness: A thirty-five-year longitudinal study. *Journal of Personality and Social Psychology, 55,* 23–27.

Pettigrew, T. F. (1979). The ultimate attribution error: Extending Allport's cognitive analysis of prejudice. *Personality and Social Psychology Bulletin, 5,* 461–476.

Piaget, J. (1932). *Moral judgment of the child.* London: Routledge and Kegan Paul.

Ross, M., & Fletcher, G. J. O. (1985). Attribution and social perception. In G. Lindzey & A. Aronson (Eds.), *The handbook of social psychology* (3rd ed., Vol. 2, pp. 73–122). Reading MA: Addison-Wesley.

Scheier, M. F., & Carver, C. S. (1985). Optimism, coping, and health: Assessment and implications of generalized outcome expectancies. *Health Psychology, 4,* 219–247.

Schlenker, B. R., & Miller, R. S. (1977). Egocentrism in groups: Self-serving biases or logical information processing? *Journal of Personality and Social Psychology, 35,* 755–764.

Secord, P. F. (1959). Stereotyping and favorableness in perception of Negro faces. *Journal of Abnormal and Social Psychology, 59*, 309–321.

Sherif, M., Harvey, O. J., White, J., Hood, W., & Sherif, C. (1961). *Intergroup conflict and cooperation: The robber's cave experiment.* Norman, OK: University of Oklahoma, Institute of Intergroup Relations.

Simon, H. H. (1990). A mechanism for social selection and successful altruism. *Science, 250*, 1665–1668.

Snyder, C. R., & Higgins, R. L. (1988). Excuses: Their effective role in the negotiation of reality. *Psychological Bulletin, 104*, 23–35.

Snyder, M. (1984). When belief creates reality. In L. Berkowitz (Ed.), *Advances in experimental social psychology* (Vol. 18, pp. 247–305). Orlando, FL: Academic Press.

Swann, W. B. (1983). Self-verification: Bringing social reality into harmony with the self. In J. Suls & A. G. Greenwald (Eds.), *Psychological perspectives on the self* (Vol. 2, pp. 33–66). Hillsdale, NJ: Lawrence Earlbaum Associates.

Swann, W. B. (1985). The self as architect of social reality. In B. Schlenker (Ed.), *The self and social life* (pp. 100–125). New York: McGraw-Hill.

Swann, W. B. (1987). Identify negotiation: Where two roads meet. *Journal of Personality and Social Psychology, 53*, 1038–1051.

Swann, W. B., & Read, S. J. (1981). Self-verification processes: How we sustain our self-conceptions. *Journal of Experimental Social Psychology, 17*, 351–370.

Sweeney, P. D., Anderson, K., & Bailey, S. (1986). Attributional style in depression: A meta-analytic review. *Journal of Personality and Social Psychology, 50*, 974–991.

Tajfel, H. (1982). *Social identify and intergroup relations.* Cambridge, England: Cambridge University Press.

Tajfel, H., & Billing, M. (1974). Familiarity and categorization in intergroup behavior. *Journal of Experimental Social Psychology, 10*, 159–170.

Tajfel, H., & Turner, J. C. (1979). An integrative theory of social contact. In W. Austin & S. Worchel (Eds.), *The social psychology of intergroup relations* (pp. 33–47). Monterey, CA: Brooks/Cole.

Taylor, S. E. (1989). *Positive illusions: Creative self-deception and the healthy mind.* New York: Basic Books.

Taylor, S. E., & Brown, J. D. (1988). Illusion and well-being: A social psychological perspective on mental health. *Psychological Bulletin, 103*, 193–210.

Tennyson, A. (1993). In G. F. Maine (Ed.), *Poems of Alfred, Lord Tennyson.* London: Collins.

Tiger, L. (1979). *Optimism: The biology of hope.* New York: Simon & Schuster.

Tooby, J., & Devore, I. (1987). The reconstruction of hominid behavioral evolution through strategic modeling. In W. G. Kinzey (Ed.), *The evolution of human behavior: Primate models* (pp. 183–237). Albany, NY: State University of New York Press.

Trivers, R. (1985). *Social evolution.* Menlo Park, CA: Benjamin/Cummings.

Trotter, R. J. (1987). Stop blaming yourself. *Psychology Today, 21*(2), 30–39.

Van den Daele, L. (1968). A developmental study of ego-ideals. *Genetic Monographs, 78*, 191–256.

Vine, I. (1992). Altruism and human nature: Resolving the evolutionary paradox. In P. M. Oliner, S. P. Oliner, L. Baron, L. A. Blum, D. L. Krebs, & M. Z. Smolenska (Eds.), *Embracing the other: Philosophical, psychological, and historical perspectives on altruism* (pp. 73–103). New York: New York University Press.

Wagstaff, G. (1982). Attitudes to rape: The "just world" strikes again? *Bulletin of the British Psychological Society, 35,* 277–279.

Walster, E., Walster, G. W., & Berscheid, E. (1978). *Equity: Theory and research.* Boston, MA: Allyn & Bacon.

Wilder, D. A. (1981). Perceiving persons as a group: Categorization and intergroup relations. In D. L. Hamilton (Ed.), *Cognitive processes in stereotyping and intergroup behavior* (pp. 213–258). Hillsdale, NJ: Lawrence Erlbaum Associates.

Winkler, J., & Taylor, S. E. (1979). Preference, expectations, and attributional bias: Two field studies. *Journal of Applied Social Psychology, 2,* 183–197.

3

Rethinking the Role of Evolution in an Ecological Model of Social Perception

Ken Springer
Diane S. Berry
Southern Methodist University

The concept of natural selection is promiscuous. Although its status as the central dogma in evolutionary biology may be due primarily to its efficacy in accounting for variation, part of its appeal lies in the fact that it seems to work so well in accounting for an endless variety of phenomena. Even when our understanding of the phenomena changes, we seem to have a knack for inventing new explanations that stem from the same set of assumptions about selection pressures. In this chapter we do not comment on the validity of natural selection as an explanatory concept in general. Instead, we are concerned with its application, as filtered through sociobiology, to a specific type of research—ecological approaches to social perception with particular attention to their predictions regarding universality and accuracy.

This approach to social perception stems from a recent alternative to traditional theories of perception, which assume that the perceptual stimuli initially available to perceivers are raw, fleeting, and essentially meaningless sensations. Through either successive or parallel stages of processing, the brain eventually produces a meaningful construction or interpretation of the stimuli. Thus, the familiar perceptual world around us is the endpoint of a set of inferential processes computed on raw sensations. In contrast, James Gibson's ecological approach to perception holds that meaningful information about objects and movement is directly and unambiguously given in invariant stimulus information (e.g., Gibson, 1979).[1] Perception is essentially the detection of *invariants*, or lawful patterns of structure or change. This is a realist view: Perception is not a matter of interpretation, but rather of detecting what is there.

---

[1] Although we focus on visual information here, the Gibsonian perspective assumes that most information is "amodal" in nature, and potentially detectable through multiple channels.

In addition to invariants, the concepts of affordance and attunement are critical components of the Gibsonian view. Invariants reveal affordances, or opportunities for action, that the environment provides to perceivers. Rather than merely detecting properties such as the flatness, density, or weight of an object, we directly perceive how these properties might allow us to interact with that object. For example, a blade of grass does not afford climbing for humans, but a good sized tree might. Gibson argued that affordances such as "climbability" are indeed what we extract from invariants—that "the 'values' and 'meanings' of things in the environment can be directly perceived" (Gibson, 1979, p. 27).

The concept of affordance involves characteristics of both the perceiver and the environment. For example, although a blade of grass does not afford climbing for a person, it well may for an ant. As this suggests, affordances do not reside purely in either the head of the perceiver or in the environment but involve the particular compatibility between perceiver and perceived (e.g., Gibson, 1966, 1979). This leads us to the concept of attunement, which involves the sensitivity of a given perceiver to a particular affordance. Attunements may be prewired, as in the example of dogs relying more on olfactory information than humans do owing to their more sophisticated chemoreceptors. Alternatively, attunements may reflect a match between perceiver and affordance that results from perceptual learning, as in the example of a person's growing sensitivity to the orthography of a first or second language. Gibson's own research focused on identifying the stimulus invariants that reveal what physical objects afford. However, he clearly recognized that these concepts are applicable to the social environment (cf. Gibson, 1979). In recent years, a number of researchers have begun to apply this approach successfully to the study of social psychological phenomena (e.g., Baron & Boudreau, 1987; Berry, 1991, 1994; Berry & Misovich, 1994; McArthur & Baron, 1983; Zebrowitz, 1990).

THE ADAPTIVE FUNCTION OF PERCEPTION

The ecological approach assumes that perception serves an adaptive function. Moreover, perception is thought to play a critical role in the survival of both species and individual (Gibson, 1979), as well as in the attainment of personal goals (Baron & Boudreau, 1987; McArthur and Baron, 1983). This is not to say that nonecological theories deny adaptive value to perception. It is extremely useful to be able to spot predators or avoid driving off cliffs, regardless of what mechanisms one believes underlie these abilities. What distinguishes the ecological approach from nonecological or "indirect" theories of perception are the particular assumptions about evolutionary pressures that it embodies. The nonecological (e.g., information processing) theorist would say that natural selection has favored the evolution of neural structures that produce adaptive interpretations of meaningless stimuli. In contrast, the ecological theorist would make the realist assumption that natural selection has favored the evolution of perceptual systems attuned to meaningful stimulus information so that adaptive responses can be made. We will see that this difference between ecological and nonecological approaches manifests in more specific assumptions about evolution as well. For example, ecologically based research has distinctive goals that include identifying specific stimulus information to which it would be adaptive to attend (e.g., Berry 1994; Cunningham, 1986; Singh, 1993) and demonstrating that people are most accurate in picking up information from stimulus dimensions of greatest adaptive value (e.g., Gangestad, Simpson, DiGeronimo, & Biek, 1992). The main purpose of this chapter is to evaluate whether ecological studies of social perception have made proper use of these and other assumptions derivable from evolutionary theories.

Among biologists, the term *adaptive* can be used to convey one or more of several related but independent meanings. The following are examples:

1. A characteristic is adaptive insofar as it helps the individual (or some larger group) adjust to its environment.
2. A characteristic is adaptive insofar as it increases the likelihood that the individual (or some larger group) will contribute to the current gene pool.
3. A characteristic is adaptive if it has been, or will be, maintained across generations through natural selection.[2]

This is a partial list, and the interrelations among the meanings are complex. Some issues, such as the role of environmental influences in the expression of adaptive abilities, do not receive much attention here. Other issues such as degrees of adaptiveness are taken up later. The critical difference among these meanings, for our purposes, is that only (3) necessarily implicates evolutionary processes.[3] This link between adaptation and natural selection receives the most attention here, because it is the aspect of evolutionary theorizing most commonly invoked in the social perception literature.

What does ecological theory have to say about the role of evolution? According to Gibson (1979), some of the affordances in the environment have been unchanged throughout our evolutionary history, and our perceptual

---

[2] The concept of natural selection assumes that (a) individuals in a species compete for resources, (b) variations in the characteristics of these individuals are heritable, (c) the characteristics of certain individuals make them especially likely to survive, and (d) these characteristics are thus more likely to be passed on to offspring and increase the individual's contribution to the future gene pool. In this way, the species gradually becomes better adapted to its environment. We are assuming here, as many do, that adaptation and natural selection are intimately related, although in certain cases it is possible for one to exist without the other.

[3] Researchers sometimes use the term *adaptation* in the very straightforward sense of an individual becoming better adjusted to his or her environment (meaning 1). In such cases it is either not clear whether an evolutionary process is intended, or clear that it is not intended. *Adaptation* is also used occasionally to implicate meaning 2 but not selection processes (e.g., cases in which an individual acquires an advantage that will not be genetically transmitted).

systems have become attuned to these invariants. The kinds of affordances Gibson focused on in this regard are mainly regularities in the physical environment, such as features of the terrain, the medium of air, and qualities of substances such as water and fire. However, he also noted the "rich and complex set of interactions, sexual, predatory, nurturing, fighting, playing, cooperating, and communicating" afforded by other human beings (Gibson, 1979, p. 128). As people have always been primarily social creatures, it follows that Gibson's commitment to the environment at large as an evolutionary influence also applies to human interaction in particular. In the Gibsonian view, we evolved not only in a physical environment but also a social one, and presumably we have adapted to fit both environments. This biologically based attunement is not the only type that occurs. Our perceptual systems also allow us to adjust to our surroundings through perceptual learning or the "education of attention" so that we can extract invariants of greatest adaptive value (Gibson, 1966). In Gibson's view, not only are certain adaptations prewired; the ability to adapt to certain kinds of affordances are prewired as well. The difference between these types of attunements are important to several of our points here.

Gibson was not very specific about the role of evolution in either the physical or the social environment. He helped formulate a distinct approach to perception as a means of adaptation, but most of his discussions emphasized the fit between perceiver and environment (or perceiver and "niche"). He had a lot to say in terms of meaning 1 but not 2 or 3. It has been Gibson's colleagues and students primarily who brought specific evolutionary considerations to bear upon the study of social perception and addressed these latter senses of adaptation.

Successfully applying evolutionary theories to the study of perception in general, or social perception in particular, requires theoretical statements that generate testable hypotheses. However, there are several ways in which the link between theory and experimental design may break down. A statement may make a great deal of sense but not be testable. For example, it seems reasonable to suppose that it is more efficient to perceive meaningful information directly than to construct it, and that natural selection should therefore favor direct perception over its indirect varieties (Shaw & Bransford, 1977). This sounds eminently plausible. At the same time, it is essential to this argument that, once upon a time, there were people who had other sorts of perceptual systems (e.g., computational ones), but who were gradually removed from the gene pool. At the moment this is an untestable idea.

A different problem arises when a theoretical statement can be readily tested, but on closer scrutiny turns out to have been formulated improperly. This is our focus. In the next two sections we discuss two types of hypotheses that were tested by way of trying to demonstrate, among other things, that social perception is adaptive (primarily in the sense that it has been subject to selection pressures—meaning 3 given earlier). The hypotheses in question are that if social perception has been molded by natural selection, then we should find evidence for both *universals* and *accuracy* in social perception. We will have more to say about various meanings of *universals* and *accuracy*. Because accuracy has received more explicit attention in the various literatures, we give it greater attention here. We first summarize the universals and accuracy approaches. In later sections we then argue that the formulation of both hypotheses has frequently been too simplistic.

THE SEARCH FOR UNIVERSALS

It is often assumed in the social perception literature that if it is adaptive to perceive certain affordances, then the ways in which people detect and respond to these affordances should be universal, owing to selection pressures. The assumption is not that people would be universally accurate, but rather that there would be universal consensus. In other words, in a given experiment there would be a normative response, regardless of its accuracy (see Funder & West, 1993, for a discussion of the differences between consensus and accuracy). Perceptual illusions would illustrate the extreme case in which perception is clearly both universal *and* inaccurate. Although universals may be clearly accurate in many cases, they will sometimes reflect only partial accuracy. For example, it appears that males unconsciously take certain facial features as indicators of fertility (e.g., Cunningham, 1986), but if even this is a universal perception, it could be accurate only to some degree of probability. Finally, it should be noted that although universals are by definition populationwide, the universals approach can be used to link consensus within large subpopulations to evolutionary pressures (e.g., sex differences in preferences for mates and sexual partners; cf. Buss & Schmitt, 1993; Kenrick, 1994).

Two brief examples of the universals approach are as follows. First, there is striking consensus both within and across cultures that individuals with immature "babyish" faces must be relatively helpless and dependent (Berry, 1994; Berry & McArthur, 1986; McArthur & Berry, 1987). Given that a tendency to perceive helplessness in configurations of neotenous features would be extremely adaptive for raising our young (Lorenz, 1943), it has been argued that natural selection has promoted this tendency. Second, the fact that adults within and across cultures tend to converge in judgments of attractiveness (e.g., Cunningham, Roberts, Barbee, Druen, & Wu, 1995; McArthur & Berry, 1987) has been accounted for in terms of the reproductive advantages accruing from the ability to detect outliers, or individuals who diverge from population means on certain characteristics that may be related to reproductive fitness, such as physical health.

Some tension exists between the universalist assumptions reflected in these studies and pure Gibsonian principles. Gibson (1979) allowed for both nonuniversal and inaccurate perception, but in his theoretical framework universal aspects of perception are always accurate. This is a consequence of Gibsonian realism: We should be attuned biologically to affordances that have been useful to us throughout our evolutionary history, and they can only be useful insofar as we usually pick them up accurately ("usually" because it is possible to be deceived or distracted by other affordances). There are in essence, then, two types of universalist approaches: the type that most researchers assume, which holds that universals arose through selection pressures, and may or may not be accurate, and the Gibsonian type, which holds that universals arose through selection pressures and must be accurate (assuming no deception or distraction). The first type is actually not distinctive to ecological theorists; any theorist could say that a given universal is the result of selection pressures. But it is important to note that an ecological theorist is more likely than a nonecological one to embrace the second type, because if perception is a matter of interpreting raw stimuli, then wrong interpretations are possible.

THE SEARCH FOR ACCURACY

The topic of accuracy and its implications for evolutionary theories has received much more explicit attention in the literature than discussions of universals. In general, the question of our degree of accuracy in person perception is a fairly old one. The question first emerged in the 1930s, was later discredited, and has recently reemerged as an alternative to the mainstream cognitive approaches that emphasize inaccuracies (Funder, 1987). It is fair to say that accuracy is more likely to be predicted from ecological theories, in which the perceiver is attuned to his or her environment, than from nonecological theories, in which the perceiver is an interpreter of impoverished stimuli, and thus fallible. Indeed, the emphasis on error and bias in the social psychology of the 1970s and 1980s is a consequence of the social cognitive viewpoint (Berry & Finch Wero, 1993).

Several senses of accuracy have been explored in the literature, including what Funder termed *consensus, self–other agreement,* and *accuracy* (Funder, 1987; Funder & West, 1993). *Consensus* refers to interobserver agreement about a target, which, as the number of observers approaches the population value, can be equated with universality. We do not address consensus here, because, as Funder points out, it does not entail accuracy. *Self–other agreement* refers to similarity in judgments made about a target person by both the target (self-judgment) and an observer who may or may not know the target (other judgment). In the literature, this is sometimes used as a criterion for accuracy. However, as Funder pointed out, *accuracy* refers to a true statement about a person, whereas "truth" is evaluated by some independent standard.

Researchers in social perception have sometimes linked accuracy—a term we will use as shorthand here for either self–other agreement or accuracy—to evolutionary processes. This type of argument has taken two forms.

First, it has been assumed that accuracy directly results from selection pressures. For example, consider the body of findings suggesting that in some cases people tend to be accurate when making certain kinds of trait attributions based on first impressions of appearance (e.g., Albright, Kenny, & Malloy, 1988; Berry, 1991; Berry, 1990; Berry & Finch Wero, 1993; Borkenau & Liebler, 1992). In general terms, this line of work demonstrated important convergences among (a) an individual's appearance, (b) behavioral propensities attributed to the individual by others on the basis of first impressions of the individual's appearance, and (c) behavioral propensities that the individual either claims to possess (e.g., Gangestad et al., 1992) or actually seems to possess according to some independent criterion (e.g., Bond, Berry, & Omar, 1994). The Gibsonian and evolutionary interpretation of such findings is that invariants specifying dispositional properties are given in perceptual stimulus information, and natural selection has favored perceivers who can successfully extract these invariants. The ability to do so is, in a word, highly adaptive.

A second and more specific way that accuracy has been linked to evolutionary theories is in the assumption that selection pressures would favor accuracy for dimensions that are more, as opposed to less, relevant to survival in a given context (e.g., Berry & Finch Wero, 1993; Gangestad et al., 1992). (Researchers who accept this assumption also by definition embrace the assumption that accuracy is causally related to natural selection.) For example, Gangestad et al. (1992) tested the prediction that accuracy pertaining to sociosexuality, defined as one's degree of "willingness to engage in sexual relations without closeness or commitment" (p. 690), should be greater than accuracy for dimensions such as social closeness. The data, about which we have more to say later, generally supported this prediction.

Finally, a comment on the independence of universality and accuracy. We have already pointed out that universality does not entail accuracy because individuals can be universally inaccurate, or partially accurate. However, if accuracy is supposed to obtain for dimensions of adaptive value, it might seem to follow that accuracy entails universality. In other words, wherever selection pressures favor accuracy, they should do so for virtually the entire population. This idea is discredited by a variety of empirical findings. For example, in Gangestad et al. (1992), males were more accurate than females in terms of attributing sociosexuality based on visual appearance, and a plausible explanation for this in terms of selection pressures has been proposed. This illustrates a case in which accuracy is especially pronounced in one group (males) but not universally. In sum, universality and accuracy may often go hand in hand, but, in principle, they must be decoupled. We turn now to a critical evaluation of the universals and accuracy approaches.

BEYOND UNIVERSALS

Once again, the universalist assumption is that universals in social perception reflect processes of natural selection. This assumption is problematic on several different levels. First, it is a misconception to assume that natural selection necessarily promotes universality (Kenrick, 1994). Selection pressures rarely produce homogeneity. In many cases, natural selection actually favors diversity, as in the example of clines, or graded variations in species linked to geographic variance (Curtis & Barnes, 1981). The implication of this point is not that the search for universals may be misguided, but rather that the assumption that universals reflect the outcome of evolutionary pressures is difficult to falsify. A given universal in social perception may well have been selected for. However, the existence of individual differences is not prima facie evidence that it has *not* been selected for.

This point has direct bearing on interpretations of the social perception data. For example, consider again the finding that individuals with babyish faces are widely considered to be relatively helpless and dependent, along with the interpretation that this is relevant to a natural tendency to nurture babies. Had it been discovered instead that a substantial proportion of people do not attribute helplessness and dependency to babyfaced individuals, it could still have been argued that evolutionary forces were at play. Specifically, it could have been argued that it is advantageous for some members of a species to be disinclined to nurturing behaviors, so that they can focus their energies on kinship- or species-sustaining behaviors such as hunting or domestic chores. Exactly this sort of argument has been proposed to account for the persistence of nonreproducing individuals in various societies, such as worker castes in insect species (Wilson, 1980).

Likewise, consider again the finding that adults and infants across cultures seem to converge on judgments of attractiveness, perhaps due to a tendency to distinguish outliers on physical dimensions. Had it been found that judgments of attractiveness are somewhat divergent, it could have been argued that this too would be advantageous, serving to reduce intraspecies conflict over sexual partners. "To each his or her own" would approximate the guiding principle. There would still be social dominance hierarchies and intrasex competition over potential mates, but the amount of energy wasted on conflict over mate selection would be lessened.

The point here is not that we need to take such arguments seriously. The point is rather that arguments can be made to fit virtually any pattern of data. This in turn tells us that the hypotheses put forth in social perception experiments have not always been falsifiable in terms of the role of evolution. As Karl Popper (1972) emphasized, however, a valid hypothesis must be falsifiable. This is a major—perhaps *the* major—problem in the application of evolutionary theorizing to the study of social perception, and it is not limited to discussion of universals. In fact, it is not limited to psychologists; even evolutionary biologists occasionally show a tendency to invoke natural selection merely on the grounds that it is consistent with a pattern of findings.

A second problem with the universals approach is that ecological theory predicts individual differences too. One of the fundamental assumptions of the ecological approach is that some of what the environment affords will differ from person to person, and perception of these affordances by a given individual will be selective (Gibson, 1979). It follows that there will be individual differences in social perception corresponding to differences in personal goals, a point that McArthur and Baron (1983) have discussed at length, but which has not always been followed up in the literature. For example, McArthur and Baron (1983) point out the following:

> Different information is essential to the behavioral goals of different people. What low-status people need to perceive in order to interact effectively with their environment may often be different from what high-status people need to perceive....Thus, the ecological approach suggests that we may discover much of interest in the domain of individual differences in social perception if we begin our investigations with a careful analysis of what it is that various individuals most need to perceive in order to interact effectively with their social environment. (pp. 219–220)

McArthur and Baron also say that affordances "are inherently specific to a particular perceiver" (p. 218), but this may be pushing a good point too far, because it implies that there are no affordances of universal importance. Although many affordances will be specific to individuals, others will have been invariant across the whole of our evolutionary history (Gibson, 1979). What this suggests is that if we want to understand the role of selection pressures in social perception we need to search for both universals and individual differences.[4]

A final problem with the universals approach arises from empirical qualifications to some of the more limited formulations. For example, since Darwin it has been widely agreed that recognizing emotional expressions is adaptive, and we believe now that, in some respects, the ability is universal (e.g., Ekman, 1982). However, there is also evidence for individual differences in sensitivity to emotion. For example, individuals raised in families who are low in expressiveness are much more sensitive to subtle nonverbal indicators of emotion than are people from highly expressive families (Halberstadt, 1986). Switching to the example of neonate abilities, it appears that although infants are prewired to discriminate among basic emotional expressions (Field, Woodson, Greenberg, & Cohen, 1982), this basic ability is modifiable through contingencies of reinforcement (e.g., Campos, Caplovitz, Lamb, Goldsmith, & Stenberg 1983).

---

[4]Of course, we need criteria for doing so because, in one sense, universals and individual differences exhaust all possible empirical outcomes. As McArthur and Baron might say, we would need to analyze carefully exactly what should be expected on the basis of evolutionary considerations. This will be difficult, though, given the idea, noted earlier, that natural selection can be shaped to fit almost any pattern of data.

In Gibsonian terms, such findings reflect perceptual learning, and they show us that it is not enough to denote recognition of basic emotions as a universal and, therefore, adaptive ability. The ability to become attuned to highly subtle emotional displays, or to detect associations between certain displays and consequences, is also adaptive; it seems just as likely a priori that these also have been subject to selection pressure. The point is that our search for universals needs to be sensitive to the interplay between biological and learned attune-ments (see Buck, 1988, Buss & Schmitt, 1993, and others for good examples). We have more to say about this in the final section of the chapter.

We present two addenda on the topic of universality. First, researchers have typically looked for universals within or across cultures. This general approach would benefit by comparisons of adults and infants (e.g., Langlois, Roggman, & Rieser-Danner, 1991). In an ecological framework, convergence between adults and children per se may be suggestive with respect to evolution, but it will not be conclusive because perceptual learning is a reasonable alternative (Springer, Berry, & Meier, 1995). However, convergence between adults and infants strongly implicates prewiring, because infants presumably will not have had sufficient time or opportunity to become appropriately socialized. This would be illustrated by the line of reasoning that because infants and adults converge in judgments of attractiveness (e.g., Langlois et al., 1990), the underlying basis for the judgment must be innate (e.g., Berry & Finch Wero, 1993).

The second addendum is, an empirical suggestion. Data will always be noisy, but in the search for universals we will be especially motivated to ignore the noise. Yet this may be exactly where we need to look. We may need to ask the following question: Who are the outliers? If we assume that a given ability is adaptive (i.e., developed under selection pressures), we might predict that in some instances the few individuals who do not conform to a standard pattern should be less likely to contribute to the gene pool. For example, if accuracy with respect to sociosexuality is particularly adaptive for males (Gangestad et al., 1992), then males who are particularly inaccurate on this dimension might be expected to have difficulties finding female romantic partners. In contrast, if we find that a substantial proportion of these males are actually quite successful with the opposite sex, we might have to either reexamine the idea that we are dealing with an evolutionarily constrained ability, or argue that alternative strategies have been selected for. When we can be reasonably sure that noise in the data is not due to the usual culprits (fatigue, failure to cooperate, etc.), then there may be a means for determining whether or not natural selection has played a role.

BEYOND ACCURACY

We suggested earlier that there are two types of accuracy assumptions: The general assumption that accuracy is produced by selection pressures, and, in some cases, the additional assumption that selection favors accuracy for the most adaptive dimensions in a given context. We now take these up in reverse order.

As we have said, a common strategy in the ecological literature is to argue that X should be more readily perceived than Y because perceiving X is more adaptive. This idea originates in Gibson's (1979) assertion that perception is selective: Particularly in social interactions, which are highly complex, we attend to the affordances of greatest personal relevance. The main problem with this idea is that solid criteria for ranking adaptiveness are either not available or debatable. An example in which criteria are not given is McArthur and Baron's (1983) contention that perceiving fear and anger should be accomplished most readily because they are most adaptive.[5] This valuation is plausible but debatable. Among humans, an inability to perceive happiness and disgust is arguably just as nonadaptive as the inability to perceive fear and anger (although not necessarily for the same reasons). Moreover, some evidence suggests that sensitivity to both anger and happiness is prewired (e.g., Field et al., 1982). In any case, the point cannot be settled in the absence of criteria for degrees of adaptiveness.

An example in which criteria for degrees of adaptiveness are provided but may be debatable is the argument in Gangestad et al. (1992) that accuracy with respect to sociosexuality is more adaptive than accuracy pertaining to dimensions such as social closeness. The proposal of Gangestad et al. has a lot to commend it. It is plausible; some of the differences in accuracy that it predicts are confirmed in empirical work; and, most important, its predictions follow from the assumption that accuracy will be greatest for dimensions relevant to reproductive outcome in certain contexts. This is a sophisticated application of natural selection, more sophisticated than the sort of argument sometimes advanced that accuracy with respect to X is "just more adaptive" than accuracy about Y, on the grounds that recognizing X is generally more likely to promote individual survival.

However, a possible problem with linking degree of adaptiveness to reproductive outcome is that people are accurate with respect to traits such as conscientiousness and extroversion (Borkenau & Liebler, 1992), which do not seem relevant to reproductive outcome. One must either argue that accuracy in such cases is linked to reproduction in some nonobvious way, or to develop the view that accuracy may be produced by mechanisms other than natural selection.

---

[5] In support of their argument McArthur and Baron point out that English contains more words for negative emotions than for positive ones. Whether this should count as evidence is debatable. In many cultures there are only two words for color (often something such as *dark* and *light*), but the individuals in these cultures have no difficulty perceiving colors such as red, and perceiving red happens to be highly adaptive when it comes to questions such as "Is this particular splotch of color blood or not?"

We turn now to the more general assumption that accuracy results from selection pressures (i.e., that when we find accuracy, we can infer a process of natural selection that favors it). The main difficulty in using accuracy as an indicator of evolutionary pressures is the falsifiability issue raised in our discussion of universals. Accuracy is consistent with selection pressures, but lack of accuracy need not rule them out. For example, consider the idea that successfully recognizing deception is highly adaptive (at least in most cases, if not always; Bond, Kahler, & Paolicelli, 1985), and the resulting hypothesis that people will be fairly good at identifying deception. This seems plausible. However, the data suggest that humans are not very good at lie detection, despite the fact that there are reliable verbal and nonverbal qualities that indeed distinguish truthful from nontruthful communication attempts (e.g., Berry & Pennebaker, 1996; DePaulo & Rosenthal, 1979; Zuckerman, DePaulo, & Rosenthal, 1981). These data may at first seem to contradict the predictions of an evolutionary approach. However, one might also convincingly argue that because it is adaptive to engage in successful deceptive strategies, poor lie detection skills on the part of perceivers are therefore to be expected. Later, we argue that this is too simplistic. The point is simply that when we consider whether humans can recognize deception, almost any pattern of data could be taken in support of selection pressures.

The issue of deception introduces a second problem with the accuracy argument that we also raised in our discussion of universals: Accuracy is not always favored by natural selection. This is starkly illustrated in cases of predators encountering deceptive strategies. Among animals, there are countless examples of prey that have developed deceptive strategies such as camouflaging to fool predators, and many of these strategies are believed to have emerged through natural selection (Wilson, 1980). Predators are sometimes fooled by these strategies, even though natural selection has presumably favored those with outstanding hunting skills, including the ability to see through deception. What seems to be critical here is that predators are fooled only some of the time. If they were fooled all of the time, they might starve. But if they were completely accurate, they might starve too, owing to excessive depletion of the food supply. However, this is not to say that accuracy itself would be attenuated by depletion of the food supply. As predators become increasingly accurate, prey may evolve new strategies for fooling them, making it necessary for the predators to catch up. For the sake of the survival of both species, it may be essential that the new strategies not be completely successful, at least not for an extended period of time. A similar argument can be made for intraspecies deception: it is more adaptive for deception to be successful sometimes as opposed to never, to guarantee the survival of the deceivers.

In summary, it would be inaccurate and oversimplified to say that natural selection favors accuracy in the case of deception. This means that psychological findings pertaining to deception will not bear on the role of evolution if we simply ask whether individuals are accurate or not. Once again, it might be fruitful to consider outliers. Natural selection rarely guarantees homogeneity in a population, so we would expect the existence of some individuals who are particularly bad at detecting lies. It might be predicted that these individuals will be less likely than the majority of the population to contribute to the gene pool, although this is not an absolutely necessary outcome.

A third problem with linking accuracy in social perception to selection pressures stems from the ecological assumption that we attend to the most relevant or adaptive information. As we noted in the introduction, Gibsonian realism holds that we usually detect relevant affordances accurately, and the attendant evolutionary assumption is that natural selection has favored accuracy. However, although it may sound paradoxical, there are a number of cases in which accurately registering the most useful information in a particular context can actually result in inaccuracies. Two prime examples are *illusory causation*, the tendency to attribute social causality to salient persons, and *illusory correlation*, the tendency to perceive correlations among highly salient event pairs. McArthur (1980) interpreted these phenomena in an ecological framework. Her general argument is that individuals are biologically attuned to detect certain co-occurrences more readily than others. In most cases, this leads to accurate judgments. When it does not, illusory causation and illusory correlation effects may result. Despite the fact that these biases may be viewed as sources of error in social perception, it is far more adaptive to occasionally make such errors of commission than to fail to detect highly relevant co-occurrences. As McArthur (1980) pointed out, it may be very "useful to be attuned to correlations between people who are rare and events which are rare ... event pairs which have been implicated in illusory correlation effects" (p. 516). If McArthur is right, then accuracy cannot be predicted in every instance of adaptive social perception. This, in turn, may once again raise the issue of falsifiability because it illustrates that in some cases, both accuracy and inaccuracy could be consistent with an evolutionary account. (See Berry, 1988, for a similar interpretation of facial stereotypes, and Funder, 1987, for a related discussion of the meaning of *errors* in social perception.) To put it crudely, it may be that what is selected for is not accuracy in every case but rather accuracy in most cases.

A final problem with accuracy, and one that is not linked to evolutionary concerns per se, is that it is still not clear when and to what extent people are accurate. For example, whether or not accuracy obtains depends to a great extent on both subject and context variables. Of particular relevance is the finding that accuracy, in the sense of self-observer agreement, improves as familiarity between observer and target increases (e.g., Funder & Colvin, 1988). Because it is assumed in ecological theories that information about "traits" is perceptually given, these approaches would interpret familiarity effects in terms of perceptual learning, as opposed to preprogrammed attunements, which would produce ceiling effects for accuracy, beginning with the first impression. However, there is no a priori guarantee that the type of perceptual learning here would involve natural selection to any meaningful extent.

There are other limitations in current applications of evolutionary theory to social perception. For example, in the literature, appeals are often made to adaptation and selection pressures, but evolutionary biology contains a wealth of other, more specific concepts that may be useful. (Sexual selection is one common exception to our point here; another example would be parental investment [Trivers, 1972], which we discuss later.) At the same time, there are many phenomena in social perception that ecological theorists have not yet considered in evolutionary terms but might find worthwhile in doing so. Some of these phenomena might amount to extremely circumscribed behaviors. For instance, dominant male hamadryas baboons produce a distinctive "swing step" when they wish to leave a group and induce subordinates to follow (Wilson, 1980). Conceivably, dominant members of human social groups indicate their plans to leave by more subtle but, nonetheless, recognizable movements. Other examples, such as regulation of group size, which have been studied extensively by sociobiologists, involve sets of complex and less clearly circumscribed behaviors. What is true of all these behaviors is that they depend critically on perceptual information, and are thus especially amenable to an ecological approach.

One ability, the adaptive value of which has not yet received much attention, is the ability to recognize individual identity across changes in appearance. For example, one implication of the literatures on craniofacial changes that result from aging (e.g., Todd, Mark, Shaw, & Pittenger, 1980) and the types of information available in point light displays (e.g., Berry, 1991; Cutting & Koslowski, 1977) is that individual identity over time and across variations in viewing conditions might be specified in perceptual information. This is an important idea, in that theoretical and empirical approaches to identity have often assumed that identity is underspecified by perceptual appearances over time and must be inferred from nonperceptual sources of information (e.g., Guardo & Bohan, 1971; Locke, 1959; Mohr, 1978). However, it has not been sufficiently noted that the existence of a perceptual invariant for identity and the ability for perceivers to detect this invariant would be extremely adaptive in terms of selection pressures that operate on the level of, say, kinship groups (e.g., kin selection).

SUGGESTIONS FOR THE USE
OF EVOLUTIONARY THEORY IN ECOLOGICAL
MODELS OF SOCIAL PERCEPTION

It is time for us to offer more positive comments. If evolutionary theories are going to play a meaningful role in ecological approaches to social perception, then we need guidelines for how and where they should be applied. In this section what we offer are not clear guidelines, but rather a set of suggestions on how to make use of some aspects of evolutionary theorizing. We begin by reiterating and, in some cases, extending several points raised earlier.

1. *Evidence for natural selection may come from carefully identifying universals and accuracy in social perception.* If falsifiable hypotheses are indeed generated, the existence of either or both universals and accuracy may implicate selection pressures.

2. *Evidence for natural selection may come from identifying individual differences, if the differences turn out to be divergences from adaptive norms associated with lower reproductive potential.* Selection pressures may be implicated if individuals who do not conform to a nearly-universal pattern of responses or show particularly poor accuracy on some dimension are also found to have especially low reproductive potential according to some reasonable criteria.

3. *Evidence for natural selection will rely on making tenable distinctions and recognizing genuine interplay between innate and learned attunements.* It will be useful for us to extend our discussion of this particular issue. Biological attunements will, by definition, usually arise through selection pressures, but it might seem that this does not necessarily hold for perceptual learning. However, it has become clear from specific research findings (e.g., Garcia & Koelling, 1966) as well as from theoretical considerations (e.g., Chomsky, 1969; Keil, 1986; Seligman, 1970) that learning is guided by biologically based constraints, and Gibson (1979) accepted this principle in the case of perceptual learning. In Gibsonian terms, we are biologically attuned to pick up certain types or classes of affordances, and in such cases learning will be facilitated. If follows that what may reflect selection pressures is not the emergence of some specific learned ability but rather sensitivity to the domain in which the ability operates. If abused children, as well as children from families low in expressiveness, become hypersensitive to certain kinds of emotional displays, what may be important from an evolutionary perspective is not the particular types of displays involved, but rather that humans have a general ability to learn how to read emotions out of facial displays, and that this ability is facilitated by a biological attunement.

4. *Further evidence for evolutionary influences may come from deployment of a wider range of concepts from evolutionary theories, and from applying such concepts to a broader range of phenomena in social perception.* For instance, we have focused primarily on claims about specific abilities or perceptions being adaptive. However, other characteristics may be adaptive, such as the tendency to respond in certain ways to social influence (e.g., Graziano, Jensen-Campbell, Shebilske, & Lundgren, 1993).

5. *Arguments for natural selection must rely on preestablished criteria for whether, and to what extent, an ability is adaptive.* The most commonly deployed criterion

is plausibility, which is, as we have indicated, a double-edged sword. Relationship to reproductive outcome also has been suggested as a criterion (e.g., Gangestad et al., 1992), which, as we said, may be debatable. Evidence for natural selection has sometimes been gathered through other approaches, including direct observation.

6. *Evidence for natural selection may depend on alternative conceptual approaches and methodologies.* We have suggested one alternative approach in the evaluation of individual differences. One other approach that we sketch here involves a typology of social perceptual abilities. An important distinction can be drawn between abilities whose basic expression is biologically preprogrammed, involves perceptual learning, or involves nonperceptual learning. For example, compare facial attractiveness and verbal deception. Recognition of each is adaptive. However, as Gibson (1979) would say, attractiveness is just one more type of optical information in the ambient array (although specified in a complex way by multiple sources, as indicated by Cunningham, 1986, and others), whereas verbal deception is a message, or form of communication, and thus the ability to detect it involves nonperceptual learning. This points to a difference between a primarily biological attunement and one that primarily involves experience, and the developmental literature indicates that whereas infants converge with adults on judgments of attractiveness, the ability to recognize lies is a later development. Although this example seems obvious, in other cases it may be more difficult to distinguish biological from perceptual and nonperceptual attunements, either across abilities or within a specific ability. For example, because deception is potentially detectable through both verbal and nonverbal information, it is not immediately obvious whether the ability to detect a particular lie is based in nonperceptual or perceptual learning. This brings us to a further suggestion.

7. *Full understanding of the role of evolution in social perception will depend on accurately characterizing the nature of genetically influenced abilities.* To illustrate our point, there is evidence that animacy, as well as specific transitory relationships among humans such as "chasing" and "fighting," are specified in dynamic stimulus information (Heider & Simmel, 1944). Berry and colleagues have presented viewers with versions of the classic Heider and Simmel film in which either motion or form is disrupted, and the data indicate that patterns of motion are the source of perceptions of animacy among adults (Berry & Misovich, 1994; Berry, Misovich, Kean, & Baron, 1992) as well as among preschoolers (Berry & Springer, 1993). It also appears that by age 5, children converge with adults in their specific perceptions of the Heider and Simmel film (Springer, Meier, & Berry, in press). However, it does not seem plausible to suppose that humans are prewired to detect something as specific as chasing behavior. A more accurate description of the competency in question is in terms of a biological attunement to certain kinds of information contained in motion (e.g., information suggesting intentionality), coupled with a learned ability to circumscribe certain events and append names to them (e.g.,

chasing). The relevance of this kind of description is in its potential for guiding specific paths of research in social perception. One would not design a study to see whether children are sensitive to chasing behavior in different contexts, but rather to see whether there is specific information in motion that consistently gives rise to impressions of intentionality, such as that involved in chasing. According to Premack (1990), this sort of information not only exists, but preschoolers are, and babies may be, sensitive to it.

Finally, another sort of suggestion about the applicability of evolutionary theory to social perception is that good models do exist, a point which may have been concealed by our essentially critical approach so far in this chapter. This is the subject of our next section.

A MODEL APPLICATION OF EVOLUTIONARY THEORY TO SOCIAL PERCEPTION

The example that we discuss here comes from some of the literatures pertaining to selection of mates and sexual partners, for which numerous sex differences have been documented within and across cultures. We begin by mentioning just a few of these differences.

Females tend to be fairly selective in choosing short-term sexual partners as wellas mates, but generally prefer older men who seem able to control and provide material resources. Mature males, in contrast, tend to be less selective than females in their choice of sexual partners, other than preferring younger women whose appearance indicates fertility and health, and to some extent promiscuity. However, in choosing mates men may be approximately equal to women in selectivity. (Findings such as these have surfaced in studies of preferences for mates or sexual partners, and in the literature on attractiveness.)

A number of evolutionary-based explanations for these and other differences have employed Trivers' (1972) analysis of parental investment (e.g., Buss & Schmitt, 1993; Kenrick, 1994; Symons, 1979). Biologically speaking, women invest much more to produce babies than men do; at the minimum, babies must grow inside their bodies and be nourished after birth, whereas men need only invest the energy required for copulation. At the same time, women stand to contribute less to the gene pool, owing to limitations imposed by the gestation period, the onset of menopause, and so fourth, whereas men can (theoretically) impregnate women on a constant basis well beyond middle age. Thus women will be more selective about mates because they need long-term support, so that in the event of pregnancy, their relatively infrequent contribution to the gene pool will be maximized. Men can afford to be less selective about sexual partners, because they have little to lose from nonoptimal impregnations, and females' selectivity will force them to attempt to establish positions in dominance hierarchies in which they compete for females. However, in the case of mate selection, in which males stand to contribute substantially to childrearing, they will be selective to the extent that females are.

The approaches alluded to here provide excellent illustrations of how evolutionary theorizing can be applied to social perception. First, the underlying claims about selection are falsifiable and extremely detailed: The typical argument is not the simplistic one that a particular behavior is adaptive, and thus selected for, but rather that what has been selected for is an entire set of behaviors, abilities, and preferences that are differentially adaptive depending on, among other things, the age, sex, and intentions (e.g., one-night stand vs. marriage) of individuals. Moreover, the set of behaviors, abilities and preferences that are supposed to have been selectively favored are not just aggregates but in some cases are interrelated or hierarchical. An example would be Buss' (1985) observation that mate selection, although influenced by the various factors just discussed as well as others, is also subject to the most basic constraint of similarity. The notion of hierarchy is illustrated by Singh's (1993) hypothesis that attractiveness judgments are made on the basis of successively applied filters, some of which are biologically preprogrammed, whereas others have a cultural basis.

There are other ways in which this literature represents a model approach. The empirical work is laudably "ecological" in that data bearing on actual selection of sexual partners and mates have been reported (e.g., Kenrick, 1994). An especially important point is that specific concepts in evolutionary theories such as parental investment are employed, and specific dimensions of stimulus information that might be relevant are examined (Cunningham, 1986; Gangestad et al., 1992; Singh, 1993). The analysis of available stimulus information, which requires extremely sophisticated approaches (e.g., Cunningham, 1986), is particularly relevant to ecological approaches.

A FINAL WORD: THE CASE OF "RECIPROCALITY"

In this section we offer one concrete but highly speculative proposal about the role of natural selection in social perception. Gibson observed that "what the male affords the female is reciprocal to what the female affords the male;...what the prey affords the predator goes along with what the predator affords the prey," and so forth for other sorts of interactions (Gibson, 1979, p. 135). This "reciprocality" that Gibson so briefly treated points to a general way in which evolutionary pressures may have influenced social perception. In brief, it may be that natural selection favors perceptual attunement to reciprocal interactions.

Notice that *reciprocal* does not necessarily mean *equitable* (e.g., you scratch my back and I'll scratch yours). Participating in an equitable interaction is something we typically learn how to do, and may or may not choose to do in specific contexts. The male–female interactions implicated by the mate selection literatures are reciprocal in the sense of inherently fitting together like pieces in a puzzle. They may or may not be equitable. To return to one of Gibson's examples, in the case of lovers, the relationship might be positive, equitable, and mutually rewarding, whereas in the case of predator and prey, the relationship is reciprocal yet wholly adversary.

It might be objected, then, that in any interaction the participants could be seen as fitting together somehow. However, one can imagine scenarios in which reciprocality is not found. For example, if both young men and young women happened to prefer older members of the opposite sex as mates, we might still be able to propagate the species, but in that case male and female mate selection would not be fully reciprocal; young males and young females would not be mutually engaged in their attempts to achieve reproductive goals.

Reciprocality might be thought of as a matter of engagement between people as they try to achieve personal goals. If men tend toward promiscuity and women tend toward long-term monogamy, then the goals of their interactions with each other may be inherently conflictual, but they will nonetheless be engaged: Men will be trying to "get something" from women, and women will be trying to "get something" from men. This is roughly analogous to the way in which predators attempt to catch their prey while the prey try to escape; both predator and prey are part of an interaction in which their mutual goals conflict. This reflects reciprocality in the Gibsonian sense, and the argument here is that it would be adaptive to be sensitive to dimensions of information relevant in such interactions. In contrast, in the fictional case in which young men and women both prefer older sexual partners, the interactions among young members of the species would not be very engaged in the sexual dimension, and it would be less likely that these individuals would be particularly accurate about each others' sexual availability.

Deception would be another example involving reciprocality. Although the ability to deceive and the ability to see through deception are adversarial, so to speak, it seems essential for both to be maintained, as we argued earlier. Natural selection may operate at the level of individual abilities or at the level of ensembles of abilities. In the case of deception, selection pressures may sustain the ensemble of being a good liar and being good at detecting lies, or, to return to the previous example, selection may favor the ensemble of female preferences for older males and older male preferences for younger females, or the ensemble of male promiscuity and female monogamy. Natural selection, therefore, would favor reciprocal interactions, and this would be part of the social world in which we evolved.

All of this brings us back to where we started: James Gibson. "We were created by the world we live in" observed Gibson (1979), and if this world includes reciprocal interactions, then natural selection might favor sensitivity to information of central importance to these interactions. One task for ecological approaches to social perception might be to empirically evaluate this idea.

ACKNOWLEDGMENT

We are grateful to Doug Kenrick and Jeff Simpson for their thoughtful comments on an earlier draft of this chapter.

REFERENCES

Albright, L., Kenny, D. A., & Malloy, T. (1988). Consensus in personality judgments at zero acquaintance. *Journal of Personality and Social Psychology, 55,* 387–395.

Baron, R. M., & Boudreau, L. A. (1987). An ecological perspective on intergrating personality and social psychology. *Journal of Personality and Social Psychology, 53,* 1222–1228.

Berry, D. S. (1988). The visual perception of people. *Journal for the Theory of Social Behavior, 18,* 345–534.

Berry, D. S. (1990). Taking people at face value: Evidence for the kernel of truth hypothesis. *Social Cognition, 8,* 343–361.

Berry, D. S. (1991). Accuracy in social perception: Contributions of facial and vocal information. *Journal of Personality and Social Psychology, 61,* 298–307.

Berry, D. S. (1991). Child and adult sensitivity to gender information in patterns of facial motion. *Ecological Psychology, 3,* 349–366.

Berry, D. S. (1994). Beyond beauty and after affect: An event perception approach to perceiving faces. In R. A. Eder (Ed), *Psychological perspectives on craniofacial problems: Insights into the function of appearance in development* (pp. 48–73). New York: Springer-Verlag.

Berry, D. S., & Finch Wero, J. L. (1993). Accuracy in face perception: A view from ecological psychology. *Journal of Personality, 61,* 497–520.

Berry, D. S., & McArthur, L. Z. (1986). Perceiving character in faces: The impact of age-related craniofacial changes on social perception. *Psychological Bulletin, 100,* 3–18.

Berry, D. S., & Misovich, S. J. (1994). Methodological approaches to the study of social event perception. *Personality and Social Psychology Bulletin, 20,* 139–152.

Berry, D. S., Misovich, S. J., Kean, K. J., & Baron, R. M. (1992). Effects of disruption of structure and motion on perceptions of social causality. *Personality and Social Psychology Bulletin, 18,* 237–244.

Berry, D. S.,& Pennebaker, J. W. (1996). *Language as lie detector: Predicting deception from variations in linguistic style.* Manuscript submitted for review.

Berry, D. S., & Springer, K. (1993). Structure, motion, and preschoolers' perceptions of social causality. *Ecological Psychology, 5,* 273–283.

Bond, C. F., Berry, D. S., & Omar, A. (1994). The kernel of truth in judgments of deception. *Basic and Applied Social Psychology, 15,* 523–534.

Bond, C. F., Kahler, K. N., & Paolicelli, L. M. (1985). The miscommunication of deception: An adaptive perspective. *Journal of Experimental Social Psychology, 21,* 331–345.

Borkneau, P., & Leibler, A. (1992). Trait inferences: Sources of validity at zero acquaintence. *Journal of Personality and Social Psychology, 62,* 645–667.

Buck, R. (1988). The perception of facial expression: Individual regulation and social coordination. In T. R. Alley (Ed.). *Social and applied aspects of perceiving faces* (pp. 141–166). Hillsdale, NJ: Lawrence Erlbaum Associates.

Buss, D. M. (1985). Human mate selection. *American Scientist, 73,* 47–51.

Buss, D. M., & Schmitt, D. P. (1993). Sexual strategies theory: An evolutionary perspective on human mating. *Psychological Review, 100,* 204–232.

Campos, J. J., Caplovitz, K. B., Lamb, M. E., Goldsmith, H. H., Stenberg, C. (1983). Socioemotional development. In M. M. Haith & J. J. Campos (Eds.), *Handbook of child psychology: Vol. 3. Infancy and developmental psychobiology* (pp. 783–915). New York: Wiley.

Chomsky, N. (1969). *The acquisition of syntax in children from five to ten.* Cambridge, MA: MIT Press.

Cunningham, M. R. (1986). Measuring the physical in physical attractiveness: Quasi-experiments on the sociobiology of female facial beauty. *Journal of Personality and Social Psychology, 50,* 925–935.

Cunningham, M. R., Roberts, A. R., Barbee, A. P., Druen, P. B., & Wu, C. (1995). "Their ideas of beauty are, on the whole, the same as ours": Consistency and variability in the cross-cultural perception of female physical attractiveness. *Journal of Personality and Social Psychology, 68,* 261–279.

Curtis, H., & Barnes, N. (1981). *Invitation to biology* (3rd ed.). New York: Worth Publishers.

Cutting, J. E., & Kozlowski, L. T. (1977). Recognizing friends by their walk: Gait perception without familiarity cues. *Bulletin of the Psychomonic Society, 9,* 353–356.

DePaulo, B. M., & Rosenthal, R. (1979). Telling lies. *Journal of Personality and Social Psychology, 37,* 1713–1722.

Ekman, P. (1982). *Emotion in the human face.* Cambridge, England: Cambridge University Press.

Field, T. M., Woodson, R., Greenberg, R., & Cohen, D. (1982). Discrimination and imitation of facial expressions by neonates. *Science, 218,* 179–181.

Funder, D. C. (1987). Errors and mistakes: Evaluating the accuracy of social judgment. *Psychological Bulletin, 101,* 75–90.

Funder, D. C., & Colvin, C. R. (1988). Friends and strangers: Acquaintenship, agreement and the accuracy of personality judgment. *Journal of Personality and Social Psychology, 55,* 149–158.

Funder, D. C., & West, S. G. (1993). Consensus, self-other agreement, and accuracy in personality judgment. *Journal of Personality, 61,* 457–476.

Gangestad, S. W., Simpson, J. A., DiGeronimo, M.,& Biek, M. (1992). Differential accuracy in person perception across traits: Examination of a functional hypothesis. *Journal of Personality and Social Psychology, 62,* 688–689.

Garcia, J. & Koelling, R. (1966). Relation of cue to consequence in avoidance learning. *Psychonomic Science, 4,* 123–124.

Gibson, J. J. (1966). *The senses considered as perceptual systems.* Boston: Houghton Mifflin.

Gibson, J. J. (1979). *The ecological approach to visual perception.* Boston: Houghton Mifflin.

Graziano, W. G., Jensen-Campbell, L. A., Shebilske, L. J., & Lundgren, S. R. (1993). Social influence, sex differences, and judgments of beauty: Putting the interpersonal back in interpersonal attraction. *Journal of Personality and Social Psychology, 65,* 522–531.

Guardo, C. J.,& Bohan, J. B. (1971). Development of a sense of self-identity in children. *Child Development, 42,* 1909–1921.

Halberstadt, A. G. (1986). Family socialization of emotional expression and nonverbal communication styles and skills. *Journal of Personality and Social Psychology, 51,* 827–836.

Heider, F., & Simmel, M. (1994). An experimental study of apparent behavior. *American Journal of Psychology, 57,* 243–259.

Keil, F. C. (1986). On the structure dependent nature of stages of cognitive development. In I. Levin (Ed.) *Stage and structure.* Norwood, NJ: Ablex.

Kenrick, D. T. (1994). Evolutionary social psychology: From sexual selection to social cognition. *Advances in Experimental Social Psychology, 26,* 75–119.

Langlois, J. H., & Roggman, L. A. (1990). Attractive faces are only average. *Psychological Science. 1,* 115–121.

Langlois, J. H., Roggman, L. A., & Rieser-Danner, L. A. (1990). Infants' differential social responses to attractive and unattractive faces. *Developmental Psychology, 26,* 153–159.

Locke, J. (1959). *An essay concerning human understanding.* New York: Dover Publications.

Lorenz, K. (1943). Die angeborenen formen moglicher erfahrung [The innate forms of potential experience]. *Zietschrift fur Tierpsychologie, 5,* 233–519.

McArthur, L. Z. (1980). Illusory causation and illusory correlation: Two epistemological accounts. *Personality and Social Psychology Bulletin, 6,* 507–519.

McArthur, L. Z., & Baron, R. M. (1983). Toward an ecological theory of social perception. *Psychological Review, 90,* 215–238.

McArthur, L. Z., & Berry, D. S. (1987). Cross-cultural agreement in perceptions of babyfaced adults. *Journal of Cross-Cultural Psychology, 18,* 165–192.

Mohr, D. M. (1978). Development of attributes of personal identity. *Developmental Psychology, 14,* 427–428.

Popper, K. R. (1972). *Objective knowledge.* Longon: Oxford University Press.

Premack, D. (1990). The infant's theory of self-propelled objects. *Cognition, 36,* 1–16.

Seligman, M. E. P. (1970). On the generality of laws of learning. *Psychological Review, 77,* 406–418.

Shaw, R., & Bransford, J. (1977). Introduction: Psychological approaches to the problem of knowing. In R. Shaw & J. Bransford (Eds.), *Perceiving, acting, and knowing* (pp. 1–39). Hillsdale, NJ: Lawrence Erlbaum Associates.

Singh, D. (1993). Adaptive significance of female physical attractiveness: Role of waist-to-hip ratio. *Journal of Personality and Social Psychology, 65,* 293–307.

Springer, K., Meier, J. A., & Berry, D. S. (in press). Nonverbal bases of social perception: Developmental changes in sensitivity to patterns of motion that reveal interpersonal events. *Journal of Nonverbal Behavior.*

Symons, D. (1979). *The evolution of human sexuality.* Oxford, England: Oxford University Press.

Todd, J. T., Mark, L. M., Shaw, R. E., & Pittenger, J. B. (1980). The perception of human growth. *Scientific American, 242,* 132–144.

Trivers, R. L. (1972). Parental investment and sexual selection. In B. Campbell (Ed.), *Sexual selection and the descent of man, 1871–1971* (pp. 136–179). Chicago: Aldine-Atherton.

Wilson, E. O. (1980). *Sociobiology.* Cambridge: The Belknap Press.

Zebrowitz, L. A. (1990). *Social perception.* Pacific Grove, CA: Brooks/Cole.

Zuckerman, M., DePaulo, B. M., & Rosenthal, R. (1981). Vebal and nonverbal communication of deception. *Advances in experimental social psychology, 14,* 1–59.

4

Perceptions of Betrayal and the Design of the Mind

Todd K. Shackelford
The University of Texas at Austin

One indication of the interpersonal significance of a trait, behavior, or event is the extent to which it is encoded into the lexicon, the natural language system (Norman, 1963; Goldberg, 1981). According to this lexical approach to identifying and describing the most significant human social events, betrayal is a momentous interpersonal occasion. *Roget's International Thesaurus (Chapman, 1977) lists dozens of synonyms for betray, including deceive, beguile, trick, hoax, dupe, gammon, bamboozle, snow, hornswaggle, take in, string along, put something over, put something across, slip one over on, pull a fast one on, delude, leave in the lurch, leave holding the bag, play one false, double-cross, cheat on, two-time, conjure, bluff, and outsmart.*

In this chapter and elsewhere (Shackelford & Buss, 1996a, 1996b), I define a *relationship betrayal* as any instance in which an expected benefit is actively withheld, awarded to parties outside of the focal relationship, or both. Every human relationship is subject to betrayal. Marital partners sometimes commit sexual and romantic infidelities. Best friends occasionally whisper secrets into unintended ears, violating trust. Parents sometimes sexually violate their own children. Betrayal is a significant element of the human "adaptive landscape" (Buss, 1991). Why do we betray those closest to us? When is betrayal likely to occur? Who is most likely to betray us? These are significant questions, yet framed as such they are unanswerable, unless we ask these questions for specific relationship contexts. The actions or events that constitute a betrayal in one relationship context may not constitute a betrayal in another such context (Buss, 1990; Shackelford & Buss, 1996a). I argue in this chapter that to understand betrayal—when, why, and to whom it occurs—we must understand the design of the human mind. To the extent that different relationship contexts posed different adaptive problems for ancestral humans, different solutions to these adaptive problems will have been selected for and thereby incorporated into the psychological apparatus of the human mind. An evolutionarily informed perspective on relationship betrayal provides a unique insight into this darker side of human nature.

PREVIOUS RESEARCH AND THEORY ON BETRAYAL HAS BEEN PREMISED ON A DOMAIN-GENERAL MIND

Much of the empirical work on human interpersonal relations has historically focused on the benefits expected, received, and exchanged within various interpersonal relationships. Little effort has been invested in elucidating the costs accompanying the relationships, or the potential of betrayal by a trusted relationship partner. Interestingly, one of the foremost distinctions between traditional Darwinian theory and the modern evolutionary psychological perspective is that the modern movement employs "cost–benefit analysis" in investigations of human social behavior (see exemplar applications in Anderson & Crawford, 1995; Daly & Wilson, 1988; Gangestad, 1993; Thornhill & Gangestad, 1993). Traditional Darwinians (Darwin, 1859/1958; see also Alexander, 1979, 1987; Betzig, 1986) focus primarily on the reproductive benefits associated with certain behaviors.[1] What little work has been done on betrayal across different relationship contexts typically defines these contexts in exceptionally broad or domain-general terms.

Standard social psychology texts (e.g., Argyle, 1992; Baum, Fisher, & Singer, 1985; Brehm & Kassin, 1990; Myers, 1993; Sabini, 1992) typically present a single chapter on "intimate relationships," in which little effort is made to delineate the different types of intimate relationships engaged in by humans. This relatively domain-general approach to a single class of close relationships is often made explicit, as in Brehm and Kassin's (1990) definition of intimate relationships: "close relationships between two adults involving at least one of the following: emotional attachment, fulfillment of psychological needs, and mutual dependence on each other's assistance" (p. 244). Although Brehm and Kassin (1990) conceded that there are a few variations of close relationships—they discuss friendships, nonmarital romantic relationships, and marriages—the implication is that each of these specific relationship contexts falls under the more general canopy of "intimate relationships."

Myriad social psychological theories purport to account for the development, maintenance, and dissolution of all interpersonal relationships within a single

---

[1]I thank Jeff Simpson for this insight.

conceptual scheme. Few such theories offer an explicit discussion of relationship betrayal (see Clark and her colleagues for one exception: Clark, 1984; Clark & Mills, 1979; Clark & Waddell, 1985). More often, these domain-general theories embed an implicit account of betrayal in their larger conceptual framework. Byrne and his colleagues (Bryne, 1971; Byrne & Clore, 1970; Byrne, Clore, & Smeaton, 1986), for example, suggest a two-stage model of relationship development. The two-stage model argues that we first avoid dissimilar others and then approach similar others. Feelings of betrayal might be produced when an otherwise similar other behaves in a dissimilar manner. The two-stage model is presumably equally applicable to all varieties of interpersonal relationships.

Murstein's (1976, 1987) stimulus–value–role theory of interpersonal relationships holds that intimate relationships, be they close friendships or romantic relationships, proceed from a *stimulus stage*, in which attraction and liking is based on external characteristics, such as physical appearance, to a *value stage*, in which attraction is a function of similarity in beliefs and values. The third and final stage of close relationships is the *role stage*, in which partners' satisfaction with and commitment to the relationship is dependent on each party performing his or her expected role. Feelings of betrayal might emerge if, in the first stage, one person attempted to deceive the other about the former's physical appearance. Once a relationship progressed to the second stage, feelings of betrayal might arise if one person acted in a manner inconsistent with a value or belief that the other person thought the two of them shared. Feelings of betrayal may arise in the third stage if one person fails to perform an expected role. No differentiation is made between the varieties of close relationships.

Social exchange theory (Blau, 1964; Homans, 1961; Thibaut & Kelley, 1959) offers a simple account for the emergence of feelings of betrayal in all interpersonal relationships: A person will feel betrayed when his or her relationship partner (close friend, lover, spouse, mother, etc.) imposes costs that exceed the benefits expected. Equity theory (Adams, 1965; Messick & Cook, 1983; Walster, Walster, & Berscheid, 1978) is an offshoot of social exchange theory, and maintains that satisfaction in any interpersonal relationship depends on the ratio of benefits derived from the relationship to the costs of participating in that relationship. According to equity theory, relationship satisfaction is produced when the members of the relationship experience the same benefit to cost ratio. Feelings of betrayal arise in both parties when this ratio is unequal. No distinction is made among the varieties of interpersonal relationships by social exchange or equity theorists.

Clark and her colleagues (Clark, 1984; Clark & Mills, 1979; Clark & Waddell, 1985) have investigated the differential dynamics of communal and exchange relationships. *Communal relationships* include all those relationships "in which the giving of a benefit in response to the need for the benefit is appropriate," whereas *exchange relationships* include all relationships "in which the giving of a benefit in response to the receipt of a benefit is appropriate" (Clark & Mills, 1979, p. 12). Thus, in the rubric of Clark and company, long-term mateships, close friendships, parent–child relationships, and sibling relationships, to name a few, all fall under the canopy of a communal relationship. Store clerk–customer relationships, temporary business partnerships, and prostitute–client relationships, for example, all are subsumed under the category of exchange relationships.

Using this highly general categorization scheme, Clark and Waddell (1985) led their subjects (all female college students) to expect involvement in either a communal relationship or an exchange relationship with a confederate (also a female college student). After introductions and a minimal amount of interaction, confederates asked the subjects for a favor, and then subsequently either offered repayment or did not offer repayment for the favor. Specifically, confederates asked subjects if they would fill out a survey for a class the confederate was taking. All subjects agreed to fill out the survey. In the exchange condition, the confederate quickly reassured the subject by saying, "I'll be able to pay you $2.00 from class funds for doing this. Just send the questionnaire back through campus mail, and I'll send you the money." In the communal condition, the confederate apologetically told the subject, "We used to have class funds to pay people for doing this, but we ran out. So I won't be able to pay you. I hope that's okay." Subjects who were under the impression that they were in a communal relationship felt betrayed by their partner's insistence on paying for the favor. In contrast, subjects who were led to believe they were interacting in an exchange relationship felt betrayed when asked if nonrepayment of the favor was acceptable.

To my knowledge, Clark and Waddell's (1985) research represents the first empirical investigation of differential perceptions of betrayal across different relationship contexts. Clark and Waddell discussed the applicability of their results to all exchange and communal relationships, however, and did not differentiate mateships, for example, from friendships or parent–child relationships. Nor did they differentiate among the varieties of exchange relationships. Rather, Clark and Waddell implied the domain-general notion that failing to offer repayment for a favor elicits feelings of betrayal within the context of any exchange relationship. Similarly, Clark and Waddell implicitly argued that an offer of repayment for a favor elicits feelings of betrayal in any communal relationship.

Within the conceptual scheme of Clark and her colleagues (Clark, 1984; Clark & Mills, 1979; Clark & Waddell, 1985; Clark, Mills, & Powell, 1986), the human psychology of betrayal is architecturally simple. Clark et al. distinguished only between betrayal in exchange relationships and betrayal in communal relationships. Implicitly, Clark and her colleagues argued for a domain-general human mind with respect to interpersonal perceptions of betrayal: There is a mechanism of the mind that is responsible for monitoring participation in exchange relationships. There is another mechanism of the mind that is responsible for monitoring participation in communal relationships.

Clark and her colleagues do not elaborate on the implications of their conceptual scheme for the design of the human mind. As social psychologists, Clark et al. study observable, interpersonal behavior, not mechanisms of the mind or the cognitive decision rules and procedures according to which these mechanisms operate. The interpersonal behavior they study, however, is generated, perceived, and interpreted by the psychological mechanisms instantiated as functional components of the mind (Buss, 1990; Kenrick, 1994; Kenrick, Sadalla, & Keefe, in press). If researchers want to understand relationship betrayal—or any other interpersonal phenomenon—they need to understand the design and functioning of the mind with respect to interpersonal behavior. Moreover, to understand the design and functioning of the human mind, we need to begin with the recognition that the mind is the cumulative product of hundreds of thousands of generations of human evolutionary history (Darwin, 1859/1958).

SOLUTIONS TO ADAPTIVE PROBLEMS ARE INSTANTIATED AS MECHANISMS IN THE MIND

Each of us represents an evolutionary success story. Each generation of our ancestors faced a recurrent set of adaptive problems. These adaptive problems included, for example, obtaining a reproductively valuable mate, reproducing with this mate, rearing healthy children to reproductive age, locating and ingesting nutritious foods, and successfully negotiating mateships, friendships, family relationships, and coalitions. Our ancestors succeeded in finding better mates, raising healthier children, locating and eating better foods, and negotiating their mateships, friendships, family relationships, and coalitional relationships better than did our ancestors' conspecifics who failed at these tasks. Our ancestors' behaviors, thoughts, and feelings were generated, in part, by evolved psychological mechanisms built into their minds. Those protohumans whose minds embodied alternative mechanisms were less likely to be our evolutionary ancestors, for they would have been out-survived and out-reproduced by those early humans—our ancestors—whose behavior, cognition, and affect was generated by the favored mechanisms. Thus, the better solutions to the adaptive problems recurrently faced by our ancestors over human evolutionary history came to be instantiated as reliable psychological mechanisms in the human mind.

Several evolutionary psychologists (Buss, 1991, 1995; Kenrick, Keefe, Bryan, Barr, & Brown, 1995; Pinker, 1994; Thornhill, 1990; Tooby & Cosmides, 1990, 1992) have struggled with the question of just what is an "evolved psychological mechanism"? How can we identify such a mechanism? How does it work? What does it do? Rather than add to the growing list of definitions of this slippery term, I shall offer two published definitions that strike me as both linguistically elegant and conceptually complete.

Buss (1991) defined an *evolved psychological mechanism* as:

> a set of processes inside an organism that (1) exist in the form they do because they (or other mechanisms that reliably produce them) solved specific problems of individual survival or reproduction; (2) take only certain classes of input, where input (a) can either be external or internal, (b) can be actively extracted from the environment or passively received from the environment, and (c) specifies to the organism the particular adaptational problem it is facing; and (3) transform that information into output through a procedure (e.g., a decision rule) where output (a) regulates physiological activity, provides information to other psychological mechanisms, or produces action, and (b) solves a particular adaptational problem. Species have evolved psychologies to the extent that they possess mechanisms of this sort. (p. 464)

Tooby and Cosmides (1992) defined *adaptation*, a more general class of evolved mechanisms that subsumes all such mechanisms, psychological and physiological.[2] According to Tooby and Cosmides, an *adaptation* or *evolved mechanism* is:

> (1) a system of inherited and reliably developing properties that recurs among members of a species that (2) became incorporated into the species' standard design because during the period of their incorporation, (3) they were coordinated with a set of statistically recurrent structural properties outside the adaptation (either in the environment or in other parts of the organism), (4) in such a way that the causal interaction of the two (in the context of the rest of the properties of the organism) produced functional outcomes that were ultimately tributary to propagation with sufficient frequency (i.e., it solved an adaptive problem for the organism). Adaptations are mechanisms or subsystems of properties crafted by natural selection to solve the specific problems posed by the regularities of the physical, chemical, developmental, ecological, demographic, social, and informational environments encountered by ancestral populations during the course of a species' or population's evolution. (1992, pp. 61–62)

Every theory of human cognitive, behavioral, and affective functioning—even the most environmentalistic—assumes that human psychology is at some basic level constructed of psychological mechanisms (Buss, 1991, 1995; Symons, 1987). If two members of a given species, or if two members of the same species are exposed to identical stimuli and respond in nonidentical ways, we must infer the operation of a mechanism or mechanisms internal to the

---

[2]All mechanisms, including "psychological" mechanisms, are fundamentally physiological mechanisms, instantiated by some process and in some form in the human brain.

organisms. These mechanisms are usefully described as *information-processing devices*. They take in certain classes of information, process that information according to a set of decision rules, and then generate some sort of output. The information accepted into the mechanism for processing may come from other psychological mechanisms internal to the organism, or it may originate in the external environment. With regard to humans, for example, an important source of external information is the multifaceted social environment comprised of other humans operating according to these same evolved mechanisms. The output generated by an evolved psychological mechanism may be in the form of information that is then channeled and processed by other psychological mechanisms internal to the organism, or the output may be in the form of manifest behavior.

The psychological mechanisms that guide human functioning have evolved because they solved the adaptive problems recurrently confronted by our ancestors. Consider the problem of which foods to ingest. For survival, certain nutrients had to be ingested and, conversely, various toxins had to be avoided. A handful of dirt would not usually be an adaptive food selection, nor would a mat of hair collected from an unwilling neighbor. A handful of berries, a bushel of vegetables, or a slab of protein-rich flesh would have carried much more nutritive value, and would therefore have been much more satisfying to our ancestral forebears. Additionally, a plethora of poisonous plants, fruits, vegetables, insects, and other organisms had to be carefully avoided (Profet, 1991). Those early humans who did not identify, procure, and ingest nutritive foods, to the exclusion of nonnutritive and toxic foods, are not likely to have been our evolutionary ancestors, for they would have been out-survived and out-reproduced by their sharper and more discriminating conspecifics.

Human psychology consists of a finite though numerous collection of evolved psychological mechanisms. The adaptive problems our evolutionary ancestors faced were many and varied: from mate selection, to food ingestion, to negotiating same-sex coalitional alliances. The solution to each of these problems evolved as a circumscribed set of decision rules that guide human thought, emotion, and behavior. The psychological mechanisms that have evolved as solutions to these multifarious adaptive problems will be as numerous and varied as the adaptive problems themselves (Buss, 1991, 1995; Symons, 1987).

Success in selecting a reproductively valuable mate has little or no direct relevance to whether one can successfully select and ingest the most nutritive foods available. Mate selection and food selection are two qualitatively different adaptive problems that will have selected for at least two qualitatively different sets of psychological mechanisms over human evolutionary history. Because the number of adaptive problems that confronted ancestral humans was finite though numerous, so, too, do evolutionary psychologists expect that the number of mechanisms instantiated in the mind is finite though numerous. These mechanisms must be (variably) domain-specific (i.e., they evolved as solutions to specific adaptive problems). Because ancestral humans did not confront a single "survive and reproduce successfully" adaptive problem, we have no reason to expect that human psychology is composed of a single "survive and reproduce successfully" mechanism that evolved as a domain-general solution (Buss, 1991, 1995; Symons, 1987; Tooby & Cosmides, 1990, 1992). Psychological mechanisms lie on a continuum from relatively more to relatively less domain-specific, according to the relatively more or less circumscribed nature of the adaptive problem that selected for their evolution and instantiation in the human mind.

The results of evolutionary psychological research in heterosexual (Buss, 1989; Buss, 1994; Greiling, 1995) and homosexual (Bailey, Gaulin, Agyei, & Gladue, 1994; Kenrick, et al., 1995) mate preferences, attraction (Langlois & Roggman, 1990), criminology and homicide (Daly & Wilson, 1988; Wilson & Daly, 1992), social exchange (Cosmides, 1989; Cosmides & Tooby, 1992; see below), psychophysical perception (Shepard, 1984, 1992), language (Pinker, 1994; Pinker & Bloom, 1992), self-deception (Lockard & Paulhus, 1988), environmental aesthetics (Kaplan, 1992; Orians & Heerwagen, 1992), and numerous other disciplines are consistent with a single, stalwart conclusion: "Human psychology involves many and complex domain-specific mechanisms, each suited to serve a particular function" (Buss, 1991, p. 462).

AN EVOLUTIONARILY INFORMED PERSPECTIVE ON BETRAYAL: COSMIDES AND TOOBY'S CHEATER-DETECTOR MECHANISM

Cosmides and Tooby (1992; see also Cosmides, 1989) argued and provided empirical evidence for the existence of an evolved "cheater-detector" mechanism that guides human social exchange. Cosmides and Tooby documented findings that humans can easily point out when someone has broken a consensual social rule (i.e., when someone has cheated). When this same problem is framed as a logic problem stripped of its social-contractual relevance, humans are amazingly poor problem solvers.

The Wason selection task (Wason, 1966) serves as the methodological backbone of Cosmides and Tooby's empirical investigations. In the Wason selection task, subjects must determine if a conditional hypothesis of the form "if A then B" has been violated by any of four instances depicted by cards. This conditional hypothesis is violated only when A is true but B is false. Consider, for example, the following problem (Cosmides & Tooby, 1992, p. 182):

Part of your new clerical job at the local high school is to make sure that student documents have been processed correctly. Your job is to make sure that the documents conform to the following alphanumeric rule: "If a person had a 'D' rating, then his documents must be marked code '3.'" You suspect the person you replaced did not categorize the students' documents correctly. The cards below have information about the documents of four people who are enrolled at this high school. Each card represents one person. One side of a card tells a person's letter rating and the other side of the card tells that person's number code. Indicate only those card(s) you definitely need to turn over to see if the documents of any of these people violate this rule.

Subjects view four cards, only one side of which is visible to them. The letter D appears on the first card, the letter F on the second card, the number 3 on the third card, and the number 7 on the fourth card. The alphanumeric rule is violated only if one of the cards has a D on one side and a number other than 3 on the reverse side. To determine whether a rule-violation had ocurred, the subject need only turn over two cards: the A card, to see if not-B appears on the reverse side, and the not-B card, to see whether A is displayed on the reverse side. In the example, these two cards display D and 7, respectively. Framed in this strictly logical format, fewer than 25% of subjects select the correct cards. Instead, most subjects elect to turn over the A card alone (D), or both the A and B cards (D and 3).

When this logical problem is framed in social-contractual terms, however, nearly 75% of subjects select the correct cards to determine whether a rule-violation—that is, cheating—has occured. Consider, for example, the Drinking Age Problem (see Cosmides & Tooby, 1992, p. 182), which is identical in logical structure to the abstract problem presented above, but presents the problem in social-contractual terms:

In its crackdown against drunk drivers, Massachusetts law enforcement officials are revoking liquor licenses left and right. You are a bouncer in a Boton bar, and you'll lose your job unless you enforce the following rule: "If a person is drinking beer, then he must be over 20 years old." The cards below have information about four people sitting at a table in your bar. Each card represnets one person. One side of a card tells what a person is drinking and the other side of the card tells that person's age. Indicate only those card(s) you definitely need to turn over to see if any of these people are breaking the law.

The four cards presented to subjects display the following information, respectively: drinking beer, drinking coke, 25 years old, 16 years old. To determine if the conditional hypothesis *if A then B* has been infracted, subjects again need only turn over two cards: the A card (*drinking beer*) and the not-B card (*16 years old*). If the person drinking beer is not over 20, or if the 16-year-old is drinking beer, then the rule stating, "If a person is drinking beer, then he must be over 20 years old," has been violated (i.e., cheating has occurred).

Cosmides and Tooby (1992) reviewed the empirical investigations using the Wason selection task, the cumulative results of which demonstrate the operation of a cheater–detector mechanism that is domain-specific insofar as it is particularly attuned to rule breaking in the context of social contracts. To test whether subjects' success with problems of logic when framed in social–contractual terms simply involves an application and extension of formal logical reasoning to social contracts rather than a demonstration of domain-specific social-contractual reasoning independent of formal logic, Cosmides (1989) presented subjects with *reversed* social contracts.

Consider the *standard* form of the social-contract problem, "If you give me your watch, I'll give you $20." This conditional rule states *if A then B*, in which *A* corresponds to "you give me your watch," and *B* corresponds to "I'll give you $20." More generally, this rule specifies that I receive the benefit (your watch) only if I pay the cost ($20). The *reversed* form of this conditional hypothesis states: "If I give you $20, then you give me your watch," that is, *if A then B*, in which *A* now refers to "I give you $20," and *B* now refers to "you give me your watch." More generally, the reversed social contract specifies that if I pay the cost ($20), I receive the benefit (your watch). The parameters of the social contract are identical when framed in the standard and reversed forms. The conditional rule is violated only when *A* is true but *B* is false.

When framed in the standard form, the logically correct selection is to turn over the *A* card to see if you give me your watch but I fail to pay you $20, and the *not-B* card to see if you fail to give me your watch although I have paid you $20. The logically correct selection happens to coincide with the adaptive selection—the selection produced by the hypothesized cheater-detector algorithm.

When this same social contract is framed in its reversed form, however, the logically correct selection differs from the selection generated according to a cheater-detector algorithm. The logically correct selection again entails turning over the *A* and *not-B* cards, because a violation is indicated only if *A* is true but *B* is false. The adaptive selection entails determining whether cheating has occurred (i.e., whether a benefit has been received at no cost, or a cost has been paid but no benefit received). Thus, the adaptive selection is to turn over the *B* card, to see if I fail to pay you the $20 cost for the watch you have given me, and the *not-A* card to see if you fail to give me your watch although I have paid you $20. Fully 75% of the subjects in the Cosmides (1989) study made the adaptive selection, consistent with the operation of a specialized cheater-detector mechanism that is functionally independent of logical reasoning.

Cosmides and Tooby's (1992) empirically substantiated conception of cheating and betrayal in social-exchange relationships was derived from a more general theory of human functioning that has successfully predicted variations in cognition, emotion, and behavior in hundreds of arenas, most of which are unrelated to cheater detection. Alternative theories of interpersonal cheating and betrayal such as social exchange theory (Blau, 1964; Homans, 1961; Thibaut & Kelley, 1959) equity theory (Adams, 1965; Messick & Cook, 1983; Walster, et al., 1978), and Clark and her colleagues' theory of communal and exchange relationships (Clark, 1984; Clark & Mills, 1979; Clark & Waddell, 1985) were not derived from any similarly empirically and theoretically substantiated metatheory of organismic functioning, much less human functioning in particular.

COSMIDES AND TOOBY'S CHEATER-DETECTOR MECHANISM MAY BE TOO DOMAIN-GENERAL

Within their larger argument for the domain-specific design of the human mind, Cosmides (1989) and Cosmides and Tooby (1992) implied the operation of a relatively domain-general cheater-detector mechanism. This mechanism is assumed to operate uniformly as long as the problem is structured as a social contract. Presumably, it matters little whether the social contract is made between long-term mates, close friends, or coalition members, for example. The results of a study I conducted with David Buss (Shackelford & Buss, 1996a; see later) suggested an important extension of Tooby and Cosmides' work on cheater detection. In that study, we demonstrated that Cosmides and Tooby's cheater-detector mechanism may be more domain-specific than their research suggests. We found that perceived betrayal varies in predictable ways with the type of social-contract relationship—mateship, friendship, or coalition—and with the domains within a given social-contract relationship.

DIFFERENT RELATIONSHIPS, DIFFERENT ADAPTIVE PROBLEMS, DIFFERENT SOLUTIONS, AND THE DESIGN OF THE MIND: SHACKELFORD AND BUSS (1996a)

Several relationship contexts are likely to have been recurrent features of the interpersonal "adaptive landscape" (Buss, 1991) over human evolutionary history. Among these significant relationship contexts are long-term mateships, close same-sex friendships, and same-sex coalitional relationships. A *long-term mateship* refers to a heterosexual, committed, presumptively monogamous, romantic, sexual relationship between two persons. A *close same-sex friendship* refers to a close, nonromantic, nonsexual relationship between two persons of the same sex. A *same-sex coalitional relationship* refers to a nonromantic, nonsexual relationship between two members of a single-sex group that is organized to accomplish specific goals. Each of these relationship contexts appears to be cross-culturally universal (Argyle & Henderson, 1984; Brown, 1991; Buss &

Schmitt, 1993; Daly & Wilson, 1988; Harcourt & deWaal, 1992). Long-term mateships, same-sex friendships, and same-sex coalitions pose a collection of adaptive problems, some of them unique to a particular relationship context, some of them shared by two or perhaps all three relationship contexts. An adaptive problem particular to mateships, for example, might be ensuring a mate's sexual and romantic fidelity. An adaptive problem posed by a close same-sex friendship might be ensuring that a friend does not reveal to others information given in confidence. Finally, an adaptive problem that likely characterized coalitional relationships throughout human evolutionary history was ensuring that each coalition member performed the tasks expected of him or her.

Along with the adaptive problems posed by each of these relationships came some set of expected benefits. Some of the benefits expected within one context may or may not have been expected within another context. Exclusive sexual access to one's relationship partner, for example, is likely to have been an expected benefit of a mateship, particularly for men, given paternity uncertainty (Buss, Larsen, Westen, & Semmelroth, 1992; Daly, Wilson, & Weghorst, 1982; Wilson & Daly, 1992). Sexual exclusivity, however, is not likely to have been a reliable and expected benefit of close same-sex friendships and coalitions. One benefit that may have been reliably associated with and expected within all three relationship contexts is that one's long-term mate, close friend, or fellow coalition member counter public derogation of one's reputation.

One way to predict which behaviors will be interpreted as betrayals of a particular relationship is to identify the benefits that might have been reliably associated with participation in a given relationship over human evolutionary history. This set of resources will come to be expected as a benefit of participating in the relationship if we assume that relationship participants incurred opportunity costs proportional to their participation in the focal relationship. The psychological mechanisms, or sets of decision rules, activated in the relationship context will have evolved a sensitivity to the provision of the resources or benefits expected within that context. Those early humans who were not sensitive to whether they received particular benefits would have suffered in fitness currencies relative to those who were sensitive (Cosmides, 1989; Cosmides & Tooby, 1992). A general proposal guiding our investigation into relationship betrayal (Shackelford & Buss, 1996a) was that humans may possess specialized psychological mechanisms that evolved to solve the adaptive problems faced by the potential diversion of resources to persons outside the focal relationship. This evolved "psychology of betrayal" might serve one of two functions: to deter the cheater from committing future betrayals, or to incline the betrayed person to terminate the relationship and search for one in which the expected benefits are not diverted.

We reasoned that the psychological mechanisms sensitive to betrayal should operate in the same manner in those domains having benefits that are common across relationships, and differently in those domains having benefits that are unique to relationship type. We investigated three interpersonal domains with regard to perceived betrayal: extrarelationship intimate involvement, intrarelationship reciprocity, and relationship commitment. We tested several hypotheses across the three relationship domains and across the three relationship contexts via perceived betrayal judgments.[3]

EXTRARELATIONSHIP INTIMATE INVOLVEMENT AND PERCEPTIONS OF BETRAYAL IN MATESHIPS, FRIENDSHIPS, AND COALITIONS

In this section, I elaborate on the hypotheses of Shackelford and Buss (1996a) about perceptions of betrayal across long-term mateships, close same-sex friendships, and same-sex coalitions with regard to one of the three domains we investigated: extrarelationship intimate involvement. Given space limitations, I have selected only this one domain because the results are so clearly suggestive of the relative domain-specificity of human betrayal psychology. The hypotheses, results, and discussion of our investigation into the remaining two domains, intrarelationship reciprocity and relationship commitment, can be found in Shackelford and Buss (1996a).

Much of the empirical research generated within an evolutionary psychological framework on the causes, consequences, and variations in human jealousy divides this conceptual realm into two fundamental dimensions of jealousy: sexual jealousy and romantic–emotional jealousy (Buss et al., 1992; Buss & Schmitt, 1993; Shackelford & Buss, 1996b; Wiederman & Allgeier, 1993; Wilson & Daly, 1992). We followed this convention in developing hypotheses about perceptions of betrayal as a function of extrarelationship intimate involvement, considering two core dimensions of such involvement: extrarelationship sexual involvement and extrarelationship romantic–emotional involvement. These two dimensions can be considered relatively independent, although not mutually exclusive, dimensions of extrarelationship intimate involvement.

---

[3]With regard to the Shackelford and Buss (1996a) study, relationship context and relationship domain are not synonymous concepts. The *context* of a relationship refers to the basic structural organization of the relationship. We examined three contexts, each with very different structural parameters, only a small subset of which are shared by two or three of the contexts. Within each context, we examined three interpersonal *domains* of relationship functioning: extrarelationship involvement, intrarelationship reciprocity, and relationship commitment. We selected these three domains for logistical reasons (we could not possibly study all relationship domains in a single investigation) but, more importantly, because these domains represent relatively independent arenas of relational discourse. Moreover, these three domains represent a gross cross-section of the variety of domains relevant to any particular relationship context.

Extrarelationship Sexual Involvement

In a presumed monogamous society, exclusive sexual access is a resource expected by both partners in a committed mateship (Buss et al., 1992; Buss & Schmitt, 1993; Wiederman & Allgeier, 1993; Wilson & Daly, 1992). Human reproduction entails fertilization internal to women. Consequently, men, but not women, over evolutionary history confronted the adaptive problem of uncertain genetic relatedness to putative offspring. A mate's sexual infidelity placed ancestral men at risk of investing in offspring to whom they were genetically unrelated. Those men who were indifferent to the sexual fidelity of their mates are less likely to have been our evolutionary ancestors, for they would have been out-reproduced by men who invested effort in retaining exclusive sexual access to their mates. A "psychology of betrayal" in a man sensitive to the sexual infidelity of his mate can be understood as a solution to the adaptive problem of threatened cuckoldry.[4]

Although women have not faced the adaptive problem of uncertain parentage, the sexual infidelity of their mate likely provides a cue to the potential or current loss of other reproductively valuable mateship-specific resources (Daly et al., 1982). A woman may fear that the resources her mate contributes will be diverted to another woman and her children (Buss & Schmitt, 1993; Daly & Wilson, 1988). A "psychology of betrayal" in a woman sensitive to the sexual infidelity of her mate can be understood as a solution to the adaptive problem of threatened loss of reproductively valuable resources (Buss et al., 1992; Buss & Schmitt, 1993).

We hypothesized that the betrayal felt in response to a mate's extrarelationship sexual involvement would be more intense when it occurs with an enemy of the mate's partner, than when it occurs with a stranger. Not only is exclusive sexual access lost; in addition, it is lost to a competitor. Even more damaging would be the case in which one's mate engages sexual relations with one's close friend. Disruption of a close reciprocal alliance would occur in addition to the loss of exclusive sexual access. Two relationships are damaged when sexual philandering occurs with a mate's close friend. A mate's sexual involvement with his or her partner's enemy similarly damages the mateship. Yet such involvement does not threaten a valuable reciprocal relationship in the process. Thus, we predicted that betrayal would be greater in response to a mate's sexual relations with a close friend than with an enemy of his or her partner.

Betrayal in Mateships

Hypothesis 1a: Extrarelationship sexual involvement will be perceived to be a greater betrayal of a mateship than of a same-sex friendship or coalitional relationship, from the perspective of both men and women.

---

[4]We are not implying conscious recognition of the threat of cuckoldry. As with the operation of most psychological processes, anticuckoldry mechanisms need not be functionally dependant on their host's immediate attention and awareness.

Hypothesis 1b: Perceived betrayal will be greater when the sexual involvement is with a *close friend* of the mate's partner than when such involvement is with *someone previously unknown* to the mate's partner.

Hypothesis 1c: Perceived betrayal will be greater when the sexual involvement is with an *enemy* of the mate's partner than when such involvement is with *someone previously unknown* to the mate's partner.

Hypothesis 1d: Perceived betrayal will be greater when the sexual involvement is with a *close friend* of the mate's partner than when such involvement is with an *enemy* of the mate's partner.

In the context of the typical close same-sex friendship or coalition, sexual involvement outside of the relationship is not likely to engender feelings of betrayal, assuming otherwise appropriate relationship participation. Exclusive sexual access is typically not a resource garnered from these relationships. If the sexual involvement of a close friend or fellow coalition member is with one's mate, however, feelings of betrayal will ensue. Also, if the extrarelationship sexual involvement is with a personal or coalitional enemy, feelings of betrayal will arise. In both relationship contexts, feelings of betrayal will be greater when the sexual involvement is with the mate of the other relationship member than when such involvement is with an enemy of the other relationship member. Loss of exclusive sexual access to a mate is likely to prove more reproductively damaging than the loss of an alliance to a personal or coalitional enemy.

Betrayal in Friendships

Hypothesis 2a: Perceived betrayal will be greater when the sexual involvement is with the *mate* of the close friend than when such involvement is with *someone previously unknown* to the close friend.

Hypothesis 2b: Perceived betrayal will greater when the sexual involvement is with an *enemy* of the close friend than when such involvement is with *someone previously unknown* to the close friend.

Hypothesis 2c: Perceived betrayal will be greater when the sexual involvement is with the *mate* of the close friend than when such involvement is with an *enemy* of the close friend.

Betrayal in Coalitions

Hypothesis 3a: Perceived betrayal will be greater when the sexual involvement is with the *mate* of the fellow coalition member than when such involvement is with *someone previously unknown* to the coalition members.

Hypothesis 3b: Perceived betrayal will be greater when the sexual involvement is with a coalitional *enemy* than when such involvement is with *someone previously unknown* to the coalition members.

Hypothesis 3c: Perceived betrayal will be greater when the sexual involvement is with the *mate* of the fellow coalition member than when such involvement is with a coalitional *enemy*.

Extrarelationship Romantic–Emotional Involvement

We predicted that extrarelationship romantic–emotional involvement would elicit feelings of betrayal in the context of a mateship, for both men and women (Buss et al., 1992; Buss & Schmitt, 1993; Wiederman & Allgeier, 1993; Wilson & Daly, 1992). A woman may fear that the resources her mate contributes to their relationship will be diverted to another woman and her children (Buss & Schmitt, 1993; Daly & Wilson, 1988). A man may fear that the romantic–emotional involvement of his mate with another man will escalate to sexual involvement, potentially rendering him a cuckold—and his investing in children fathered by another man (Buss & Schmitt, 1993; Daly et al., 1982). Betrayal felt in response to a mate's extrarelationship romantic involvement might be particularly intense when it occurs with an enemy or rival of the mate's partner. Not only are reproductively valuable resources threatened; in addition, they are threatened to be lost to a competitor. More damaging would be the case in which a mate becomes romantically involved with his or her partner's close friend: Reproductively valuable resources are threatened, and a close reciprocal alliance is undermined in the process. Two relationships are damaged when romantic involvement occurs with a mate's close friend. A mate's romantic entanglement with their partner's rival similarly damages the mateship. Such involvement does not, however, threaten a valuable reciprocal relationship in the process. Thus, we predicted that betrayal will be greater in response to a mate's romantic–emotional relations with a close friend than with an enemy of his or her partner.

Betrayal in Mateships

Hypothesis 4a: Extrarelationship romantic–emotional involvement will be perceived to be a greater betrayal of a mateship than of a same-sex friendship or coalitional relationship, from the perspective of both men and women.

Hypothesis 4b: Perceived betrayal will be greater when the romantic–emotional involvement is with a *close friend* of the mate's partner than when such involvement is with *someone previously unknown* to the mate's partner.

Hypothesis 4c: Perceived betrayal will be greater when the romantic–emotional involvement is with an *enemy* of the mate's partner than when such involvement is with *someone previously unknown* to the mate's partner.

Hypothesis 4d: Perceived betrayal will be greater when the romantic–emotional involvement is with a *close friend* of the mate's partner than when such involvement is with an *enemy* of the mate's partner.

In the context of the typical close same-sex friendship or coalition, romantic involvement outside of the relationship is not likely to elicit feelings of betrayal, assuming otherwise appropriate relationship participation. Exclusive romantic–emotional investment is not usually a resource garnered from these relationships. If, however, the romantic involvement of a close friend or fellow coalition member is with the other relationship member's mate, feelings of betrayal will arise. Also, if the extrarelationship romantic involvement is with a personal or coalitional enemy, feelings of betrayal will result. In both relationship contexts, these feelings of betrayal are predicted to be greater when the romantic–emotional involvement is with the mate of the other relationship member than when such involvement is with an enemy of the other relationship member. The loss of resources garnered within a mateship is likely to prove more reproductively damaging than is the loss of an alliance to a personal or coalitional enemy.

Betrayal in Friendships

Hypothesis 5a: Perceived betrayal will be greater when the romantic–emotional involvement is with the *mate* of the close friend than when such involvement is with *someone previously unknown* to the close friend.

Hypothesis 5b: Perceived betrayal will be greater when the romantic–emotional involvement is with an *enemy* of the close frien than when such involvement is with *someone previously unknown* to the close friend.

Hypothesis 5c: Perceived betrayal will be greater when the romantic–emotional involvement is with the *mate* of the close friend than when such involvement is with an *enemy* of the close friend.

Betrayal in Coalitions

Hypothesis 6a: Perceived betrayal will be greater when the romantic–emotional involvement is with the *mate* of the
fellow coalition member than when such involvement
is with *someone previously unknown* to the coalition
members.

Hypothesis 6b: Perceived betrayal will be greater when the romantic–emotional involvement is with a coalitional *enemy*
than when such involvement is with *someone previously
unknown* to the coalition members.

Hypothesis 6c: Perceived betrayal will be greater when the romantic–emotional involvement is with the *mate* of the fellow
coalition member than when such involvement is with a
coalitional *enemy*.

Sex Differences in Perceived Betrayal

We predicted that both sexes would feel betrayed by the sexual or romantic–emotional infidelity of a long-term mate. Studies that do not disassociate
sexual from romantic mate infidelity find no sex differences in what are
effectively global measures of betrayal (Wiederman & Allgeier, 1993). When
sexual and romantic infidelity are disassociated, however, men display greater
psychological, physiological, and behavioral distress while imagining a mate's
sexual infidelity, whereas women manifest greater distress while imagining a
mate's romantic–emotional infidelity (Buss et al., 1992). The pressing adaptive
problem for mated men is the threat of cuckoldry associated with a mate's
sexual infidelity. The pressing adaptive problem for mated women is the
threatened loss of reproductively valuable resources associated with a mate's
romantic–emotional involvement with another woman. For the mated
woman, the adaptive concern confronted is not the sexual infidelity of her
mate per se, but, rather, it is the threatened diversion of his time and resources
to this other woman in a bartering effort to gain, and perhaps retain, sexual
access to her.

Hypothesis 7a: Men will perceive greater betrayal in the *sexual* infidelity
of a mate, regardless of the identity of the third party
(not previously known, close friend, or enemy of the
man).

Hypothesis 7b: Women will perceive greater betrayal in the *romantic–emotional* infidelity of a mate, regardless of the identity
of the third party (not previously known, close friend, or
enemy of the woman).

METHOD

Subjects

Subjects were 204 undergraduates (89 men, 115 women) at a large Midwestern university fulfilling a requirement for their introductory psychology course. Of these subjects, 82% were Caucasian. The mean age of the subjects was 18.76 years, with ages ranging from 17 to 27 years.

Materials

The questionnaire opened with several biographical questions, followed by a section on the subjects' romantic relationship history. Next, subjects responded to three sections of 22 items each (66 items, in total). Each of the items within a given section depicted a different interpersonal interaction between the target person and his or her relationship partner. For each section of items, a different relationship context was depicted. Thus, subjects responded to three sets of 22 items in which the target person interacted with his or her mate, close same-sex friend, and same-sex fellow coalition member, respectively. A given item was presented across the three relationship contexts in identical form, except for the substitution of the words *partner*, *close friend*, and *fellow coalition member*. The sex of the target person matched the sex of the respondent, so that men responded to items about John, whereas women responded to items about Mary. Because some of the items dealt with the target's romantic partner across all three contexts, the target was described as currently involved in a committed, romantic, sexual relationship.

The instructions given to women for the *mateship* section read:

> Mary S., a female college student, is involved in a committed, romantic, sexual relationship. The man Mary is involved with is referred to as her "partner." The statements below contain acts that might occur in Mary's committed, romantic, sexual relationship.

Male subjects read similar instructions, except that "Mary" was replaced by "John," in addition to other sex-specific word substitutions.

The instructions given to women for the *friendship* section read:

> Mary S., a female college student, is involved in a close friendship with another female. Mary is also involved in a committed, romantic, sexual relationship. The man she is involved with is referred to as her "partner." The statements below contain acts that might occur in Mary's close friendship.

Again, male subjects read similar instructions, except that "Mary" was replaced by "John," in addition to other sex-specific word substitutions.

The instructions given to women for the *coalition* section read:

Mary S., a female college student, is a member of a coalition. A coalition is a group of people who work together to accomplish specific goals. In the following statements, Mary's fellow coalition member is a female. Importantly, Mary is NOT close friends with this particular coalition member. Mary is also involved in a committed, romantic, sexual relationship. The man Mary is involved with is referred to as her "partner." The statements below are acts that might occur in Mary's coalition.

Male subjects read similar instructions, except that "Mary" was replaced by "John," in addition to other sex-specific word substitutions.

Subjects were instructed to rate, for each item, how betrayed the target person would feel if the particular incident were to occur. The scale provided ranged from $0 = $ not at all betrayed and $8 = $ unbearably betrayed. Finally, subjects responded to a small set of forced-choice questions: 11 for the mateship context, and 8 each for the friendship and coalitional contexts. The forced-choice questions were included to provide an alternative means of testing several of the hypotheses. The two items comprising a given forced-choice question were identical to two of the items presented in the main section of the survey. Subjects were asked to select which one of the two situations would lead the target person to experience greater feelings of betrayal.

Procedures

Subjects from a larger departmental subject pool were randomly scheduled to participate in an experiment entitled "Relationship Betrayal," with approximately 30 subjects of the same sex participating in each session.

RESULTS

Ipsatization of Betrayal Ratings

Regardless of the content of the item, we found subjects imputing greater betrayal when a given act occurred within a mateship than when the act occurred in a friendship and coalition. According to the sign test for two groups (Hays, 1988, pp. 139–141), the probability that all 66 rated survey items would elicit greater betrayal in mateships than in friendships, assuming an equal likelihood that either context might elicit the greater betrayal, is $p = 1.36 \times 10^{-20}$. Additionally, all 66 items elicited greater betrayal in the mateship and friendship contexts than in the coalitional context, $p = 1.36 \times 10^{-20}$ in each case. Also, across contexts, women offered higher betrayal ratings than men for 56 of the 66 items. According to the sign test for two groups, $p = 7.1 \times 10^{-15}$.

To accurately test the hypotheses, it was necessary to ipsatize betrayal ratings for all items, which effectively standardized ratings within relationship context and within sex. Our interest was in determining whether differences in perceived betrayal occur as a function of the relationship context in which the particular act is embedded over and above the general tendency for a particular act to elicit greater perceived betrayal of a mateship than of a friendship or coalition, and of a friendship than of a coalition. Similarly, we predicted several sex differences in perceived betrayal over and above the general tendency for women to report greater perceived betrayal than men for a given act, regardless of the context in which that act was embedded. Although we had not fully anticipated such striking effects of relationship context and sex of subject, it nevertheless would have been inappropriate to test our hypotheses on the nonipsatized betrayal ratings. Because many (a priori) t tests were conducted, and because we ipsatized betrayal ratings post hoc, we adopted the conservative strategy of (a) resetting alpha from .05 to .01, unless otherwise indicated, and (b) employing two-tailed alpha regions to reduce the probability of Type I error.

Extrarelationship Sexual Involvement

Table 4.1 displays ipsatized betrayal ratings as a function of the relationship context. As shown in Table 4.1, betrayal ratings given in response to discovering a relationship partner *in flagrante delicto* with someone previously unknown to the target are highest when the relationship partner is the target's mate, relative to when the relationship partner is the target's close friend or fellow coalition member. Thus, Hypothesis 1a was supported.

Table 4.2 displays ipsatized betrayal ratings as a function of the identity of the third party, and is organized according to type of extrarelationship involvement. Table 4.2 shows that betrayal ratings are highest when the target discovers his or her mate having sex with the target's best friend, relative to when the third party is the target's enemy or someone previously unknown to the target. Betrayal ratings given when the target discovers his or her mate having sex with an enemy are equivalent to those given when the third party was someone previously unknown to the target. Thus, hypotheses 1b and 1d were supported, whereas hypothesis 1c was not.

Also shown in Table 4.2, perceived betrayal ratings are highest when the target discovers a close friend or fellow coalition member having sex with the target's mate, relative to when the third party is the target's personal or coalitional enemy or someone previously unknown to the target. Also, betrayal ratings are higher when the third party is a personal or coalitional enemy than when the third party is someone previously unknown to the target. Thus, hypotheses 2a through 3c were supported.

TABLE 4.1

Ipsatized Betrayal Ratings as a Function of Relationship Context

	Relationship Context		
	Mateship	Friendship	Coalition
	Mean	Mean	Mean
Item (Male Form)	(SD)	(SD)	(SD)
Extrarelationship Sexual Involvement			
Hypothesis 1a			
John witnessed his [relationship partner] having sex with someone previously unknown to John.	1.08[a] (.37)	-.94b (.78)	-.82[c] (.54)
Extrarelationship Romantic–Emotional Involvement			
Hypothesis 4(a)			
John's [relationship partner] fell in love with someone previously unknown to John.	0.58[a] (.44)	-1.25b (.63)	-.80[c] (.49)

Note. $N = 204$. Mean ipsatized ratings with different superscript letters differ at $p \leq .001$, correlated means t-tests (two-tailed). In the survey completed by subjects, *relationship partner* is substituted with *partner, close friend,* or *fellow coalition member.*

TABLE 4.2

Ipsatized Betrayal Ratings as a Function of Identity of Third Party (Nonrelationship Member)

	Identity of Third Party			
	Previously Unknown	Best Friend	Enemy	Mate
	Mean	Mean	Mean	Mean
Item (Male Form)	(SD)	(SD)	(SD)	(SD)
Extrarelationship Sexual Involvement **Hypotheses 1b through 1c**				
John witnessed his partner having sex with . . .	1.08[a] (.37)	1.18[b] (.36)	1.10[a] (.35)	N/A
Hypotheses 2a through 2c				
John witnessed his close friend having sex with . . .	-.94[a] (.78)	N/A	-.15[b] (.83)	1.70[c] (.57)
Hypotheses 3a through 3c				
John witnessed his coalition member having sex with . . .	-.82[a] (.54)	N/A	-.07[b] (.78)	2.06[c] (.81)
Extrarelationship Romantic–Emotional Involvement				
Hypotheses 4b through 4c				
John's partner fell in love with . . .	.58[a] (.44)	.91[b] (.43)	.78[c] (.40)	N/A
Hypotheses 5a through 5c				
John's close friend fell in love with . . .	-1.24[a] (.64)	N/A	-.46[b] (.68)	1.13[c] (.67)
Hypotheses 6a through 6c				
John's coalition member fell in love with . . .	-.80[a] (.49)	N/A (.61)	-.40[b]	1.08[c] (.80)

Note. $N = 204$. Mean ipsatized ratings with different superscript letters differ at $p \leq .001$, correlated means t tests (two-tailed).

Extrarelationship Romantic–Emotional Involvement

As shown in Table 4.1, betrayal ratings for the case in which the relationship partner falls in love with someone previously unknown to the target are highest when the relationship partner is the target's mate, relative to when the relationship partner is the target's close friend or fellow coalition member. Thus, hypothesis 4a was supported. Table 4.2 reveals that betrayal ratings are highest when the target's mate falls in love with the target's best friend, relative to when the third party is the target's enemy or someone previously unknown to the target. Also, higher betrayal ratings are given when the target's mate falls in love with the target's enemy than when he or she falls in love with someone previously unknown to the target.

Table 4.2 also shows that subjects perceive the greatest betrayal when the target's close friend or fellow coalition member falls in love with the target's mate, relative to when the third party is the target's personal or coalitional enemy, or someone previously unknown to the target. Moreover, subjects perceive greater betrayal when the target's close friend or coalition member falls in love with a personal or coalitional enemy than when the third party is someone previously unknown to the target. Thus, hypotheses 4b through 6c were supported.

Sex Differences in Perceived Betrayal

Table 4.3 displays ipsatized betrayal ratings as a function of sex. Table 4.3 shows that men perceive greater betrayal than women when a mate is sexually unfaithful, regardless of the identity of the third party. Women perceive somewhat greater betrayal than men when a mate falls in love with the target's best friend. The sexes perceive equal betrayal when a mate falls in love with an enemy of the target and with someone previously unknown to the target. Thus, for the rated items, hypothesis 7a was supported, whereas hypothesis 7b was not supported.

Table 4.4 displays betrayal ratings given in response to a mate's extrarelationship sexual versus romantic involvement as a function of sex. When subjects are forced to select which situation would elicit greater betrayal, men select sexual infidelity more often than do women, whereas women select romantic emotional infidelity more often than do men. This sex difference is true regardless of the identity of the third party. Thus, hypotheses 7a and 7b were supported by the forced-choice probes.

DISCUSSION

Extrarelationship Sexual Involvement

The extrarelationship sexual involvement of a mate, relative to that of a friend or coalition member, elicited greater feelings of betrayal regardless of the

TABLE 4.3

Ipsatized Betrayal Ratings as a Function of Sex

	Sex		
	Male Mean	Female Mean	
Item	(SD)	(SD)	Significance[a]
Extrarelationship Sexual Involvement (Mateship)			
Mary/John witnessed her/his partner having sex with someone she/he didn't know.	1.16 (.40)	1.02 (.32)	$p = .005$
Mary/John witnessed her/his partner having sex with her/his best friend.	1.24 (.36)	1.13 (.36)	$p = .023$
Mary/John witnessed her/his partner having sex with someone she/he considered an enemy.	1.18 (.39)	1.04 (.31)	$p = .004$
Extrarelationship Romantic–Emotional Involvement (Mateship)			
Mary's/John's partner fell in love with her/his best friend.	.85 (.48)	.96 (.38)	$p = .076$
Mary's/John's partner fell in love with someone she/he considered an enemy.	.84 (.38)	.74 (.42)	n.s.
Mary's/John's partner fell in with someone she/he didn't know.	.63 (.46)	.54 (.43)	n.s.

Note. N = 204.

[a]As per independent means t test (two-tailed).

identity of the third party, thus supporting hypothesis 1a. In a presumed monogamous society, exclusive sexual access is a primary benefit expected of and by both persons involved in a committed mateship (Buss, 1994; Wiederman & Allgeier, 1993; Wilson & Daly, 1992). Because human fertilization is internal to women, men have faced the adaptive problem of uncertain paternity. A woman's sexual infidelity placed her mate at risk of investing in another man's children. Presumably, those men who were not concerned with the sexual fidelity of their mate would have been out-reproduced by those men—our evolutionary ancestors—who were so concerned.

Although women have not faced the adaptive problem of uncertain maternity, the sexual infidelity of a mate likely foreshadowed the potential loss of reproductively valuable mateship-specific resources to another woman and her children (Daly et al., 1982). Those women who were sensitive to this possibility are likely to have been more successful in the reproductive realm, and hence are more likely to be our ancestors (Buss et al., 1992; Buss & Schmitt, 1993).

Hypothesis 1c held that the betrayal elicited by a mate's sexual infidelity would be more intense when it occurs with an enemy of the mate's partner than when it occurs with someone unknown to the mate's partner, because not only would sexual exclusivity be lost, it would be lost to a rival. In fact, the relevant betrayal ratings were equivalent, failing to support hypothesis 1c. Subjects in our study were not imputing greater betrayal to a mate for sexual involvement

TABLE 4.4

Perceived Betrayal in Response to Mate's Sexual vs. Romantic Infidelity as a Function of Sex

Third (Nonmate) Party: Someone Not Previously Known to Mate

Endorsed as a Greater Betrayal	*Sex*		
	Male Count (%)	*Female* Count (%)	
Sexual Infidelity	68 (77.3%)	72 (62.6%)	
Romantic Infidelity	20 (22.7%)	43 (37.4%)	Significance:[a] $p = .02$

Third (Nonmate) Party: Mate's Best Friend

Endorsed as a Greater Betrayal	*Sex*		
	Male Count (%)	*Female* Count (%)	
Sexual Infidelity	66 (74.2%)	71 (61.7%)	
Romantic Infidelity	23 (25.8%)	44 (38.3%)	Significance:[a] $p = .06$

Third (Nonmate) Party: Mate's Enemy

Endorsed as a Greater Betrayal	*Sex*		
	Male Count (%)	*Female* Count (%)	
Sexual Infidelity	67 (75.3%)	68 (59.1%)	
Romantic Infidelity	22 (24.7%)	47 (40.9%)	Significance:[a] $p = .01$

Note. $N = 204$.
[a]As per Pearson Chi-square Test for Independence.

with their partner's enemy, relative to when such involvement occurs with someone previously unknown to the partner. This is true for male and female subjects when considered independently. Although we suspect the operation of a ceiling effect, the failure of hypothesis 1c is as yet inexplicable. It is unclear why subjects do not differentiate extramateship sexual involvement with a partner's enemy from like involvement with someone unknown to the partner. We found strong support for hypotheses 1b and 1d, which held that betrayal in response to a mate's sexual infidelity would be most intense when one mate is involved with the other's close friend. In this situation, a close reciprocal alliance is disrupted in addition to the loss of exclusive sexual access.

Extrarelationship sexuality was not predicted to engender feelings of betrayal in the typical close same-sex friendship or coalition, as sexual exclusivity is not an expected benefit of these relationships. Hypotheses 2a, 2b, 3a, and 3b held that significant feelings of betrayal would arise, however, in the event a close friend or fellow coalition member became sexually involved with a relationship partner's mate, or with a personal or coalitional enemy. We found solid support for these hypotheses, as we did for hypotheses 2c and 3c, which held that the greatest betrayal would be elicited when a friend or coalitional member became involved with a mate because the loss of sexual exclusivity is likely to prove more reproductively damaging than is the loss of an ally to a personal or coalitional enemy.

Extrarelationship Romantic–Emotional Involvement

The extrarelationship romantic involvement of a mate, relative to that of a friend or coalition member, elicited greater feelings of betrayal regardless of the identity of the third party, supporting hypothesis 4a. A woman may fear that her mate's investment in her and their children will be channeled to another woman and her children should her mate develop an emotional attachment to his mistress (Buss & Schmitt, 1993; Daly & Wilson, 1988; Wilson & Daly, 1992). A man, too, might feel betrayed by his mate's romantic–emotional infidelity, fearing that such involvement might escalate to sexual relations, decreasing his certainty that he is the father of his mate's children (Buss et al., 1992; Buss & Schmitt, 1993; Daly & Wilson, 1988; Daly et al., 1982).

Hypothesis 4c held that the betrayal felt by a mate's romantic infidelity would be particularly intense when the infidelity occurs with a rival of the mate's partner: not only would reproductively valuable investment be lost; it would be lost to an intrasexual competitor. We found strong support for this hypothesis, as we did for hypotheses 4b and 4d, which held that betrayal elicited by a mate's romantic infidelity would be most intense when one mate is involved with the other's close friend. In this scenario, a close reciprocal alliance is disrupted in addition to the loss of reproductively valuable resources.

Extrarelationship romantic emotional involvement was not predicted to elicit feelings of betrayal in the friendship and coalitional contexts, because romantic exclusivity is not an expected benefit of these relationships. Hypotheses 5a, 5b, 6a and 6b held that significant feelings of betrayal would arise, however, in the event that a close friend or fellow coalition member became emotionally involved with a relationship partner's mate, or with a personal or coalitional enemy. We found solid support for these hypotheses. Also supported were hypotheses 5c and 6c, which held that the greatest betrayal would be elicited when a friend or coalitional member became involved with a mate, because the loss of romantic–emotional exclusivity is likely to prove more reproductively damaging than is the loss of an ally to a personal or coalitional enemy.

Sex Differences in Perceived Betrayal

Studies that do not distinguish sexual from romantic–emotional mate infidelity do not find sex differences in global betrayal or jealousy (Wiederman & Allgeier, 1993). When the two types of unfaithfulness are disassociated, men manifest greater upset in response to a mate's sexual infidelity, whereas women are more distraught by a mate's romantic–emotional infidelity (Buss et al., 1992). Over evolutionary history men have faced the adaptive problem of uncertain paternity, a problem which is fundamentally exacerbated by a mate's sexual infidelity per se. Accordingly, Hypothesis 7a held that men would experience more betrayal when a mate was sexually unfaithful, regardless of the identity of the third party. Hypothesis 7a was solidly supported. For a mated woman, the adaptive concern is not the sexual infidelity of her partner per se, but the associated diversion of his commitments and investments to another woman and that woman's children. Correspondingly, hypothesis 7b held that women would experience more betrayal when a mate was romantically unfaithful, regardless of the identity of the third party. Hypothesis 7b was supported by the forced-choice items. This hypothesis was not, however, supported by the rating data.

The wording of the sexual and romantic infidelity items may not have been comparable in the intensity of infidelity presented. The sexual infidelity scenario had the mate walking in on his or her partner having sex with someone else. The romantic infidelity scenario presented one mate telling the other that he or she had fallen in love with someone else. A comparable romantic infidelity situation might entail one mate discovering the other investing extensive resources in another person. For example, one mate could find an expensive piece of jewelry together with an affectionate card addressed to an illicit lover. Were such an item employed, we might find that women perceive greater betrayal than do men in the romantic infidelity of a mate.

THREE RELATIONSHIP CONTEXTS, THREE CHEATER-DETECTOR ALGORITHMS

The results of the Shackelford and Buss (1996a) study are inconsistent with the operation of a single, context-general, cheater-detector algorithm. We investigated perceptions of betrayal associated with the extrarelationship intimate involvement of a long-term mate, a close same-sex friend, and a fellow coalition member. Perceptions of betrayal vary across these three relationship contexts. The extrarelationship intimate involvement of a mate elicits striking perceptions of betrayal, regardless of the identity of the third party with whom this involvement occurs. A friend or fellow coalition member's extrarelationship involvement elicits similarly high levels of betrayal only when it occurs with a mate or enemy of the other relationship member. That a long-term mate's, but not a friend or coalition member's, intimate involvement with a stranger elicits intense feelings of betrayal is consistent with the operation of at least two cheater-detector mechanisms. One such mechanism appears to be activated in the long-term mateship context, whereas the other appears to be specific to the two nonmateship contexts.

For the domain of extrarelationship intimate involvement, perceptions of betrayal do not vary significantly across the friendship and coalitional contexts. In our original empirical article, however, we present evidence consistent with the operation of two cheater-detector algorithms, one of which is specific to friendships, the other apparently activated only in the coalitional context. We find, for example, that a relationship partner's insistence on immediately returning a favor elicits significant betrayal in the close friendship context, whereas no such betrayal is manifested in the coalitional context. We also document that a friend's failure to share his or her intimate feelings produces a tremendous sense of betrayal. A fellow coalition member's reluctance to share his or her feelings, however, does not elicit feelings of betrayal. Thus, although the results presented in this chapter are consistent with the operation of two cheater-detector mechanisms, the results of our original study suggest the operation of three such mechanisms, one each for the three relationship contexts we investigated.

More central to this chapter, the results of our research cannot be accounted for by a single cheater-detector mechanism that operates uniformly across mateships, friendships, and coalitions. Working from an evolutionary psychological paradigm, we successfully predicted contextual variations in perceived betrayal associated with a relationship partner's extrarelationship intimate involvement. Moreover, we successfully predicted within-context variation in perceived betrayal associated with a relationship partner's extrarelationship intimate involvement. This predictable within-context variation provides additional information about the design features of the cheater-detector mechanism(s) operative in each context. Utilizing an evolutionary psychological framework, we accurately predicted that a long-term mate's infidelity would elicit greater betrayal when this involvement occurs with a close friend of the duped partner than when such involvement occurs with someone the betrayed considers an enemy. Our results suggest that cheater detection is not a unidimensional enterprise. Rather, the algorithms guiding cheater detection may be more specialized, having evolved as specific solutions to specific adaptive problems posed by different relationship contexts over human evolutionary history.

Finally, this chapter presents data suggesting that the cheater-detector mechanism activated in the long-term mateship context is not identical in design and functioning across the sexes. Instead, employing an evolutionary psychological framework, we successfully predicted that this cheater detection algorithm is more attuned specifically to sexual infidelity when couched within male psychology, but particularly sensitive to romantic–emotional infidelity when it is operative within the female psychological architecture.

CONCLUDING REMARKS

Our evolved "psychology of betrayal" appears to be more domain-specific than previous work indicated. Even the relatively domain-specific theory of cheater detection (Cosmides, 1989; Cosmides & Tooby, 1992) may be too domain-general to account for the context-specific forms of relationship violations that occur across and within mateships, friendships, and coalitions. In our full empirical report, we document that whether a particular action or event is considered a betrayal is a function of both the particular interpersonal domain represented by the action or event and the relationship context in which the action or event occurred. This chapter presents the results of our investigation into the domain of extrarelationship intimate involvement. The picture painted by these results is not of a single, domain-general cheater-detector mechanism uniformly operative across mateships, friendships, and coalitions. Instead, these results are consistent with the functioning of two, relatively domain-specific cheater-detector algorithms, one of which is activated only in the long-term mateships context, the other operative only in the two nonmateship contexts. Additionally, evidence suggests a cheater-detector mechanism that, consistent with our evolutionary psychological predictions, appears to function differently in the long-term mateship psychologies of men and women.

Utilizing an evolutionary psychological framework, Shackelford and Buss (1996a) generated 10 specific predictions about perceptions of betrayal across three relationship contexts, seven of which are presented in this chapter. Alternative theories of interpersonal relationships such as social exchange theory (Blau, 1964; Homans, 1961; Thibaut & Kelley, 1959), equity theory (Adams, 1965; Messick & Cook, 1983; Walster et al. , 1978), and Clark and her colleagues' theory of communal and exchange relationships (Clark & Mills, 1979; Clark & Waddell, 1985) are theories of general relationship functioning. These alternative theories tend not to make specific, testable predictions about perceptions of betrayal across particular relationship contexts. Nor do these alternative theories offer specific and testable predictions about perceived betrayal in particular domains of interpersonal functioning within and across different relationship contexts. Additionally, the predictions we tested were generated a priori from a cogent theory of human psychological functioning. Although alternative theories of interpersonal functioning might be able to explain some of our findings, they can only do so post hoc. Moreover, these alternative theories were not derived from an empirically substantiated metatheory of general organismic functioning, much less from a metatheory of particularly human psychological functioning.

Our research could be extended in a variety of directions to reveal what we suspect is an even more domain-specialized "psychology of betrayal" than our investigation suggests. The research of Shackelford and Buss (1996a) could be extended by investigating other relationship contexts. Do perceptions of betrayal vary across different parent–child contexts such as genetic, step-, and adoptive? Do certain actions elicit betrayal in the context of a same-sex friendship, but not in the context of an opposite-sex friendship? Where do relationship contexts such as doctor–patient and therapist–client fit in, given that they entail elements of close friendships, business arrangements, and parent-child relationships? How might the evolved psychological mechanisms that are sensitive to betrayal process these relationships? What sorts of actions or events might be considered betrayals in these relationships?

Perceived betrayal could also be investigated with respect to other relationship domains and subdomains. Intrarelationship sexuality would be a fruitful domain to examine. Within this general domain, researchers could investigate perceived betrayal associated with presumptions of sexual access, or with an insistence on sexual relations. Might there be sex differences in perceived betrayal in response to an opposite-sex close friend's insistence on sexual relations?

Another future direction for research is identifying the tactical, behavioral, and emotional output of the evolved psychological mechanisms that are attuned to betrayal. Once a perceived betrayal has occurred, what is the betrayed party's reaction? For example, paternity uncertainty may have selected for a more intense and more immediate reaction by men relative to women in response to a long-term partner's real or suspected sexual infidelity. What determines decisions to terminate the relationship in response to a perceived betrayal? What actions deter future betrayals in particular relationship contexts and domains? Are there sex differences in what is considered a reasonable or appropriate response to a particular betrayal in a certain relationship context? For example, paternity uncertainty may have provided selective pressures on male (but not female) psychology such that men perceive a violent—even homicidal—retaliation for a long-term partner's infidelity to be a more reasonable, justifiable response than do women. Future research might also investigate the emotional reactions to a perceived betrayal in a given context and interpersonal domain, such as outrage, sadness, and humiliation. Paternity uncertainty, for example, may have selected for the angry, rageful, even pseudopsychotic reaction more characteristic of men than of women in response to a partner's sexual infidelity (Buss, 1994; Buunk, 1995; Daly & Wilson, 1988; Shackelford & Buss, 1996b; Wiederman & Allgeier, 1993; Wilson & Daly, 1992).

Another important domain of inquiry is how a betrayal is detected. That is, in dyadic contexts, what actions or events lead one relationship partner to suspect that the other has betrayed him or her? These cues to betrayal will, of course, vary with the relationship domain and across relationship contexts. Cues to a long-term mate's sexual infidelity, for example, may differ from cues to a long-term mate's romantic–emotional infidelity. David Buss and I (Shackelford & Buss, 1996b) recently completed a series of studies on this topic. In study 1, we identified 170 cues to a partner's infidelity. In two subsequent studies, for each of the 170 acts we obtained estimates of the probability of infidelity given that an act has been observed (act diagnosticity), and of the likelihood of the act's occurrence given that an infidelity has occurred (act likelihood). We found evidence that within a particular relationship context—long-term mateships—and with regard to a particular domain of interpersonal behavior—sexual and romantic–emotional extrarelationship involvement—the psychological structure of the mind is recognizably domain-specific. There does not appear to be a general infidelity-detection mechanism. Rather, the results of this series of studies are consistent with the operation of one or more mechanisms dedicated to detecting sexual infidelity, and an independent (although functionally related) set of mechanisms designed to identify romantic–emotional infidelity.

Shackelford and Buss (1996a) provide an initial sketch of several design characteristics of the human mind as revealed by perceptions of betrayal within and across mateships, friendships, and coalitions. Betrayal occurs in other relationship contexts, and in relationship domains not explored in our research to date. Betrayal is a probabilistic element of every relationship. Betrayal is a fundamental interpersonal phenomenon that reveals the evolved design of the human mind.

ACKNOWLEDGMENT

I am grateful to the editors and to David Buss for many helpful comments and suggestions on an earlier version of this chapter.

REFERENCES

Adams, J. S. (1965). Inequity in social exchange. In L. Berkowitz (ed.), *Advances in experimental social psychology* (Vol. 2, pp. 267–299). New York: Academic Press.

Alexander, R. D. (1979). *Darwinism and human affairs.* Seattle: University of Washington Press.

Alexander, R. D. (1987). *The biology of moral systems.* Hawthorne, NY: Aldine de Gruyter.

Anderson, J. L., & Crawford, C.B. (1995, July). *Female infanticide: A stochastic simulation model to assess fitness costs and benefits.* Poster presented at the Seventh Annual Convention of the Human Behavior and Evolution Society, Santa Barbara, CA.

Argyle, M. (1992). *The social psychology of everyday life.* New York: Routledge.

Argyle, M., & Henderson, H. (1984). The rules of friendship. *Journal of Social and Personal Relationships, 1,* 211–237.

Bailey, J. M., Gaulin, S., Agyei, Y., & Gladue, B. A. (1994). Effects of gender and sexual orientation on evolutionarily-relevant aspects of human mating psychology. *Journal of Personality and Social Psychology, 66,* 1081–1093.

Baum, A., Fisher, J. D., & Singer, J. E. (1985). *Social psychology.* New York: Random House.

Betzig, L. (1986). *Despotism and differential reproduction: A Darwinian view of history.* Hawthorne, NY: Aldine de Gruyter.

Blau, P. M. (1964). *Exchange and power in social life.* New York: Wiley.

Brehm, S. S., & Kassin, S. M. (1990). *Social psychology.* Boston: Houghton Mifflin.

Brown, D. E. (1991). *Human universals.* New York: McGraw-Hill.

Buss, D. M. (1989). Sex differences in human mate preferences: Evolutionary hypotheses tested in 37 cultures. *Behavioral and Brain Sciences, 12,* 1–49.

Buss, D. M. (1990). Evolutionary social psychology: Prospects and pitfalls. *Motivation and Emotion, 14,* 265–286.

Buss, D. M. (1991). Evolutionary personality psychology. *Annual Review of Psychology, 42,* 459–491.

Buss, D. M. (1994). *The evolution of desire: Strategies of human mating.* New York: Basic Books.

Buss, D. M. (1995). Evolutionary psychology: A new paradigm for psychological science. *Psychological Science, 6,* 1–30.

Buss, D. M., Larsen, R. J., Westen, D., & Semmelroth, J. (1992). Sex differences in jealousy: Evolution, physiology, and psychology. *Psychological Science, 3,* 251–255.

Buss, D. M., & Schmitt, D. P. (1993). Sexual strategies theory: An evolutionary perspective on human mating. *Psychological Review, 100,* 204–232.

Buunk, B. P. (1995). Sex, self-esteem, dependency and extradyadic sexual experience as related to jealousy responses. *Journal of Social and Personal Relationships, 12,* 147–153.

Byrne, D. (1971). *The attraction paradigm.* New York: Academic Press.

Byrne, D., & Clore, G. L. (1970). A reinforcement model of evaluative processes. *Personality: An International Journal, 1,* 103–128.

Byrne, D., Clore, G. L., & Smeaton, G. (1986). The attraction hypothesis: Do similar attitudes affect anything? *Journal of Personality and Social Psychology, 51,* 1167–1170.

Chapman, R. L. (Ed.). (1977). *Roget's International Thesaurus* (4th ed.). New York: Harper & Row.

Clark, M. S. (1984). Record keeping in two types of relationships. *Journal of Personality and Social Psychology, 47,* 549–557.

Clark, M. S., & Mills, J. (1979). Interpersonal attraction in exchange and communal relationships. *Journal of Personality and Social Psychology, 37,* 12–24.

Clark, M. S., Mills, J., & Powell, M. (1986). Keeping track of needs in communal and exchange relationships. *Journal of Personality and Social Psychology, 51,* 333–338.

Clark, M. S., & Waddell, B. (1985). Perceptions of exploitation in communal and exchange relationships. *Journal of Social and Personal Relationships, 2,* 403–418.

Cosmides, L. (1989). The logic of social exchange: Has natural selection shaped how humans reason? Studies with the Wason selection task. *Cognition, 31,* 187–276.

Cosmides, L., & Tooby, J. (1992). Cognitive adaptations for social exchange. In J. Barkow, L. Cosmides, & J. Tooby (Eds.), *The adapted mind: Evolutionary psychology and the generation of culture* (pp. 163–228). New York: Oxford University Press.

Daly, M., & Wilson, M. (1988). *Homicide.* Hawthorne, NY: Aldine de Gruyter.

Daly, M., Wilson, M., & Weghorst, S. J. (1982). Male sexual jealousy. *Ethology and Sociobiology, 3,* 11–27.

Darwin, C. (1859/1958). *The origins of species by means of natural selection or the preservation of favoured races in the struggle for life.* New York: New American Library.

Gangestad, S. W. (1993). Sexual selection and physical attractiveness: Implications for mating dynamics. *Human Nature, 4,* 205–235.

Goldberg, L. (1981). Language and individual differences: The search for universals in personality lexicons. In L. Wheeler (Ed.), *Review of personality and social psychology* (Vol. 2, pp. 141–165). Beverly Hills, CA: Sage.

Greiling, H. (1995). *Women's mate preferences across contexts.* Paper presented at the Seventh Annual Convention of the Human Behavior and Evolution Society, Santa Barbara, CA.

Harcourt, A. H., & deWaal, F. B. M. (Eds.). (1992). *Coalitions and alliances in humans and other animals.* New York: Oxford University.

Hays, W. L. (1988). *Statistics* (4th ed.). Chicago: Holt, Rinehart & Winston.

Homans, G. C. (1961). *Social behavior.* New York: Harcourt, Brace, & World.

Kaplan, S. (1992). Environmental preference in a knowledge-seeking, knowledge-using organism. In J. Barkow, L. Cosmides, & J. Tooby (Eds.), *The adapted mind: Evolutionary psychology and the generation of culture* (pp. 581–598). New York: Oxford University Press.

Kenrick, D. T. (1994). Evolutionary social psychology: From sexual selection to social cognition. In M. P. Zanna (Ed.), *Advances in experimental social psychology* (Vol. 26, pp. 75–122). San Diego, CA: Academic Press.

Kenrick, D. T., Keefe, R. C., Bryan A., Barr, A., & Brown, S. (1995). Age preferences and mate choice among homosexuals and heterosexuals: A case for modular psychological mechanisms. *Journal of Personality and Social Psychology, 69,* 1166–1172.

Kenrick, D. T., Sadalla, E. K., & Keefe, R. C. (in press). Evolutionary cognitive psychology: The missing heart of modern cognitive science. In C. B. Crawford & D. Krebs (Eds.), *Evolution and human behavior: Ideas, issues, and applications.* Mahwah, NJ: Lawrence Erlbaum Associates.

Langlois, J. H., & Roggman, L. A. (1990). Attractive faces are only average. *Psychological Science, 1,* 115–121.

Lockard, J. S., & Paulhus, D. L. (1988). *Self-deception: An adaptive mechanism?* Englewood Cliffs, NJ: Prentice Hall.

Messick, D. M., & Cook, K. S. (Eds.). (1983). *Equity theory: Psychological and sociological perspectives.* New York: Praeger.

Murstein, B. I. (1976). *Who will marry whom?: Theories and research in marital choice.* New York: Springer.

Murstein, B. I. (1987). A clarification and extension of the SVR theory of dyadic pairing. *Journal of Marriage and the Family, 49,* 929–933.

Myers, D. G. (1993). *Social psychology* (4th ed.). New York: McGraw-Hill.

Norman, W. T. (1963). Toward an adequate taxonomy of personality attributes: Replicated factor structure in peer nominations personality ratings. *Journal of Personality and Social Psychology, 66,* 574–583.

Orians, G. H., & Heerwagen, J. H. (1992). Evolved responses to landscapes. In J. Barkow, L. Cosmides, & J. Tooby (Eds.), *The adapted mind: Evolutionary psychology and the generation of culture* (pp. 555–580). New York: Oxford University Press.

Pinker, S. (1994). *The language instinct: How the mind creates language.* New York: William Morrow.

Pinker, S., & Bloom, P. (1992). Natural language and natural selection. In J. Barkow, L. Cosmides, & J. Tooby (Eds.), *The adapted mind: Evolutionary psychology and the generation of culture* (pp. 451–494). New York: Oxford University.

Profet, M. (1991). The function of allergy: Immunological defense against toxins. *The Quarterly Review of Biology, 66,* 23–62.

Sabini, J. (1992). *Social psychology.* New York: Norton.

Shackelford, T. K., & Buss, D. M. (1996a). Betrayal in mateships, friendships, and coalitions. *Personality and Social Psychology Bulletin, 22,* 1151–1164.

Shackelford, T. K., & Buss, D. M. (1996b). Cues to Infidelity. Manuscript submitted for publication.

Shepard, R. N. (1984). Ecological constraints on internal representation: Resonant kinematics of perceiving, imagining, thinking, and dreaming. *Psychological Review, 91,* 417–447.

Shepard, R. N. (1992). The perceptual organization of colors: An adaptation to regularities of the terrestrial world? In J. Barkow, L. Cosmides, & J. Tooby (Eds.), *The adapted mind: Evolutionary psychology and the generation of culture* (pp. 495-532). New York: Oxford University Press.

Symons, D. (1987). If we're all Darwinians, what's the fuss about? In C. Crawford, M. F. Smith, & D. Krebs (Eds.), *Sociobiology and psychology: Ideas, issues, and applications* (pp. 121–146). Hillsdale, NJ: Lawrence Erlbaum Associates.

Thibaut, J. W., & Kelley, H. H. (1959). *The social psychology of groups.* New York: Wiley.

Thornhill, R. (1990). The study of adaptation. In M. Berkoff & D. Jamieson (Eds.), *Explanation, evolution, and adaptation* (pp. 31–61). Boulder, CO: Westview.

Thornhill, R., & Gangestad, S. W. (1993). Human facial beauty: Averageness, symmetry, and parasite resistance. *Human Nature, 4,* 237–269.

Tooby, J., & Cosmides, L. (1990). On the universality of human nature and the uniqueness of the individual: The role of genetics and adaptation. *Journal of Personality, 58,* 17–67.

Tooby, J., & Cosmides, L. (1992). The psychological foundations of culture. In J. Barkow, L. Cosmides, & J. Tooby (Eds.), *The adapted mind: Evolutionary psychology and the generation of culture* (pp. 19–136). New York: Oxford University Press.

Walster, E., Walster, G. W., & Berscheid, E. (1978). *Equity: Theory and research.* Boston: Allyn & Bacon.

Wason, P. (1966). Reasoning. In B. M. Foss (Ed.), *New horizons in psychology*. Harmondsworth: Penguin.

Wiederman, M. W., & Allgeier, E. R. (1993). Gender differences in sexual jealousy: Adaptationist or social learning explanation? *Ethology and Sociobiology, 14,* 115–140.

Wilson, M., & Daly, M. (1992). The man who mistook his wife for chattel. In J. Barkow, L. Cosmides, & J. Tooby (Eds.), *The adapted mind: Evolutionary psychology and the generation of culture* (pp. 289–322). New York: Oxford University Press.

III

Interpersonal Attraction

5

Angels, Mentors, and Friends: Trade-offs among Evolutionary, Social, and Individual Variables in Physical Appearance

Michael R. Cunningham
Perri B. Druen
Anita P. Barbee
University of Louisville

The human face is a product of evolution that serves multiple functions. The face houses the organs for the major senses such as sight and smell and also transmits emotional states through nonverbal expressions. The face conveys the individual's demographic group, such as age and gender and serves as a primary clue for perceivers to recognize individuals from a distance. Most important for a Darwinian analysis, the face is a potent stimulus for social attraction and mating behavior.

Each configuration of features may be thought of as an experiment on the social and physical environment. Each face and body provides an opportunity for natural and sexual selection to increase or decrease the success of the individual conveying that appearance. Success, however, comes in many forms. From an evolutionary perspective, sexual success is the most important, indicated by genetic fitness (the number of reproducing offspring) and inclusive fitness (the number of reproducing kin).

Among humans, several alternate forms of success may serve as mediators of sexual success. Social success, as indicated by status compared to competitors, and by the number of resource-providing friendship alliances, may directly contribute to sexual success. Physical success, including the accumulation of resources for self and kin, hardiness against weather and disease, and longevity that permits continued investment in offspring, also may contribute to genetic and inclusive fitness. Different ecologies, species, groups, and individuals may select different strategies to achieve success, such as being relatively aggressive versus relatively timid, or having a large versus a small number of offspring (Cunningham & Barbee, 1991). Each approach to success confers certain advantages and disadvantages.

A trade-off is a compromise that results from such multiple options, each with its own benefits and costs. Several levels of trade-offs are involved in physical appearance. At the first level, trade-offs are evident in physical form and in the amount of space that each of the various physical organs can occupy on the limited amount of territory available on the face. Nostrils cannot point upward because of rainfall. Eyes cannot be too wide without impinging on the nose. If teeth are too long the jaw cannot be lifted.

The second type of trade-off exists because a single facial or bodily feature cannot be two different shapes or colors simultaneously, and thus cannot convey two opposing qualities that are based upon size or color. Both dominance and submissiveness, for example, may have adaptive value in different environments. But a face perfectly constructed to frighten competitors to gain status and resources may not be ideally configured to attract friends. If both goals are to be accomplished, trade-offs must be made in the form of physical features. One adaption is a face that is neutral at rest, but capable of both frowning and smiling at different times.

The third type of trade-off is involved in social perception and relationship selection. When we choose friends and romantic partners, we would like them to be wonderful in every way, and a perfect fit to our every need. We recognize, however, that no person has every desirable quality and that trade-offs must be made, therefore, in the qualities that we seek in partners. Individual perceiver motives play a role in determining which qualities will be traded off for others. For example, we may seek a dominant partner who can defend us, but also may want a compliant partner who will not dominate us. Our motives, personalities, and needs will determine the extent to which dominance or compliance will prevail in the choices of partners that we make.

In the following sections, we describe how facial and bodily features convey various types of fitness, ways in which trade-offs arise, and how the trade-offs are decided based on the motives of the perceiver.

TRADE-OFFS RESULTING
FROM PHYSICAL SELECTION PRESSURES

The realities of genetics and physical ecologies impose initial constraints on facial and bodily appearance. Ecological forces such as temperature, gravity, and the need for periodic intakes of nutrition produce pressure against extreme values of traits in a species; these are termed stabilizing selection pressures (Plomin, 1981). The human nose, for example, cannot grow too small without interfering with respiration and air filtration, or too large without interfering with vision or eating and the human gene pool lacks the DNA to produce the elephant's trunk. The human iris cannot be too translucently blue without losing its ability to protect the retina from sunlight. The human chin cannot be too small without losing the capacity to bite effectively, nor can it be so large that it becomes too heavy to be supported by the rest of the face.

Average Qualities

The phenotype that is most common in a given ecology may be well adapted to it. Langlois and Roggman (1990) extended this analysis by arguing that appearances close to the population mean will be attractive to conspecifics. Langlois and Roggman (1990) and Langlois, Roggman, and Musselman (1994) created composite images by digitizing a series of faces, and then computer averaging the gray scale values of the pixels in roughly the same places on the different faces. The creation of such composites entails the risk of numerous artifacts, such as exceptional symmetry and freedom from blemishes; this topic was discussed elsewhere (Alley & Cunningham, 1991; Pittenger, 1991). Langlois and Roggman reported that computer averaging produced composite faces that were rated as higher in attractiveness than the mean of the 32 to 96 individual faces used to create the composites. Yet the mean of the female composites was not very high in absolute terms (Langlois & Roggman, 1990, M = 3.25 vs. 2.43; Langlois, et al., 1994, M = 3.46 vs. 2.30 on a 5-point scale). In fact, some individual faces that contributed to the composite were more attractive than the best composite (4.05 vs. 3.26; Langlois & Roggman, 1990), suggesting that beauty requires more than averageness.

The conclusion that averaged faces might be attractive, but not highly attractive, was given further support by Perrett, May, and Yoshikawa (1994). If averageness is the most important quality in beauty, then the more faces that contribute to a composite, the more attractive the composite should be. By contrast, Perrett, et al., (1994), demonstrated that a composite created from the 15 highest rated faces from a group of 60 was rated higher than a composite created from the average of all 60 faces. Furthermore, a *high* + *50%* face was created by exaggerating by a factor of 50% the shape differences between the high attractiveness composite and the average. The *high* + *50%* face was preferred over the high composite and the average composite, both by college students in the United Kingdom and by those in Japan. Thus, moderately extreme shapes enhanced attractiveness.

Extreme Qualities

An average value of a trait may be physically adaptive and moderately socially desirable, but intrasexual competition may favor values of traits that differ from the population mean. Depending on the trait, the most successful, or desirable, male or female morphology may not be equivalent to the average phenotype. Biologists have documented numerous examples of directional selection for extreme characteristics, such as dramatic peacock feathers, large deer antlers, and vivid coloration (Wickler, 1973). Such extreme traits may be of direct benefit in intrasexual competition or may be of indirect benefit in conveying such qualities as parasite resistance to members of the opposite sex (Thornhill & Gangestad, 1994).

Directional and stabilizing selection forces may be in conflict. A trait such as large antlers in deer, or tallness in male humans, may convey advantages in intrasexual competition. But a truly extreme trait, such as gargantuan height, may be maladaptive because of problems with gravity, and in obtaining sufficient nutrition. If the extreme trait is based on a combination of recessive genes, the risks may multiply. Sexual selection pressures also can suppress extreme variations within a subpopulation. A large chin, for example, has been found to be attractive in men, but unattractive in women (Cunningham, Barbee, & Pike, 1990; Keating, 1985). If a man's chin is too small, he may not attract a mate, or his son may experience a disadvantage to his reproductive success. If the man's chin is too large, his daughter may seem unattractive.

The contending advantages of averageness and extremity may result in a trade-off that favors a point of optimal discrepancy from the population mean. The compromise point may be discrepant enough from the average to provide competitive benefits, but close enough to the average to avoid increased costs. The point of optimal discrepancy that produces maximum attractiveness is based both on the effects of physical ecology and on the attunements of perceivers, and may vary from feature to feature.

Physical Limitations and the Capacity
of Features to Communicate Qualities

The Multiple Fitness Model (Cunningham, Roberts, Barbee, Druen, & Wu, 1995, see Fig. 5.1) integrates both evolutionary and social cognitive dynamics to address the complex trade-offs inherent in facial appearance, and in the perception of physical attractiveness and other attributes. The multiple fitness model outlines five categories of physical attractiveness features in the face and body, and links each category of feature to a particular quality that the features convey. The five basic qualities that physical features may communicate, which often are mutually incompatible, include fitness to be a beneficiary,

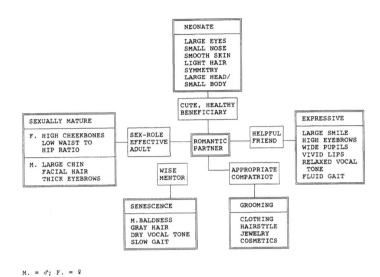

FIG. 5.1. The Multiple Fitness Model's configural specification of attractiveness.

fitness to meet adult challenges, fitness to be a wise mentor, fitness to serve as a nurturant friend, and fitness to be a compatriot.

It might be noted that the term fitness has three overlapping meanings in this model. A target who is perceived to be highly fit on a dimension, such as fit to meet challenges, may actually have greater capability to adapt to ecological and social demands, and, thus, may possess greater genetic and inclusive fitness compared to targets who are perceived to be less fit. In addition, a target who is perceived to be highly fit on a dimension, such as fit as a beneficiary, may stimulate the perceiver to engage in behaviors that increase the target's genetic and inclusive fitness, thus inciting a sort of self-fulfilling prophecy. Finally, a target who is accurately seen as fit on a dimension may "fit" the perceiver's needs, and thereby increase the perceiver's genetic and inclusive fitness. Effectively perceiving the various fitness qualities in perceptual targets may improve social choices, longevity, and reproductive success. Perceptions of fitness may not be veridical, of course, if perceivers rely on nondiagnostic features or if targets engage in deception (Rowatt, Cunningham & Druen, 1995).

Social perception and attraction may involve multiple hurdles, with the fitness of the body evaluated from a distance; the fitness of the face becoming visible a bit closer; and finally, the desirability of the person's voice, sense of humor (Lundy, Cunningham, & Lister, 1995), integrity (Graves & Cunningham, 1995), smell, texture and taste (Cunningham & Rauscher, 1994) becoming apparent with increased time and familiarity. Although this chapter will concentrate on the face, mention will be made of the other dimensions.

Neonate Features: Fitness as a Beneficiary.   Parents who have more children than resources, or relatives with more than one niece, nephew, grandchild, or cousin, must decide in whom to invest scarce benefits. The child with the greatest perceived likelihood of survival and reproduction would be the most likely to increase the inclusive fitness of the kin, and therefore would be the most fit beneficiary for the investment of resources.

Neonatenous features such as large eyes, smooth skin, and a small nose may convey such fitness as a beneficiary and suggest such desirable qualities as youthful vivaciousness, openness, and agreeableness (Berry & McArthur, 1985, 1986). Several studies found that such features are linked to judgments of cuteness and youth, and thus may lead perceivers to believe that the target is entitled to nurturance, caretaking, and other resources (Alley, 1983). Hildebrandt and Fitzgerald (1978), for example, measured the size of various facial features of infants, and found that eye height and width, forehead height and cheek width were positively correlated with cuteness ratings, whereas nose width, ear height, and mouth height were negatively associated with perceived cuteness. Parents who were particularly responsive to the wide eyes and the cute noses of their healthy offspring may have provided better care for them, and thus left more surviving and reproducing children (Alley, 1983), who in turn could have perpetuated responsiveness to neonate cues.

Hair color may also serve as a cue to neoteny. Natural blonde hair color is genetically determined but developmentally linked. It is much more frequent in White babies than adults, suggesting that it may be a neonate cue conveying childlike vulnerability, hedonism, and suitability as a beneficiary. We conducted an experiment in which the same 21 women were portrayed either as blondes or as brunettes by artistically retouching photographs in two stimulus sets, which contained a balanced number of targets with both hair colors. When the targets were blonde, they were rated as more attractive, feminine, emotional, and pleasure seeking. When the same women were brunette, they were seen as more intelligent, suggesting that brunettes are seen as having fewer neonatenous qualities than blonds (Cunningham, Went, Rodenhiser, Roberts, & Richardson, 1989). Brunette hair is also preferred over blonde hair for males (Feinman & Gill, 1978), consistent with the observation that males are expected to be older and more mature than females (Kenrick, Groth, Trost, & Sadalla, 1993). Thus, it is possible that social stereotypes were built on perceptions of children, such that blonde hair suggests both the positive and negative qualities of youthfulness.

The Multiple Fitness Model suggests that what seems cute to adults when they look at babies also looks desirable to adults when they look at other adults. Neonatenous features may be important in the perception of female attractiveness, both because healthy youthfulness is important for female reproductive fitness, and because the elicitation of nurturant responses may help to elicit greater care-taking and resource investment from males. But it is important to recognize that females may wish to nurture and invest in males, and male longevity may benefit from such female social support (Barbee et al., 1992). Furthermore, males who display neonatenous features may seem less threatening, and be more capable of eliciting resources from other resource-rich people. As a consequence, it may be adaptive for females to be attracted to males who possess some neonatenous qualities. Based on careful facialmetric measurements, our lab determined that White adults rated both women's and men's faces as more attractive if they possessed neonatenous large eyes and a relatively small nose (Cunningham 1981, 1986).

Perceptions of the attractiveness of features does not vary greatly across cultures. When we asked Asian and Hispanic students who had recently arrived in America, as well as White American students, to rate the attractiveness of Asian, Hispanic, Black, and White photographed women, we obtained extremely high correlations in attractiveness ratings across groups ($r = .77$ to .99). Not only did these three populations rate the targets with high levels of consistency, but they were similarly influenced by many facial features (see Table 5.1). The stability of facialmetric assessments and attractiveness judgments across ethnic groups was further confirmed in a study conducted in Taiwan, and in another study using American Blacks (Cunningham et al., 1995).

Sexual Maturity Features: Fitness to Meet Adult Challenges

Increased levels of hormones during puberty produce distinct sexual maturations in certain parts of the body, including breasts, and hips, as well as the development of bodily hair in females (Singh, 1993), and muscle mass, genitals, larynx, as well as the development of facial and body hair in males. The face also is transformed: Cheekbones become more prominent and cheeks become thinner in both genders, and male facial and eyebrow hair coarsens (Enlow, 1990; Farkas, 1987; Tanner, 1978).

The Multiple Fitness Model suggests that sexual maturity features convey fitness to meet adult challenges, including such specific tasks as demonstrating virility or fertility, forcefulness and readiness for competition, and strength for parenting. Keating, Mazur, and Segall (1981) found that faces rating high on sexual maturity features were perceived to convey strength, dominance, status, and competency to perceivers of both genders. Some maturity features, such as

TABLE 5.1

Correlations of Female Facialmetrics with Attractiveness Ratings Reported by Six Groups of Judges

Neonatal Features	White S = 48	Hispanic S = 48	Taiwanese S = 20	Black S = 29	Gay Male S = 50	Lesbian S = 50
Eye Area	.57[a]	.48[a]	.45[c]	.57[a]	.38[b]	.22
Eye Height	.62[a]	.55[a]	.48[c]	.44[c]	.25[d]	.12
Eye Width	.51[b]	.39[b]	.41[d]	.64[a]	.44[b]	.31[c]
Nose Area	-.41[c]	-.36[c]	-.40[d]	-.40[c]	-.16	-.05
Nose Length	.18	.15	.08	.21	.35[b]	.21
Nose Tip Width	.12	.08	-.48[c]	-.18	-.04	.08
Nostril Width	.34[c]	.28[d]	.20	.12	.24[d]	.16
Sexual Maturity Features						
Features That Are Appropriate For Females and Males						
Cheekbone Prominence	.29[c]	.36[b]	.17	.40[c]	.33[c]	.19[c]
Cheek Thinness	.37[c]	.44[b]	.36	.24	.20	.36[c]
Facial Narrowness	.43[b]	.45[b]	-.07	.09	-.03	.03
Features That Are Appropriate For Males						
Eyebrow Thickness	-.48[b]	-.48	-.50[c]	-.49[b]	-.12	.02
Chin Area	-.46[b]	-.44[b]	-.59[c]	-.24	-.39	-.33[c]
Chin Length	-.57[a]	-.54[a]	-.38[d]	.02	-.26[d]	.12
Chin Width	-.19	-.17	-.58[c]	-.43[c]	-.29[c]	-.36[b]
Expressive Features						
Eyebrow Height	.56[c]	.44[a]	.17	.43[c]	.09	.09
Smile Area	.46[a]	.30[b]	.35	.43[c]	.35[b]	.35[b]
Smile Height	.41[c]	.38[b]	.26	.35[c]	.23[d]	.23[c]
Smile Width	.36[b]	.29[c]	.42[d]	.38[c]	.55[b]	.55[b]
Upper Lip Width	.08	-.09	.06	.05	-.06	.02
Lower Lip Width	.36[c]	.30[b]	.29	.08	.30[c]	.14
Grooming Features						
Hair Width	.40[a]	.40[b]	.49[a]	.48[b]	.14	.10
Hair Length	.01	.05	.09	.56[b]	.27[d]	.22

Correlations Among Columns of 22 Facialmetric–Attractiveness Relations

	White	Hispanic	Taiwanese	Black	Gay Male	Lesbian
Hispanic	.98[a]					
Taiwanese	.87[a]	.85[a]				
Black	.78[a]	.77[a]	.88[a]			
Gay Male	.78[a]	.76[a]	.87[a]	.80[a]		
Lesbian	.58[b]	.57[b]	.75[a]	.70[a]	.86[a]	

Note. Judges: White, $N = 46$; Hispanic, $N = 13$; Taiwanese, $N = 38$; Black, $N = 29$; Gay Male, $N = 50$; Lesbian, $N = 50$

[a]$p < .001.$ [b]$p < .01.$ [c]$p < .05.$ [d]$p < .10.$

adult jaw profiles, also may convey functional adaptation to the physical environment (Carello, Grosofsky, Shaw, Pittenger, & Mark, 1989).

Sexual maturity features may be subdivided into those that are appropriate for adult males, and those that are appropriate for adult females. Earlier investigators suggested that the development of facial and bodily features that are sex-typed as appropriate for a gender may enhance the appearance of attractiveness (Bar-Tal & Saxe, 1976; Gillen, 1981; Goffman, 1979; Nakdimen, 1984; Singh, 1993). The Multiple Fitness Model further noted that sex-typed features from the other gender may reduce attractiveness. We found that sexual maturity cues predicted to be desirable for females, such as prominent cheek-bones, were associated with greater female attractiveness ratings by Asian, Hispanic, and White American college students. By contrast, sexual maturity features possessed by females that were predicted to be desirable for males, such as thicker eyebrows and longer chins, were associated with diminished attractiveness in the females (Cunningham, et al., 1995).

The Distinction Between Neoteny and Sexual Maturity. All facial and bodily features initially display a neonatal form and then grow with age. The difference between neonatal cues and sexual maturity cues may seem confusing until the distinction between linear development and nonlinear sexual maturation is recognized (Enlow, 1990). Features such as the male chin display one pattern of growth from birth to early adolescence, and a second pattern of transformation from callow adolescence to mature, bearded masculinity. Thus, sexual maturity features indicate postpubescent status more clearly than do diminished levels of neonate cues.

The sexual maturity category also emphasizes the sexually dimorphic nature of development: An androgynous infant appearance develops into a qualitatively different masculine or feminine appearance. An adult female face may appear less mature than an adult male face in some respects (Friedman & Zeibrowitz, 1992), but one continuum of maturation seems insufficient for all comparisons. An adult female's rounded breasts and hips do not seem less sexually mature than an adult male's hairy muscular chest. Female sexual maturity cues, however, may better convey fitness to meet traditionally feminine tasks of reproduction and nurturance, whereas the male chest may better convey fitness for tradition-ally masculine tasks such as establishing dominance and acquiring resources.

Whereas sexual maturity features, and to a lesser extent neonate features, convey fitness for reproduction, other categories of features may convey fitness for various forms of social relationships with the perceiver.

Senescence Features: Fitness as a Mentor

Physical maturation does not end with sexual maturation (Enlow, 1990). Whereas thick hair is more sexually mature than baby-fine hair, it does not seem more mature than baldness. This perplexity can be resolved by classifying thick hair as a sexual maturity feature and baldness as a senescence feature. Grey hair and male pattern baldness are genetically determined and might have adaptive significance for the inclusive fitness of the target and the genetic fitness of the perceiver. The relation of senescence features to kin selection altruism and fitness as a mentor will be described in more detail later.

Expressive Features: Fitness as a Supportive Friend

The physical features that facilitate nonverbal expressions also contribute to the attractiveness of the face. Some expressive features are sexually dimorphic; females tend to display slightly higher set eyebrows and slightly larger lips than males (Tanner, 1978). Because expressive features are influenced by emotions and are controllable by the target, however, they may have less importance as gender cues and may be more influential in conveying positive motivational dispositions to the perceiver.

Expressive cues, such as a larger than average smile, could suggest happiness and congeniality (Keltner, 1995; Lanzetta & Orr, 1986; McGinley, McGinley & Nicholas, 1978); highly arched eyebrows could suggest nonthreatening interest and social approachability (Eibl-Eibesfeldt, 1989; Izard, 1971; Moore, 1985); and vivid lips and dilated pupils could suggest excitement and arousal (Hess, 1965). Thus, expressive features may suggest fitness as a warm, responsive, supportive friend. Attractive expressive features also might accent the effects of other facial attributes, making neonatal features appear more submissive, or sexual maturity features appear more confident (cf. Forgas, 1987).

Friendliness and supportiveness may be necessary in a mate, both for nurturing offspring and for long-term individual health and well-being (Barbee, et al., 1993; Barbee, Yankeelov, & Druen 1992). Individuals generally interpret facial expressions of social motivation similarly around the world (Ekman, et al., 1987, Fridlund, 1991; Scherer & Wallbott, 1994). Cross-cultural consistency in the perception of facial expressions such as smiles could contribute to consistencies in ratings of physical attractiveness.

Other nonverbal channels, such as voice (Raines, Hechtman & Rosenthal, 1990) and gait (Montepare & Zebrowitz-McArthur, 1988) may contribute to the perception of positive expressiveness, although such channels also may convey information about maturity and age.

Grooming Features: Fitness as a Compatriot

Many of the features in the face and body that influence attractiveness judgments are modifiable by individual grooming to enhance or subdue their biological meanings (Liggett, 1974). Given such plasticity, culture may build on evolutionary dynamics by specifying grooming attributes that indicate successful adaptation and fitness (Low, 1979). Males may shave their beards, for example, to appear more youthful and less threatening. Cosmetics may be used to enhance the size of the eyes, smoothness of the skin, prominence of the cheekbones, and fullness of the lips, thereby conveying neoteny, sexual maturity and expressiveness (cf. Cash, Dawson, Davis, Bowen, & Galumbeck, 1989; Maron, 1994).

Although people have a great deal of control over grooming, they may be unaware of the meanings conveyed (cf. Wolf, 1991). Some women and men may have their hair colored to alter their apparent youthfulness. Full, shiny, well-kept hair, for example, may convey both neonatenous vitality and a sexually mature interest in attracting a partner. It might be observed that other species also evaluate fitness based on hair: Rams prefer ewes with full coats of wool over recently shorn ewes (Tilbrook & Cameron, 1989).

The perception of grooming cues may vary across individuals, cultures, and eras (Goffman, 1979; Steele, 1985). A number of grooming features, including clothing (Ashmore, Solomon & Longo, 1990; Wheeler & Eghrari, 1987), suntanning (Miller, Ashton, McHoskey & Gimbel, 1990), teeth form, jewelry, tattooing, as well as hair style, makeup, and body weight, can be used to convey status, group membership, or similarity in values to the perceiver. For example, American women prefer men wearing suits to men wearing tee shirts (Cunningham, et al., 1989).

The foregoing suggests that, beyond mimicking other fitness categories, grooming features may convey fitness as a compatriot by communicating that the target fits into a specific group that may be valued by the perceiver (i.e., healthy young professional, adventurous motorcyclist, conservative Muslim, Indian Brahmin). Being seen as a compatriot may convey the impression that the target is appropriate for a helping alliance. Such an alliance may entail kin selection altruism if genetic similarity to the self is seen as strong (Rushton, Russell & Wells, 1985) or it may be based on the expectation of reciprocal altruism. Happy, expressive people may do more than others to enhance their appearance through their clothing, hair, and jewelry (Diener, Wolsie, & Fujita, 1995). Attractiveness as a compatriot may be conveyed in nearly as many ways as there are human groups.

TRADE-OFFS APPARENT IN THE INTERACTIONS
OF CATEGORIES OF FEATURES

At the level of physical form trade-off, an individual feature, such as a nose, cannot simultaneously convey neoteny, sexual maturity, senescence, expressiveness, and grooming. In addition, each category of feature may convey both positive and negative attributes. At the level of the face as a whole, there may be a virtue in a trade-off of features from different categories to convey an optimally adaptive mixed message. The next section describes facial configurations based on interactions of feature categories.

The Trade-off of Neonatenous and Sexual Maturity Features

By conveying vitality, openness, and youthfulness, neonatal features may suggest desirability as a mate. Yet other qualities suggested by neonatenous features, such as irresponsibility, naivete, and sexual immaturity, seem less desirable. To make a face more romantically attractive, some compensation seems necessary for the negative aspects of neoteny.

One way that the negative implications of neoteny might be reduced would be for development to modify all neotenous features in the direction of greater maturity by changing roundness into angularity (Aronoff, Woike & Hyman, 1992) on all features. Midway through such a developmental process, the face would convey intermediate age, the optimal stage for mating (Symons, 1979). Yet there are grounds for concluding that the appearance of intermediate development may not be sufficient to induce the perception of romantic attractiveness. The low neoteny appearance of small eyes and a large nose could indicate young adulthood, but that configuration may convey only the loss of the desirable qualities of youth, and not the clear gain of the desirable qualities of sexual maturity (Cunningham, 1986).

The Multiple Fitness Model suggests that a romantically ideal face entails a trade-off that includes the retention of some noticeable neonate features, combined with the acquisition of prominent sexual maturity features. Consistent with that prediction, faces rated as highest in physical attractiveness included pronounced neonate features in the center of the face, such as large eyes, and pronounced sexual maturity features at the periphery, such as prominent cheekbones for women (Cunningham, 1986; Johnston & Franklin, 1993) and a large chin for men (Cunningham, et al., 1990).

All interpretations about evolutionary outcomes are necessarily speculative, but it is conceivable that the ideal combination of specific neonate and sexual maturity features identified by the Multiple Fitness Model was not random. The prototype for a healthy infant includes large eyes and a small nose (Alley, 1983), whereas small eyes and a broad nose are seen in a number of genetic and prenatal disorders, including Down's syndrome and fetal alcohol syndrome (Smith, 1982). Furthermore, large eyes may be more effective than other neonate cues, such as round cheeks, in conveying the desirable qualities of youthfulness (Terry, 1977). This may be the case because a target's eyes draw a disproportionate amount of attention during facial scanning (Hess; 1965, McKelvie, 1976). To complete the picture, a small nose does not obscure attractive eyes, and allows maturity features, such as prominent cheekbones, to be clearly evident on the periphery of the face.

Prominent secondary sexual characteristics may convey macrocompeten-cies, such as physical strength or reproductive capacity. Males are seen as attractive with somewhat more sexual maturity in their faces than females (Freidman & Zeibrowitz, 1992; Keating, 1985), whereas females are seen as attractive with more sexual maturity in the hips than males (Singh, 1993). Prominent secondary sexual characteristics also may suggest micro-competen-cies. Because sexual hormones compete with immune system functioning and lower resistance to disease, prominent sexual maturity characteristics may indicate an effective immune system and resistance to parasites (Gangestad & Buss, 1993; Hamilton & Zuk, 1982; Moller, 1990). The impression of immuno-competence may be enhanced, however, if sexual maturity characteristics were combined with the retention of healthy neonatal characteristics, such as baby-smooth skin that is unblemished by acne.

Trade-offs in the Romantic Ideal Configuration

The male and female romantic ideal of facial physical attractiveness is based on specific combinations of the five categories, shown in Fig. 5.1:

Female romantic beauty: attractiveness as a mate = high neoteny + high feminine sexual maturity + high expressiveness + high grooming + low senescence

Male romantic handsomeness: attractiveness as a mate = medium neoteny + high masculine sexual maturity + high expressiveness + high grooming + low to medium senescence.

The Multiple Fitness Model's predictions about the components of the romantically ideal female face have been tested across a variety of samples, including Whites, Blacks, Hispanics, and Asian and Taiwanese samples. The correlations in attractiveness ratings of females, including target females who vary in race and sexual orientation, have been quite high across groups; White–Black, $r = .94$; White–Hispanic, $r = .94$; White–Taiwanese, $r = .87$; Taiwanese–Hispanic, $r = .94$ (Cunningham, et al., 1995).

Homosexuals vary from social norms in their partner choices but may not differ from heterosexuals in their attractiveness criteria (Donovan, Hill, & Jankowiak, 1989; Howard, Blumstein, & Swartz, 1987). We had heterosexual males and females and gay males and lesbians evaluate 100 randomly organized photos of male and female heterosexuals and gay male and lesbian targets. These determinations were based on self-reports not given to the raters. The correla-tions in attractiveness ratings averaged $r = .80$ across the four groups. There were strong consistencies in judging female targets, for example, heterosexual males–heterosexual females, $r = .91$; heterosexual males–gay males, $r = .83$; heterosexual males–lesbians, $r = .67$; gay males–lesbians, $r = .84$ (DeHart & Cunningham, 1993).

Correlations were also calculated between the columns of facialmetric–attractiveness relationships to determine if the same features were given the same weight by different groups (see Table 5.1). As the lower portion of Table 5.1 indicates, the same neonatal, sexual maturity, expressive, and grooming features generally were associated with perceived attractiveness across groups. Facialmetrics stipulated by the multiple fitness model effectively predicted the ratings of all four groups, suggesting that perceptual judgments about male and female attractiveness may be independent of sexual orientation. Facialmetric–attractiveness relations were a bit weaker for lesbians than for other groups, but it may be more important to note that all groups were significantly correlated in the weight that they placed on the 22 facialmetrics.

Although most of our research has investigated the romantic ideal configuration of features, dozens of other feature combinations are also possible and provide the basis for other types of attraction. Selection of an individual for a long-term mating relationship, for example, may stimulate interest in high expressiveness, but that may not be required for a short-term sexual relationship (cf. Cashdan 1993; Kenrick et al., 1993). Different configurations of features may be involved in attraction to a supportive friend or to a sage mentor, as will be discussed later.

The Trade-off of Sexual Maturity and Senescence Features

Both male facial hair and male pattern baldness are genetically based. Although such features may have occurred randomly or as a by-product of other qualities that were under direct selection pressure, it also is possible that they contributed to adaptation. Male facial beardedness is associated with the sexual maturation stage and has been hypothesized to signal aggressive dominance (Freedman, 1979; Guthrie, 1976). Although some studies found that facial hair increased a male's physical attractiveness (Pellegrini, 1973; Reed & Blunk, 1990), most studies found that those with substantial facial hair were perceived as less attractive than those with clean shaven faces (Cunningham et al., 1990; Feinman & Gill, 1978; Wogalter & Hosie, 1991).

It is possible that beards evolved through a process of secondary sexual selection. In secondary selection, beards were valuable as a threat signal during direct male versus male competition for dominance and resources. Because females were more likely to mate with dominant males having abundant resources than with males who were less successful competitors, the frequency of beards increased in the population. Currently, however, success may depend more on social qualities such as cooperation, alliance formation and control of aggressiveness, rather than on physical intimidation (cf. Graves & Cunningham, 1995; Harcourt & deWall, 1992; Jensen-Campbell, Graziano, & West,

1995). As a consequence, contemporary Western grooming standards empha-size a clean-shaven face that conveys appeasement, rather than threat.

Male pattern baldness, by contrast, is associated with senescence, the next stage of physical maturation, and has been found to be associated with de-creased perceptions of physical attractiveness (Cash, 1990; Roll & Verinis, 1971; Synnott, 1987; Wogalter & Hosie, 1991). As a consequence, balding may have evolved through tertiary selection. In tertiary selection, baldness may not have aided in sexual attraction, but instead contributed to the postpartum male's survival against male competitors, by signaling nonthreatening social maturity, wisdom and nurturance. This appearance also may have given the bald males more opportunity to engage in nurturance of their offspring and relatives. Such outcomes may have increased the fitness of kin who carried the gene for baldness (cf. Bernstein, 1976).

We tested the social perception hypotheses using experimentally modified male facial stimuli (Muscarella & Cunningham, 1996). We presented faces with two levels of facial hair (beard with moustache, and clean-shaven) and three levels of cranial hair (full, receding, and bald). Targets with facial hair were perceived as more aggressive, less appeasing, less attractive, older, and lower on the nurturant form of social maturity than those with clean-shaven faces. These results are consistent with the evolutionary hypothesis that facial hair signals male sexual maturity and dominance.

A decrease in the amount of cranial hair also was associated with increased perceptions of social maturity and appeasement, and decreased perceptions of attractiveness and aggressiveness, consistent with the ethological speculation that male pattern baldness serves as an appeasement signal. Perceived age increased 7.25 years from the full cranial hair and clean shaven configuration to the bald and facial hair configuration. Perceived attractiveness did not change when age was removed as a covariate, suggesting that baldness itself, rather than its association with age, affects the perception of attractiveness. It remains possible, however, that bald males may be uniquely attractive to perceivers who have felt victimized or disappointed by aggressive or domineer-ing males. Such possibilities should be tested in future studies.

The Trade-off of Sexual Maturity and Expressive Features

As noted earlier, each category of facial feature may convey both a positive and a negative quality. Berry and Zeibrowitz-McArthur (1988) for example, found that those with pronounced maturity features were seen as less likely than those with babyfaced features to have committed a negligent act, but more likely to have committed an act of deliberate fraud. Similarly ambivalent outcomes were obtained for expressive features. Heilman and Saruwatari (1979) found that attractive women with pleasant smiles were more likely than unattractive women to be hired for low-level clerical positions, but less likely to be hired for upper-level managerial positions. The combination of high sexual maturity and high expressiveness is part of the romantically ideal face, and may appear to offer both competency and friendliness. Individual perceiver motives, such as affective state, may determine the trade-off when the choice of one category precludes the other, as will be discussed later.

TRADEOFFS MADE BY PERCEIVERS

Thus far, we have detailed the five categories of physical attractiveness features and the qualities that they signal. We have also discussed how it is impossible for a single feature to simultaneously signal several qualities, especially when they are mutually incompatible, but also how individuals make trade-offs and compromises among all the features in order to appear maximally fit across multiple fitness dimensions. We are finally ready to discuss the third type of trade-off—the trade-off faced by individuals when evaluating features and their meanings when individuals are trying to select a romantic or other partner.

The Multiple Fitness Model suggests that the shape, color and other qualities of the features of an individual's face and body are scanned by perceivers to determine the extent to which the target "fits" the perceiver's affective needs and cognitive expectations. The model suggests that attractiveness is a characteristic of the target of perception, but such perceptions are intrinsically linked to the motives of the perceivers. The perceiver must have a need, or attunement, for a target of perception to serve as an affordance to fit that need (McArthur & Baron, 1983), and such perceptual tendencies are the product of evolutionary, sociocultural, and individual forces.

Perceiver Motives

The Multiple Fitness Model suggests that attraction is influenced both by the perceived fitness qualities of the target and by the perceiver's internal motives that create responsivity to such qualities. When specifying the determinants of attractiveness, the question must eventually be asked: attractiveness for what?

Although it has been repeatedly found that judges show high concordance in their ratings of physical attractiveness (Berscheid & Walster, 1974; Jackson, 1992), research on physical attractiveness has seldom clarified the type of attraction under scrutiny, or the motives of the perceiver. It appears, however, that participants asked to rate the global attractiveness of various portraits generally respond in terms of the perceived mating desirability of the targets. Using data from Cunningham (1986), ratings by men of the attractiveness of

16 photographed women correlated with the men's expressed preferences of women for a dinner date, $r = .73$; for sexual intercourse, $r = .70$; and for having children, $r = .72$. In Cunningham, et al., (1990), women were initially asked to rate a series of male faces for attractiveness and then rerate the same faces for sexiness; the two ratings correlated, $r = .90$. Women may be less inclined than men to have sex with someone after a brief acquaintance (Kenrick, et al., 1993), but when members of either gender evaluate the attractiveness of a potential partner, they appear to be using the criteria of sexual attractiveness.

Yet despite such consistency across judges, it is reasonable for attractiveness ratings to reflect some variabiliy as a function of perceiver motives. After all, even succulent food is not very appealing if hunger is not a salient motive because one has just eaten. Similarly, when the need for a sexual partner has been satisfied because the individual already is in a committed relationship, the desirability of highly attractive targets may be reduced (Johnson & Rusbult, 1989; Simpson, Gangestad & Lerma, 1990). Conversely, people tend to rate their own friends higher on attractiveness than other perceivers rate the same targets, and to rate individuals from one's own race higher than members of other races (Cross & Cross, 1971; Cunningham, et al., 1995).

Such outcomes should not be narrowly interpreted to mean that beauty is solely in the eye of the beholder and that judgments of attractiveness are unpredictable. The impact of individual needs and motives, and of social, historical, and ecological dynamics on responses may vary depending on the category of perception under consideration. We will describe some examples in which variation from the norm in the perceiver's motive, personality, or ecology influenced attraction responses.

Affect and the Trade-off Between Expressiveness and Sexual Maturity Features

Whereas some of the criteria for attractiveness judgments may be universal, the Multiple Fitness Model also predicts areas of individual variability. The model suggests that men, regardless of their affective states, will be most attracted to a woman possessing both sexually mature and expressive facial features, because they convey multiple types of fitness, such as nubility and high status, plus warmth and acceptance. Yet faces can be moderately attractive by possessing just a portion of the ideal combination, such as high sexual maturity but low expressiveness. Given the fact that most potential dates fall short of perfection, a suitor is faced with a trade-off situation, and must decide which qualities are most important. If the romantic ideal is unavailable, men's affective states might determine such decisions (cf. Kiesler & Baral, 1970; Walster, 1970).

Positive affect may have evolved to signal favorable adaption to the social and physical ecology (Cunningham, 1987). Men experiencing elated affect may experience an increased aspiration level, plus an optimistic view of their capacity to handle the personality of a challenging potential date (Cunningham, 1988a, 1988b; Cunningham, Steinberg, & Grev, 1980; Isen, 1984). Men in a positive mood might be willing to approach a woman who possessed high sexual maturity but low expressiveness features, although such a woman may be both attractive and potentially rejecting. By contrast, men experiencing induced negative affect might prefer warmth and nurturance over the risks associated with trying to attain a high-status, sexy date. Consequently, we expected that induced negative affect would be associated with a choice of a date with features conveying low sexual maturity but high expressiveness.

The matching hypothesis (Hatfield & Sprecher, 1985; Walster, Aronson, Abrahams, & Rottman, 1966) or the similarity-attraction hypothesis (Byrne, Griffitt & Stefaniak, 1967; Locke & Horowitz, 1990) might predict that negative affect males would choose a female with low sexual maturity and low expressiveness. Such an appearance would require the lowest aspiration level and might suggest an emotional state similar to that of the perceiver.

These speculations were tested in a series of studies (Wong & Cunningham, 1990). In the first experiment, men watched positive, neutral, or negative affect-inducing movies. Afterward, they selected a date from a set of female targets who varied in facial type. When the choices included women with both sexual maturity features and expressive features, colloquially termed an "angel," the men overwhelmingly chose such women (see Table 5.2). In a second condition, none of the targets had both categories of positive features. Men experiencing positive affect chose women with sexually mature but inexpressive features, whereas men experiencing negative affect chose women with expressive but sexually immature features. These findings were replicated in a second experiment using written descriptions of the women. Such results demonstrate that physical attractiveness involves a combination of qualities, but individual affective states and motives may influence partner selections when trade-offs must occur.

Personality and the Trade-off Between Neonatal and Sexual Maturity Features

We extended the findings from our mood study, which used men as perceivers, to a study of personality effects in female perceivers. In two studies, we found that females preferred the ideal neonatal and sexually mature configuration when judging males, but if they were forced to choose from among less-than-ideal faces, then trade-offs based on the rater's personalities were evident (Pike & Cunningham, 1989).

Negative emotionality may indicate poor adaptation to the social and physical environment, and signal that changes are necessary (Cunningham,

TABLE 5.2

Effects of Induced Affect and Stimulus. Set on Percentage Selecting Female Dates

Stimulus Set	Ideal Present	Ideal Absent
Positive Affect		
High Mature + High Expressive	.92	—
High Mature + Low Expressive	—	.83
Low Mature + High Expressive	.08	.17
Low Mature + Low Expressive	.00	.00
Neutral Affect		
High Mature + High Expressive	.92	—
High Mature + Low Expressive	—	.50
Low Mature + High Expressive	.08	.50
Low Mature + Low Expressive	.00	.00
Negative Affect		
High Mature + High Expressive	.83	—
High Mature + Low Expressive	—	.08
Low Mature + High Expressive	.17	.92
Low Mature + Low Expressive	.00	.00

Note. N = 12 per condition.

1987, 1988a, 1988b). As a consequence, neurotic, emotionally unstable individuals were expected to prefer partners whose appearance suggested that they might be dissimilar to the perceiver and consequently capable of fulfilling unmet, complementary needs (Leonard, 1975; Reik, 1944; Seyfried & Hendrick, 1973).

By contrast, emotionally stable individuals may feel well-adapted, and may choose partners whose appearance suggests a personality that is similar to their own (cf. Berry & Brownlow, 1989).

We found a similarity–attraction relationship for emotionally stable individuals. Females who scored as nonneurotic on the Eysenck Neuroticism Scale, and as masculine on the Spence–Helmreich Personality Attributes Questionnaire, preferred more sexually mature and masculine male faces, whereas nonneurotic feminine females chose more babyfaced and feminine appearing males. The pattern was reversed for neurotic females. Neurotic feminine females preferred mature-faced males who might protect them. Neurotic masculine females, by contrast, chose babyfaced males who might serve as beneficiaries rather than competitors. Thus, mate selection for neurotic individuals may be based on personality complementarity rather than personal or genetic similarity (cf. Hinsz, 1989; Rushton et al., 1985).

Race, Ecology, and Tradeoffs in Body Appearance

Bodies may be just as important as faces in judgments of global physical attractiveness (Alicke, Smith & Klotz, 1987; Franzoi & Herzog, 1987). The Multiple Fitness Model proposed that social selection pressures may operate more heavily on the face, and that physical selection pressures may operate more heavily on the body. That is, human faces are primarily selected to enhance interactions with other people, whereas human bodies are primarily selected for survival in the ecology. Because the same human species may live in a wide variety of ecologies, more variation across groups is expected in preferred bodies than in preferred faces (Ford & Beach, 1951). When the ecology offers a surfeit of food, for example, slenderness may be desirable, but when the ecology offers few resources, a body weight that displays stored capital may be seen as attractive (Anderson, Crawford, Nadeau, & Lindberg, 1992).

We found that Black men gave highly similar attractiveness ratings to female faces compared to White men, $r = .94$. If Black judgments of facial attractiveness were determined by White social norms, then consistency should also be expected in judgments of bodies. Yet Blacks specified a body weight for the ideal date that was seven pounds heavier than that specified by Whites (Cunningham, et al., 1995), perhaps reflecting group differences in perceptions of access to ecological resources (Furnham & Alibhasi, 1983)

Consistency has been found in judgments of the sexual maturity feature of female waist-to-hip ratio (Singh, 1993), but preferences for other aspects of the body, such as preferred height and silhouette, have been found to display a great deal of individual variability (Beck, Ward-Hull and McLear, 1976; Lippa, 1983; Maier & Lavrakas, 1984; Wiggins, Wiggins & Conger, 1968). Traditionally feminine women, for example, showed a preference for stereotypically masculine tapering "V" physiques (Lavrakas, 1975). It is conceivable that feminine women perceive the ecology to be more challenging than do other women, requiring the resources of a strong man. Group and individual differences in the perception of ecologies seem to warrant further investigation.

The Trade-off of Ideal Romantic Attractiveness, Personality, and Resources

The trade-offs involved in the physical development of features and between categories of features in social perception are only part of the process. Besides the trade-offs inherent in choosing one face over another in terms of the target's physical attractiveness, individuals are faced with trade-offs when choosing physical attractiveness in the context of other desirable or undesirable aspects of a dating or marriage partner.

A prospective partner might possess only one, or only two, of such positive attributes as physical attractiveness, wealth, or a desirable personality (i.e., understanding, loyal, good sense of humor). Under such circumstances, a trade-off must be made based on personal values or needs. Snyder, Berscheid and Glick (1985) reported that high self-monitors were more influenced by physical attractiveness, whereas low self-monitors were more influenced by personality when evaluating prospective dates. Survey investigators also found that women rated earning potential as more important and physical attractiveness as less important than did men, consistent with parental investment theory (Buss, 1994; Feingold, 1992).

Experimental tests of trade-offs in choosing a dating partner have been relatively rare (Feingold, 1990). Sprecher (1989) experimentally manipulated written descriptions of an opposite sex target's physical attractiveness, earning potential, and expressiveness. She found that women and men were both most affected by physical attractiveness in judging the attractiveness of the person, but females perceived that they were more affected by earning potential and expressiveness.

Our lab extended Sprecher's (1989) study by using photographs to represent physical attractiveness, by collecting separate estimates of dating and marriage interest in the target, and by causing some participants to choose from among three targets, each of whom had only one of the three positive attributes of physical attractiveness, desirable personality, or wealth obtained through good luck. In a second condition, participants chose from among three targets, each of whom had two desirable qualities. We found that men were significantly more likely than women to choose as a date the target who possessed only physical attractiveness, and women were more likely to choose to date a man who possessed only a pleasing personality (Cunningham, et al., 1989; see Table 5.3). Both genders placed more emphasis on personality in the choice of a marriage partner than a date, similar to earlier findings (Kenrick et al., 1993). When the trade-off involved two of three positive attributes, males and females were similar in selecting the mix of physical attractiveness and pleasing personality over the mix of physical attractiveness and wealth or the mix of pleasing personality and wealth.

Such results suggest that wealth obtained through good luck, such as winning a lottery, may not be as attractive to women as other indicators of resources such as a high-status job (cf. Townsend & Levy, 1990), which may convey good genes and desirable social alliances. The trade-offs considered by college women may not be displayed in more impoverished ecologies and will be subjected to further investigation.

PROXIMAL MEDIATORS OF SOCIAL PERCEPTIONS

Evolutionary interpretations of behavior emphasize distal variables such as ecology and parental investments on long-term population dynamics. Tradi-

TABLE 5.3

Proportion of Males and Females Selecting Dates and Marriage Partners
as a Function of Physical Attractiveness, Personality and Wealth

Context	Date		Marriage	
	Males	Females	Males	Females
Participants	28	28	28	28
One Positive Attribute				
Physical Attractiveness	.64	.32	.39	.29
Personality	.29	.46	.50	.50
Wealth	.07	.21	.11	.21
	Males	Females	Males	Females
Participants	24	26	24	26
Two Positive Attributes				
Physical Attractiveness and Personality	.77	.78	.71	.81
Physical Attractiveness and Wealth	.13	.08	.13	.08
Personalty and Wealth	.13	.15	.17	.12

Note. N = 28, all conditions.

tional social psychology tends to emphasize more proximal variables such as social roles and learning. Each perspective is useful in highlighting the short-comings of the other.

Evolutionary accounts are weak in accounting for proximal dynamics and in specifying cognitive and motivational mediators of behaviors. It has been demonstrated that infants display physical attractiveness preferences similar to those of adults (Langlois, et al., 1987), but it is not clear how the genes tell the forebrain to focus on a big smile or large eyes. Perhaps responses to some visual patterns entail an underlying predisposition or prototype, in the same way that the tongue's response to sugar is positive in our species, and the response to quinine is not. How individuals integrate all of the features of an individual face into a configuration that can be evaluated is a question that has puzzled investigators of facial recognition and identification (Dunning & Stern, 1994) and requires further research.

Beauty Across Societies and Historical Periods

Social structures may build on genetic predispositions to optimize the fit between human needs and physical realities. Cultures and social norms can adjust to a changing ecology faster than can human genes. The experiences and expressed attitudes of other people may convey valuable survival information about the adaptation. Social conformity pressure can cause attractiveness ratings to shift a few points (Graziano, Jensen-Campbell, Shebilsky, & Lundgren, 1993). New social norms can encourage women to select men who are expressive and inclined to cuddle and nurture their offspring rather than men who are extreme in masculine sexual maturity. In inhospitable climates

where reproduction must be carefully considered, a face that shows less feminine sexual maturity and a little less expressiveness may be seen as more
acceptable than it is elsewhere (Cunningham, et al., 1995).

Yet, there may be limits in the extent to which social forces may alter
judgments that have evolutionary significance. No matter how often your
mother told you that taking bitter medicine was good for you, your tongue still
did not enjoy it. In the same way, it does not appear that such pressure can turn
the appearance of facial beauty into judgments of ugliness, or vice versa. It may
seem unfair, but we know of no culture in which the essence of female romantic
attractiveness consists of tiny eyes, a large nose, pitted skin, low cheekbones, a
small smile, unsymmetrical teeth, and a negative waist-to-hip ratio (Cunningham, et al., 1995).

Radical feminists, by contrast, have suggested that beauty ideals were created
by patriarchal forces to demoralize women. Wolf (1991), for example, asserted:

> There is no legitimate historical or biological justification for the beauty myth; it
> is. . . . a result of. . . . the need of today's power structure, economy, and culture
> to mount a counteroffensive against women (p. 13). . . . Women's identities must
> be premised upon our "beauty" so that we will remain vulnerable to outside
> approval (p. 14). . . . Since the women's movement had successfully taken apart
> most other necessary fictions of femininity, all the work of social control. . . . had
> to be reassigned. . . . This reimposed onto liberated women's faces and bodies all
> the limitations, taboos and punishments of the repressive laws, religious institu-
> tions, and reproductive enslavement that no longer carried sufficient force (p.
> 16). . . . The beauty myth was institutionalized in the last two decades . . . (p. 20).

We recognize that self-image and judgments of partners, can suffer by
comparison with idealized models of beauty (Kenrick & Gutierres, 1980,
Kenrick, Gutierres, & Goldberg, 1989), and we deplore any use of beauty
standards to demoralize or control women. At the same time, such sociopolitical
concerns should not foreclose questions about the origin and content of such
ideals. If ideals of beauty are truly recent social constructions, then beauty
ratings should show substantial inconsistency from the "second-wave" surge of
the women's movement in the 1970s to the present time of backlash efforts to
reimpose constraints on women (Wolf, 1991).

We examined the temporal stability of attractiveness judgments by comparing the ratings made by college students of female photos in 1976 with ratings
made of the same photos 17 years later, in 1993 (Cunningham, et al., 1995).
The stability in beauty ratings across time was substantial, $r = .92$, suggesting
that images of facial beauty are not recent social constructions. As noted above,
however, body attractiveness may show greater instability.

We also compared the facial measurements of female film stars from the
1930s through the 1950s with those of the stars of the 1990s (Cunningham, et
al., 1989). Vintage stars had more highly arched eyebrows than did contempo-

raries, but both groups had more ideal neonatal and mature features than did a random sample of female faces. Such results suggest that attractiveness judgments have been quite stable within the 20th century. Facialmetric research is currently being conducted on faces in ancient art and on the faces of American Presidents (Wong & Cunningham, 1995).

CONCLUSIONS

The Multiple Fitness Model is an integrative position that specifies a variety of trade-offs that are made in the perception of faces, bodies, and romantic partners. An essential point is that physical attractiveness is not a unitary dimension. Instead, there are multiple dimensions of personal fitness, and those dimensions can create multiple forms of attractiveness, each of which may fit different perceiver needs.

Romantic attractiveness may convey the widest array of fitness qualities, but that schema should not override more systematic processing of an individual's fitness to meet a specific need. Quite simply, the fact that an individual looks as though he or she has the qualities of a good mate does not mean that this truly is the case, nor does it mean that the individual possesses the fitness required for other roles, such as the sociable fitness to be a receptionist, for example, or the intellectual fitness to be a good attorney.

In the same way, signs of age, such as wrinkles or baldness, cannot be seen as the essence of romantic attractiveness, but they can be seen as attractive for the wisdom and noncompetitiveness they may represent. Fortunately, individuals who are androgynous (Anderson & Bem, 1981; Moore, Graziano, & Millar, 1987) and anyone for whom the decision is seen as important (Petty & Cacioppo, 1986) can ignore physical attractiveness when making judgments.

We all may be attractive in our own unique way, but that fact should not preclude study of the evolutionary forces played out on our faces and bodies. It is only when we understand our biological underpinnings that we can strive to place them in their proper context and overcome the perceptual inaccuracies and discriminatory behavior they may breed. This brief overview cannot stipulate all of the qualifiers and caveats for our findings, but it should provide further illustration of the heuristic value of combining evolutionary and traditional social and personality variables.

REFERENCES

Alicke, M. D., Smith, R. H., & Klotz, M. L. (1987). Judgments of physical Attractiveness: The role of faces and bodies. *Personality and Social Psychology Bulletin, 12,* 381–389.

Alley, T. R. (1983). Infantile head shape as an elicitor of adult protection. *Merrill-Palmer Quarterly, 29,* 411–427.

Alley, T. R., & Cunningham, M. R. (1991). Averaged faces are attractive, but very attractive faces are not average. *Psychological Science, 2,* 123–125.

Anderson, J. L., Crawford, C. B., Nadeau, J., & Lindberg, T. (1992). Was the Duchess of Windsor right? A cross-cultural review of the socioecology of ideals of female body shape. *Ethology and Sociobiology, 13,* 197–227.

Anderson, S. M., & Bem, S. L. (1981). Sex-typing and androgyny in dyadic interaction: Individual differences in responsiveness to physical attractiveness. *Journal of Personality and Social Psychology, 41,* 74–86.

Aronoff, J., Woike, B. A., & Hyman, L. M. (1992). Which are the stimuli in facial displays of anger and happiness? Configurational bases of emotion recognition. *Journal of Personality and Social Psychology, 62,* 1050–1066.

Ashmore, R. D., Solomon, M. R., & Longo, L. C. (1990, August). *Thinking about physical attractiveness: A single psychological dimension or multiple content-specific continuum?* Paper presented at the annual meeting of the American Psychological Association, Boston, MA.

Barbee, A. P., Cunningham, M. R., Winstead, B., Derlega, V., Gulley, M. R., Yankeelov, P. A., & Druen, P. B.(1993). The effects of gender role expectations on the social support process. *Journal of Social Issues, 49,* 175–190.

Barbee, A. P., Yankeelov, P. A., & Druen, P. B. (July, 1992). *Social support as a mechanism for relationship maintenance.* Conference of the International Society of the Study of Personal Relationships, Orono, ME.

Bar-Tal, D., & Saxe, L. (1976). Physical attractiveness and its relationship to sex-role stereotyping. *Sex Roles, 2,* 123–133.

Beck, S. P., Ward-Hull, C. I., & McLear, P. M . (1976). Variables related to women's somatic preferences of the male and female body. *Journal of Personality and Social Psychology, 34,* 1200–1210.

Bernstein, I. S. (1976). Dominance, aggression and reproduction in primate societies. *Journal of Theoretical Biology, 60,* 459–472.

Berry, D. S., & Brownlow, S. (1989). Were the physiognomists right? Personality correlates of facial babyishness. *Personality and Social Psychology Bulletin, 15,* 266–279.

Berry, D. S., & McArthur, L. Z. (1985). Some components and consequences of a babyface. *Journal of Personality and Social Psychology, 48,* 312–323.

Berry, D. S., & Zeibrowitz-McArthur, L. (1986). Perceiving character in faces: The impact of age-related craniofacial changes on social perception. *Psychological Bulletin, 100,* 3–18.

Berry, D. S., & Zeibrowitz-McArthur, L. (1988). What's in a face: Facial maturity and the attribution of legal responsibility. *Personality and Social Psychology Bulletin, 14,* 23–33.

Berscheid, E., & Walster, E. (1974). Physical attractiveness. In L. Berkowitz (Ed.), *Advances in experimental social psychology* (pp. 157–215). New York: Academic Press.

Buss, D. M. (1994). *The evolution of desire.* New York: Basic Books.

Byrne, D., Griffitt, W., & Stefaniak, D. (1967). Attraction and similarity of personality characteristics.*Journal of Personality and Social Psychology, 5,* 82–90.

Carello, C., Grosofsky, A., Shaw, R. E., Pittenger, J. B., & Mark., L. S. (1989). Attractiveness of facial profiles is a function of distance from archetype. *Ecological Psychology, 1*, 227–251.

Cash, T. F. (1990). Losing hair, losing points? The effects of male pattern baldness on social impression formation. *Journal of Applied Social Psychology, 20*, 154–167.

Cash, T. F., Dawson, J., Davis, P., Bowen, M., & Galumbeck, C. (1989). Effects of cosmetic use on the physical attractiveness and body image of American college women. *Journal of Social Psychology, 129*, 349–355.

Cashdan, E. (1993). Attracting mates: Effects of paternal investment on mate attraction strategies. *Ethology and Sociobiology, 14*, 1–23.

Cross, J. F., & Cross, J. (1971). Age, sex, race and the perception of facial beauty. *Developmental Psychology, 5*, 433–439.

Cunningham, M. R. (1981). Sociobiology as a supplementary paradigm for social psychological research. In L. Wheeler (Ed.), *Review of personality and social psychology, Vol 2*, (pp. 69–106). Beverly Hills: Sage.

Cunningham, M. R. (1986). Measuring the physical in physical attractiveness: Quasi–experiments on the sociobiology of female facial beauty. *Journal of Personality and Social Psychology, 50*, 925–935.

Cunningham, M. R. (1987). Levites and brother's keepers: A sociobiological perspective on prosocial behavior. *Humboldt Journal of Social Relations, 13*, 35–36.

Cunningham, M. R. (1988a). Does happiness mean friendliness? Mood, and heterosexual self-disclosure. *Personality and Social Psychology Bulletin, 14*, 283–297.

Cunningham, M. R. (1988b). What do you do when you're feeling blue? Affect, motivation, and social behavior. *Motivation and Emotion, 12*, 309–331.

Cunningham, M. R., & Barbee, A. P. (1991). Differential K-selection versus ecological determinants of race differences in sexual behavior. *Journal of Research in Personality, 25*, 205–217.

Cunningham, M. R., Barbee, A. P., & Pike, C. L. (1990). What do women want? Facialmetric assessment of multiple motives in the perception of male facial physical attractiveness. *Journal of Personality and Social Psychology, 59*, 61–72.

Cunningham, M. R., & Rauscher, S. (1994). *The role of various sensory qualities at two points in dating relationships.* Unpublished Honors Thesis, University of Louisville, KY.

Cunningham, M. R., Roberts, R., Barbee, A. P., Druen, P. B., & Wu, C., (1995). "Their ideas of beauty are, on the whole, the same as ours": Consistency and variability in the cross-cultural perception of female physical attractiveness. *Journal of Personality and Social Psychology, 68*, 261–279.

Cunningham, M. R., Shaffer, D. R., Barbee, A.P., Wolff, P. L., & Kelley, D. J. (1990). Separate processes in the relationship of elation and depression to altruism: Social versus personal concerns. *Journal of Experimental Social Psychology, 26*, 13–33.

Cunningham, M. R., Steinberg, J., & Grev, R. (1980). Wanting to and having to help: Separate motivations for positive mood and guilt induced helping. *Journal of Personality and Social Psychology, 38*, 181–192.

Cunningham, M. R., Wong, D. T., Rodenhiser, J. M., Roberts, R. A., & Richardson, T. (1989, October). *Facialmetric analyses of physical attractiveness.* Paper presented at the Kentucky Psychological Association, Louisville, KY.

DeHart, D. D., & Cunningham, M. R. (1993, April). *Perceptual correlates of attractiveness judgments and judgments of homosexuality by heterosexual and homosexual males and females.* Paper presented at the Southeastern Psychological Association, Atlanta GA.

Diener, E., Wolsie, B., & Fujita, F. (1995). Physical attractiveness and subjective well-being. *Journal of Personality and Social Psychology, 69,* 120–129.

Donovan, J. M., Hill, E., & Jankowiak, W. R. (1989). Gender, sexual orientation, and truth-of-consequences in studies of physical attractiveness. *Journal of Sex Research, 26,* 264–271.

Dunning, D., & Stern, L. B. (1994). Distinguishing accurate from inaccurate eyewitness identification via inquires about decision processes. *Journal of Personality and Social Psychology, 67,* 818–835.

Eibl-Eibesfeldt, I. (1989). *Human Ethology.* New York: Aldine DeGruyter.

Ekman, P., Friesen, W. V., O'Sullivan, M., Chan, A., Diacoyanni-Tarlatzis, I., Heider, K., Krause, R., LeCompte, W. A., Pitcairn, T., Ricci-Bitti, P., Scherer, K., Tomita, M., & Tzavaras, A. (1987). Universals and cultural differences in the judgment of facial expressions of emotion. *Journal of Personality and Social Psychology, 53,* 712–717.

Enlow, D. M. (1990).*Handbook of facial growth* (3rd ed). Philadelphia: W. B. Saunders.

Farkas, L. (1987). *Anthropometric facial proportions in medicine.* Springfield, IL: C. C. Thomas.

Feingold, A. (1990). Gender differences in effects of physical attractiveness on romantic attraction: A comparison across five research paradigms. *Journal of Personality and Social Psychology, 59,* 981–993.

Feingold, A. (1992). Gender differences in mate selection preferences: A test of the parental investment model. *Psychological Bulletin, 112,* 125–139.

Feinman, S., & Gill, G. W. (1978). Sex differences in physical attractiveness preferences. *Journal of Social Psychology, 105,* 43–52.

Ford, C. S., & Beach, F. A. (1951). *Patterns of sexual behavior.* New York: Harper.

Forgas, J. P. (1987). The role of physical attractiveness in the interpretation of facial expression cues. *Personality and Social Psychology Bulletin, 13,* 478–489.

Forgas, J. P. (1991). Affective influences on partner choice: Role of mood in social decisions. *Journal of Personality and Social Psychology, 61,* 708–720.

Franzoi, S. L., & Herzog, M. E. (1987). Judging physical attractiveness: What body aspects do we use? *Personality and Social Psychology Bulletin, 13,* 19–33.

Freedman, D. G. (1979). *Human sociobiology: A holistic approach.* New York: Free Press.

Fridlund, A. J. (1991). Evolution and facial action in reflex, social motive, and paralanguage. *Biological Psychology, 32,* 3–100.

Friedman, H., & Zeibrowitz, L. A. (1992). The contribution of typical sex differences in facial maturity to sex role stereotypes. *Personality and Social Psychology Bulletin, 18,* 430–438.

Furnham, A., & Alibhasi, N. (1983) Cross-cultural differences in perception of female body shapes. *Psychological Medicine, 13,* 829–837.

Gangestad, S. W., & Buss, D. M. (1993). Pathogen prevalence and human mate preference. *Ethology and Sociobiology, 14,* 89–96.

Gillen, B. (1981). Physical attractiveness: A determinant of two types of goodness. *Personality and Social Psychology Bulletin, 7,* 277–281.

Goffman, E. (1979). *Gender advertisements.* New York: Harper & Row.

Graves, C. R., & Cunningham, M. R. (1995). *The effects of wealth, competitiveness, and dishonesty on the heterosexual attractiveness of males.* Unpublished manuscript, University of Louisville, KY.

Graziano, W. G., & Jensen-Campbell, L., Shebilske, L., & Lundgren, S. (1993). Social influence, sex differences and judgments of beauty. Putting the "interpersonal" back in interpersonal attraction. *Journal of Personality and Social Psychology, 65,* 522–531.

Guthrie, R. D. (1976). *Body hotspots.* New York: Van Nostrand Reinhold.

Hamilton, W. D., & Zuk, M. (1982). Heritable true fitness and bright birds: A role for parasites? *Science, 218,* 384–387.

Harcourt, A. H., & deWaal, F. B. M. (1992). *Coalitions and alliances in humans and other animals.* Oxford: Oxford University Press.

Hatfield, E., & Sprecher, S. (1985). *Mirror, mirror . . . The importance of looks in everyday life.* New York: State University of New York Press.

Heilman, M., & Saruwatari, L. R. (1979). When beauty is beastly: The effects of appearance and sex on evaluations for managerial and non-managerial jobs. *Organizational Behavior and Human Performance, 23,* 360–372.

Hess, L. H. (1965). Attitude and pupil size. *Scientific American, 212,* 46–54.

Hildebrandt, K. A., & Fitzgerald, H. E. (1978). Adults' responses to infants varying in perceived cuteness. *Behavioral Processes, 3,* 159–172.

Hinsz, V. B. (1989) Facial resemblance in engaged and married couples. *Journal of Social and Personal Relationships, 6,* 223–229.

Howard, J. A., Blumstein, P., & Schwartz, P. (1987). Social or evolutionary theories? Some observations on preferences in human mate selection. *Journal of Personality and Social Psychology, 53,* 194–200.

Isen, A. M. (1984). Toward understanding the role of affect in cognition. In R. Wyer & T. Srull (Eds.), *Handbook of social cognition* (Vol. 3, 129–178). Hillsdale, NJ: Lawrence Erlbaum Associates.

Izard, C. E. (1971). *The face of emotion.* New York: Appleton-Century Crofts.

Jackson, L. A. (1992). *Physical appearance and gender.* Albany: State University of New York Press.

Jensen-Campbell, L. A., Graziano, W. G., & West, S. G. (1995). Dominance, prosocial orientation, and female preferences: Do nice guys really finish last? *Journal of Personality and Social Psychology, 68,* 427–440.

Johnson, D. L., & Rusbult, C. E. (1989). Resisting temptation: Devaluation of alternate partners as a means of maintaining commitment in close relationships. *Journal of Personality and Social Psychology, 57,* 967–980.

Johnson, V. S., & Franklin, M. (1993). Is beauty in the eye of the beholder? *Ethology and Sociobiology, 14,* 183–199.

Keating, C. F. (1985). Gender and the physiognomy of dominance and attractiveness. *Social Psychology Quarterly, 48,* 61–70.

Keating, C. F., Mazur, A., & Segall, M. H.(1981). A cross-cultural exploration of physiognomic traits of dominance and happiness. *Ethology and Sociobiology, 2,* 41–48.

Keltner, D. (1995). Signs of appeasement: Evidence for the distinct displays of embarrassment, amusement and shame. *Journal of Personality and Social Psychology, 68,* 441–454.

Kenrick, D. T., Groth, G. E., Trost, M. R., & Sadalla, E. K. (1993). Integrating evolutionary and social exchange perspectives on relationships: Effects of gender, self-appraisal, and involvement level on mate selection criteria. *Journal of Personality and Social Psychology, 64*, 951–969.

Kenrick, D. T., & Gutierres, S. E. (1980). Contrast effects and judgments of physical attractiveness. *Journal of Personality and Social Psychology, 38*, 131–140.

Kenrick, D. T., Gutierres, S. E., & Goldberg, L. L. (1989). Influence of popular erotica on judgments of strangers and mates. *Journal of Experimental Social Psychology, 25*, 159–167.

Kiesler, S. B., & Baral, R. L. (1970). The search for a romantic partner: The effects of self-esteem and physical attractiveness on romantic behavior. In K. J. Gergen & D. Marlowe (Eds.), *Personality and social behavior* (pp. 155–165). Reading, MA: Addison-Wesley.

Langlois, J. H., & Roggman, L. A. (1990). Attractive faces are only average. *Psychological Sciences, 1*, 115–121.

Langlois, J. H., Roggman, L. A., Casey, R. J., Ritter, J. M., Rieser-Danner, L. A., & Jenkins, V. Y. (1987). Infant preferences for attractive features: Rudiments of a stereotype? *Developmental Psychology, 23*, 363–369.

Langlois, J. H., Roggman, L. A., & Musselman, L. (1994). What is average and what is not average about attractive faces? *Psychological Sciences, 5*, 214–220.

Lanzetta, J. T., & Orr, S. P.(1986). Excitatory strength of expressive faces: Effects of happy and fear expressions and context on the extinction of a conditioned fear response. *Journal of Personality and Social Psychology, 50*, 190–194.

Lavrakas, P. J. (1975). Female preferences for male physiques. *Journal of Research in Personality, 9*, 324–34.

Leonard, R. L., Jr. (1975). Self-concept and attraction for similar and dissimilar other, *Journal of Personality and Social Psychology, 31*, 926–929.

Liggett, J. (1974). *The human face.* New York: Stein & Day.

Lippa, R. (1983). Sex typing and the perception of body outlines. *Journal of Personality, 51*, 667–682.

Locke, K. D., & Horowitz, L. M. (1990). Satisfaction in interpersonal Interactions as a function of similarity in level of dysphoria. *Journal of Personality and Social Psychology, 58*, 823–31.

Low, B. S. (1979). Sexual selection and human ornamentation. In N. A. Chagnon & W. Irons, *Evolutionary biology and human social behavior* (pp.462–486). North Scituate, MA: Duxbury Press.

Lundy, D. E., Cunningham, M. R., & Lister, S. C. (1995). *The humanizing effect of humor: The role of opening gambit and physical attractiveness on romantic interest.* Manuscript submitted for publication.

Maier, R. A., & Lavrakas, R. A. (1984). Attitudes toward women, personality rigidity, and idealized physique preferences in males. *Sex Roles, 11*, 425–433.

Maron, M. (1994). *Makeover miracles.* New York: Crown Publishers.

McArthur, L. Z., & Baron, R. M. (1983). Toward an ecological theory of social perception, *Psychological Review, 90*, 215–238.

McGinley, H., McGinley, P., & Nicholas, K. (1978). Smiling, body position and interpersonal attraction. *Bulletin of the Psychonomic Society, 12*, 21–24.

McKelvie, S. J. (1976). The role of the eyes and mouth in the memory of a face. *American Journal of Psychology, 2*, 311–323.

Miller, A. G., Ashton, W. A., McHoskey, J. W., & Gimbel, J. (1990). What price attractiveness? Stereotype and risk factors in suntanning behavior. *Journal of Applied Social Psychology, 20,* 1272–1300.

Moller, A. P. (1990). Parasites and sexual selection: Current studies of the Hamilton and Zuk hypothesis. *Journal of Evolutionary Biology, 3,* 319–328.

Montepare, J. M., & Zeibrowitz-McArthur, L. (1988). Impressions of people created by age-related qualities of their gaits. *Journal of Personality and Social Psychology, 34,* 537–542.

Moore, J. S., Graziano, W. G., & Millar, M. G. (1987). Physical Attractiveness, sex role orientation, and the evaluation of adults and children. *Personality and Social Psychology Bulletin, 13,* 95–102.

Moore, M. M. (1985). Nonverbal courtship patterns in women: Context and consequences. *Ethology and Sociobiology, 6,* 237–248.

Muscarella, F., & Cunningham, M. R. (1996). The evolutionary significance and social perception of male pattern baldness and facial hair. *Ethology and Sociobiology, 17,* 99–117.

Nakdimen, K. A. (1984). The physiognomic basis of sexual stereotyping. *American Journal of Psychiatry, 14,* 499–503.

Pellegrini, R. J. (1973). Impressions of the male personality as a function of beardedness. *Psychology, 10,* 29–33.

Perrett, D. I., May, K. A., & Yoshikawa, S. (1994). Facial shape and judgments of female attractiveness, *Nature, 368,* 239–242.

Petty, R. E., & Cacioppo, J. T. (1986). *Communication and persuasion: Central and peripheral routes to attitude change.* New York: Springer-Verlag.

Pike, C. L., & Cunningham, M. R. (1989). *Differences in perceived male facial attractiveness as a function of masculinity-femininity.* Unpublished manuscript, University of Louisville, KY.

Pittenger, J. B. (1991). On the difficulty of averaging faces: Comments on Langlois & Roggman. *Psychological Science, 2,* 351–357.

Plomin, R. (1981). Ethnological behavioral genetics and development. In K. Immelmann, G. W. Barlow, L. Petrinovich, & M. Main (Eds.), *Behavioral development: The Bielefeld Inter-disciplinary Project* (pp. 252–276). Cambridge, England: Cambridge University Press.

Raines, R. S., Hechtman, S. B., & Rosenthal, R. (1990). Physical attractiveness of face and voice: Effects of positivity, dominance and sex. *Journal of Applied Social Psychology, 20,* 1558–1578.

Reed, J. A., & Blunk, E. M. (1990). The influence of facial hair on impression formation. *Social Behavior and Personality, 18,* 169–175.

Reik, T. (1944). *A psychologist looks at love.* New York: Farrar & Rinehart.

Roll, S., & Verinis, J. S. (1971). Stereotypes of scalp and facial hair as measured by the semantic differential. *Psychological Reports, 28,* 975–980.

Rowatt, W. C., Cunningham, M. R., & Druen, P. B. (1995). *Lying to get a date: Self-Monitoring and deceptive self-presentation in the initiation of romantic Relationships.* Manuscript submitted for publication.

Rushton, P., Russell, R. J. H., & Wells, P. A. (1985). Genetic similarity theory: Beyond kin selection. *Behavior Genetics, 14,* 179–193.

Scherer, K. R., & Wallbott, H. G. (1994). Evidence for universality and cultural variation of differential emotion response patterning. *Journal of Personality and Social Psychology, 66*, 310–328.

Seyfried, B. A., & Hendrick, C. (1973). When do opposites attract? When they are opposite in sex and sex role attitudes. *Journal of Personality and Social Psychology, 25*, 15–20.

Simpson, J. A., Gangestad, S. W., & Lerma, M. (1990). Perception of physical attractiveness: Mechanisms involved in the maintenance of romantic relationships. *Journal of Personality and Social Psychology, 59*, 1192–1201.

Singh, D. (1993). Adaptive significance of female physical attractiveness: Role of waist-to-hip ratio. *Journal of Personality and Social Psychology, 59*, 1192–1201.

Smith, D. W. (1982). *Recognizable patterns of human malformation.* Philadelphia: W. B. Saunders.

Snyder, M., Berscheid, E., & Glick, P. (1985). Focusing on the exterior and the interior: Two investigations of the initiation of personal relationships. *Journal of Personality and Social Psychology, 48*, 1427–1439.

Sprecher, S. (1989). The importance to males and females of physical attractiveness, earning potential, and expressiveness in initial attraction. *Sex Roles, 21*, 591–607.

Steele, V. (1985). *Fashion and erotiicsm: Ideals of feminine beauty from the Victorian era to the jazz age.* New York: Oxford University Press.

Symons, D. (1979). *The evolution of human sexuality.* New York: Oxford Press.

Synnott, A. (1987). Shame and glory: A sociology of hair. *The British Journal of Sociology, 38*, 381–413.

Tanner, J. M. (1978). *Foetus into man: Physical growth from conception to maturity.* London: Open Books.

Terry, R. (1977). Further evidence on components of facial attractiveness. *Perceptual and Motor Skills, 45*, 130.

Thornhill, R., & Gangestad, S. W. (1993). Human facial beauty: Averageness, symmetry and parasite resistance. *Human Nature, 4*, 237–269.

Thornhill, R., & Gangestad, S. W. (1994). Human fluctuating asymmetry and sexual behavior. *Psychological Science, 5*, 297–302.

Tilbrook, A. J., & Cameron, A. W. N. (1989). Ram mating preferences for woolly rather than recently shorn ewes. *Applied Animal Behavior Science, 24*, 301–312.

Townsend, J. M., & Levy, G. D. (1990). Effects of potential partner's physical attractiveness and socioeconomic status on sexuality and partner selection. *Archives of Sexual Behavior, 371*, 149–164.

Walster, E. (1970). The effect of self-esteem on liking for dates of various social desirabilities. *Journal of Experimental Social Psychology, 6*, 248–253.

Walster, E., Aronson, V., Abrahams, D., & Rottman, L. (1966). Importance of physical attractiveness in dating behavior. *Journal of Personality and Social Psychology, 4*, 508–516.

Wheeler, L., & Eghrari, H. (1987). *Sexy, sophisticated, or wholesome: Perceptions of different types of attractive females.* Rochester, NY: University of Rochester.

Wickler, W. (1973). *The sexual code.* New York: Anchor.

Wiggins, J. S., Wiggins, N., & Conger, J. C. (1968) Correlates of heterosexual somatic preference. *Journal of Personality and Social Psychology, 10*, 82–90.

Wogalter, M. S., & Hosie, J. A. (1991). Effects of cranial and facial hair on perceptions of age and person. *The Journal of Social Psychology, 31*, 589–591.

Wolf, N. (1991). *The beauty myth: How images of beauty are used against women*. New York: Morrow.

Wong, D. T., & Cunningham, M. R. (1990, April). *Interior versus exterior beauty: The effects of mood on dating preferences for different types of physically attractive women*. Paper presented at the Southeastern Psychological Association, Atlanta, GA.

Wong, D. T., & Cunningham, M. R. (1995). *All the president's faces: Facialmetric analyses of personality profiles and greatness ratings in U.S. leaders*. Unpublished manuscript, University of Louisville.

6

Interpersonal Attraction from an
Evolutionary Psychology Perspective:
Women's Reactions to Dominant
and Prosocial Men

William G. Graziano
Texas A&M University
Lauri A. Jensen-Campbell
Florida Atlantic University
Michael Todd
Texas A&M University
John F. Finch
Arizona State University

Preference on the part of the women, steadily acting in any one direction, would ultimately affect the character of the tribe; for the women would generally choose not merely the handsomest men, according to their standard of taste, but those who were at the same time best able to defend and support them. Such well-endowed pairs would commonly rear a larger number of offspring than the less favoured. (Darwin, 1871, chapter XX, p. 585.)

Ideas are like organisms in evolutionary theory. Both are typically in the process of adaptation and development. At any given point in time, ideas seem to be perfectly embedded in a tangled bank of other contemporary concepts and ideas, as well as an immediate intellectual context. If we take a longer time frame, however, we can see that ideas often develop from more primal concepts. In the process of adapting to marauding criticisms and other ideas, they are transformed from their ancestral shape. Like contemporary organisms, contemporary ideas are hardly the final word. Efforts to hold ideas constant, to retain a pristine orthodoxy, are probably doomed to failure, just as are efforts to keep organisms from changing in response to changing environments.

We can apply this thinking to evolutionary psychology. The evolutionary theories published by Darwin in 1859 and 1871 are monumental human achievements that set a solid foundation for subsequent thinking and research in the life sciences. Nonetheless, Darwin himself would have objected to the idea that his theories were the final word on the evolution of life, much less on human psychic life. Darwin (1859), said: "In the future, I see open fields for far more important researches. Psychology will be securely based. . . " (p. 243; see also Darwin, 1871, p. 590). After all, Darwin's work was published before modern scientific psychology emerged as a distinct discipline.

According to Buss (1995), current evolutionary theory is probably best regarded as a loosely federated, hierarchically organized nomological network-in-progress, with basic foundational principles (e.g., evolution by natural selection) at the most general level. To expand the theory to nontraditional content areas such as interpersonal interaction, we will almost certainly need to confront ambiguities in foundational concepts and fill gaps with new assumptions and additional considerations not anticipated by Darwin more than a century ago. We will probably need to work at multiple levels of abstraction, using ideas and methods not available to Darwin in his time (Brewer & Caporael, 1990; Buss, 1990).

In this chapter, we adapt ideas from evolutionary theory to shed light on human interpersonal attraction. Interpersonal attraction is a topical apple that is not as far from the evolutionary tree as are other aspects of human sociality; since Darwin (1871), evolutionary theory has been concerned with factors affecting mating choice. There seems to be no shortage of critical commentaries on evolutionary approaches to interpersonal attraction, but most of these are "opinion pieces" unincumbered by data. Empirical work that directly evaluates evolutionary hypotheses about interpersonal attraction is now emerging (Buss, Larsen, Westen, & Semmelroth, 1992; Cunningham, Barbee, & Pike, 1990; Sadalla, Kenrick, & Vershure, 1987). We build on the new evolutionary psychology work, but we confront several conceptual ambiguities and deploy some ideas and methodological tools from social, personality, and developmental psychology that might suggest new lines of research in evolutionary psychology. In this chapter, we focus on women's attraction to men.

This chapter is organized into three sections. First, we consider theoretical issues surrounding evolved mechanisms as proximal mediators of behavior and cognition in current evolutionary psychology. In this discussion, we consider how these mechanisms might be linked to human interpersonal attraction. In particular, we consider the role that prosocial, altruistic behavior might play in attraction. Second, we present results of a program of research concerned with psychological processes underlying women's attraction to men. Finally, we show how results from this program of research may suggest avenues for future research within the framework of evolutionary psychology. In the process, we show ways in which some revisions in current evolutionary psychology theory may lead to a better fit with the empirical research literature on human interpersonal attraction.

EVOLVED MECHANISMS
AND INTERPERSONAL ATTRACTION

Buss (1995) noted that all psychological theories imply an evolved mechanism underlying manifest behavior. According to Buss, the core of the debate between evolutionary and nonevolutionary psychologists is not the evolved part but the mechanism part of the implication. That is, the key issue is not about evolution's role in shaping mechanisms that underlie behaviors, but about "the nature of the psychological mechanism that evolution by selection has fashioned" (p. 5). The central task for evolutionary psychology is to discover, describe, and explain the nature of these mechanisms (Tooby & Cosmides, 1990).

Buss (1995) also noted that evolved psychological mechanisms are coordinated sets of processes inside an organism that have three characteristics. First, the mechanisms exist in the form that they do because they solved specific problems of individual survival or reproduction over human evolutionary history. They are "mental organs" dedicated to solving specific psychological tasks, just as body organs such as the liver are dedicated to the specific task of detoxifying poisons, and the heart is dedicated to pumping blood (Tooby & Cosmides, 1990). Second, each mechanism is dedicated to certain classes of information or input. For example, just as the frog's retina contains a mechanism with "bug detectors," so the mechanism underlying men's attraction to women is reputed to be especially responsive to the input of women's waist-to-hip ratio (Singh, 1993). The mechanism underlying women's sexual attraction to men is reputed to be especially responsive to dominance cues from men (Sadalla et al., 1987; but see Jensen-Campbell, Graziano, & West, 1995). Third, the mechanism transforms the input into output through a specific decision rule leading to the regulation of physiological activity, provides input to other mechanisms, or produces manifest action. We assume that the transformation mechanism and its associated output has solved a particular adaptive problem, at least in the evolutionary past.

From these statements, we can see several lines for constructive theoretical development. First, there is a certain Gouldian, ScienceB, "backward look" to this discussion in the view that evolved mechanisms solved the problems of the historical past, but that they are not guaranteed to solve current problems (Dahlstrom, 1991; Gould, 1986). Perper (personal communication, October 10, 1995) noted that Paleolithic humans probably had an evolved mechanism that allowed them to work with stone and flint. Do we retain this mechanism now? It seems that the answer depends on the specific nature of that mechanism and how it could be pressed into service to solve new and different problems as conditions change. It is not possible, of course, to know in any direct way what the original mechanism might have been. We can only infer from other observations about its characteristics.

Second, even if we could directly learn their nature, it will still be no small task to uncover the potentially vast number of specific, dedicated mechanisms that serve a narrow adaptive function. We may debate the adaptiveness of a particular output behavior, but such debate is probably a tangent to a larger, more important issue.

Given the vastness of the task, we probably need to assign priorities in our research agendas. For example, given the larger perspective of evolutionary theory, our hunch is that most evolutionists would probably assign a higher priority to uncovering mechanisms underlying female preferences in mates than to uncovering the mechanism underlying fear of snakes. Both processes have implications for survival, but the former has implications for reproductive fitness, too. One complication is that such a priority system can lead to narrow coverage of human sociality, apparent reductionism, and minimalist analyses of human social behavior (Cantor, 1990). Another complication is that lower priority behaviors such as snake phobias are probably more tractable for an evolutionary analysis because they are specific, and thus likely to show the dedicated pattern. It is probable that more complex, higher-priority behavior, especially human interpersonal behavior, involves more than one mechanism (cf. Cunningham et al., 1990; Cunningham, Druen, & Barbee, chapter 5, this volume).

This line of thinking leads to our third point. Even if we focus on single mechanisms, we need a precise specification of the transformation process—the decision rule—that leads an organism or biobehavioral system from one state into another. Of the three desiderata for a mechanism outlined by Buss (1995), this third one is probably the most important (Cosmides & Tooby, 1987; cf. Hendrick, 1995). Specifying transformation processes is harder than pointing to a behavior, speculating on its function during the environment of evolutionary adaptation, and labeling it a mechanism.

A great deal hinges on the way we conceptualize evolved mechanisms, especially for psychological theory and research on interpersonal attraction. Several writers (Buss, 1995; Hendrick, 1995) argued that a major difference between sociobiology and evolutionary psychology lies in the nature of the mechanisms presumed to underlie behavior. In sociobiological analyses, mechanisms that mediate behavior are distal and appear not to require psychological processes, at least as we conventionally conceptualize them. If there is psychological mediation in sociobiological explanation, it is akin to the idea of innate releasing mechanisms. The sociobiological approach may be a reasonable approach for dealing with Formicinae, but it is limiting in dealing with complex human social behavior such as interpersonal attraction. In evolutionary psychology, however, the focus is on proximate mediation, specifically with processes occurring within a specific context in individual organisms.

The shift in focus in recent evolutionary psychology is more than a change from distal to proximal. It is a blueprint for a new pattern of alliances between evolutionary theory and the cognitive sciences. According to Cosmides and Tooby (1987), proximate mediating mechanisms for behavior are "most closely allied with the cognitive level of explanation than with any other level of proximate causation. This is because the cognitive level seeks to specify a psychological mechanism's function, and natural selection is a theory of function" (p. 284).

If we accept the admittedly controversial distinction between the sociobiological approach and the evolutionary approach, then we can discuss the corresponding differences in conceptualizations of human interpersonal attraction. Sociobiological theory appears to bypass (or at least underplay) the role of psychological mediation, implicitly making proximal processes of attraction noncausal effect variables. Beyond the theoretical difficulties (Buss, 1995), the classical sociobiological approach appears to overlook proximal variables such as perceived physical attractiveness, perceived probability of rejection, and the availability of other options (e.g., Cl_{alt}) all systematically affect interpersonal attraction. The bulk of research by psychologists on interpersonal attraction uses a quasi-Lewinian phenomenological framework in which cognition sits near the center of the stage (Berscheid & Walster, 1978; Kelley, 1980; Rusbult, Verette, Whitney, Slovik, & Lipkus, 1991). This is not to say that the attraction literature does not contain motivational accounts (Berscheid & Graziano, 1979; Cunningham et al., chapter 5, this volume; Graziano, Jensen-Campbell, & Hair, 1996; Simpson, Gangestad, & Lerma, 1990), but that these accounts usually focus on motivational effects on cognitive aspects of attraction. Building links to sociobiology from interpersonal attraction would be a much more difficult task than building links to evolutionary psychology.

It is theoretically possible, of course, that these proximal variables are noncausal by-products of the more basic, distal evolutionary mechanisms, just as it is possible that cognition is an epiphenomenon of other more basic biobehavioral processes. We could fight this metatheoretical "existence fight," or we could concentrate on evaluating more testable propositions about links among different kinds of variables from evolutionary psychology. If propositions are not yet available, then we could amend applicable evolutionary theories to add the missing proximal psychological mechanisms, testing to see what functions (if any) the mechanism might play in mate choice. That is, we can check directly to see how well the more distal predictor acts with a proximal attraction mediator in place (Baron & Kenny, 1986). Incidentally, with some conspicuous exceptions (Kenrick, Groth, Trost, & Sadalla, 1993) evolution-oriented writing about human mate selection is often poorly informed about relevant empirical findings from the psychology-based, human interpersonal attraction literature.

Evolutionary accounts of human mate selection will be less comprehensive than they aspire to be if they do not accommodate themselves to the attraction literature and coordinate themselves with evolutionary variables and mechanisms.

Trivers' (1972) theory of parental investment offers a good illustration. Buss (1995) explicitly used this theory to illustrate how propositions from a midlevel evolutionary theory can be tested and possibly falsified. Trivers' theory appears to make far-reaching predictions about patterns of human mate choices, but it is noncognitive and sociobiological in cast. Trivers (1972) said, "The pattern of relative parental investment in species today seems strongly influenced by the early evolutionary differentiation into mobile sex cells fertilizing immobile ones, and sexual selection acts to mold the pattern of relative investment" (p. 173). Trivers' theory is based largely on the behavior of nonhuman animals, mostly birds. The absence of a proximal psychological mechanism, much less a cognitive mechanism, is striking to persons interested in applying this theory to human attraction. We assume that the proximal psychological mechanism in humans would involve the phenomenology of attraction. Presumably, this is the kind of evolved cognitive mechanism Tooby and Cosmides (1992) had in mind.

One consequence of Trivers' noncognitive, nonproximal approach is that his theory seems to draw fire from critics more for its distal causation than for substantive issues in its predictions (Graziano, Jensen-Campbell, Shebilske, & Lundgren, 1993; Jackson, 1992; Jensen-Campbell, et al., 1995; Nisbett, 1990). In fairness to Trivers' theory, this work was published nearly 25 years ago as part of a doctoral dissertation when the field of evolutionary psychology did not yet exist. Recent theoretical refinements (Buss, 1992; Buss & Schmitt, 1993) make Trivers' theory more amenable to psychological analysis, but further theoretical development is still needed (Perper, 1989).

A third line of theoretical development is related to the second. Given the role that cognition plays in evolved mechanisms, we need to know more about what evolutionary psychologists such as Buss (1995) mean by the term *cognition*. Tooby and Cosmides (1992) asserted that *cognition* can mean many different things, but that for them, "terms such as 'cognitive' and 'information processing' refer to a language or level of analysis that can be used precisely to describe any psychological process: Reasoning, emotion, motivation, and motor control can all be described in cognitive terms, whether the processes that give rise to them are conscious or unconscious, simple or complex" (p. 65).

The advantage of this kind of definition is that it expands the range of phenomena that fit under the evolutionary explanatory umbrella. There are, however, several disadvantages. First, the claim that evolved mechanisms are "more closely allied with cognitive levels of explanation than with any other level of proximate causation" becomes true by definition, not by empirical analysis. If virtually any biobehavioral process can be cognitive, then any intraorganismic process that is functionally responsive to external input and affects self-regulatory processes or behavior seems to qualify as a cognitive mechanism. Any mediation of behavior, and by implication, any evolved mechanism, is cognitive by definition. In this decade following the "cognitive revolution," such a definitional stance probably would not raise many eyebrows, yet, the definition may induce researchers to overlook interpersonal attraction processes and variables that do not match the cognition prototype well (Hendrick, 1995; cf. Zajonc & Markus, 1984). We discuss this matter in more detail in our mediated structural equation analysis of women's attraction to men.

ATTRACTION FROM AN EVOLUTIONARY PSYCHOLOGY PERSPECTIVE

Individuals prefer the company of some people more than others. One especially important form of selectivity in humans involves choice of conjugal or romantic partners (Buss, 1989, 1992; Ellis, 1992). Such choices influence the kinds of physical and psychological environments the chooser will experience, and the potential reproductive consequences are powerful. Despite its importance, however, the mechanisms underlying social selectivity are poorly understood. Evolutionary theory provides one comprehensive explanation for social selectivity in terms of Darwin's (1871) notion of sexual selection. In intrasexual selection, individuals of one sex compete for mating opportunities with individuals of the other sex. In intersexual selection, individuals of one sex prefer individuals of the other sex who have certain attributes (e.g., in birds, colorful plumage). Darwin called this later form of selection "female choice" because he believed that females of many species were more selective in choosing mating partners than were males (cf. Small, 1992).

The theoretical term *female choice* is unfortunate in several respects. First, an overt choice in one context may not reflect a generalized preference (Slovic, 1995). Overt mate choices made in one context may reflect social coercion, not unconstrained preferences. Darwin himself recognized this complication in the operation of sexual selection (Darwin, 1871, chapter 20, in the section entitled "Early Betrothals and Slavery of Women"). Constraint on female mate choices can take many forms, and these probably vary along a continuum of coercion (Burbank, 1995). If a woman were to choose a dominant aggressor for a sexual partner, she may be choosing to stay alive, not expressing a personal preference for dominant men.

From the perspective of social psychological theory and research on human interpersonal attraction, current evolutionary psychology is unsophisticated when it equates choices with preferences. More generally, current evolutionary psychology is underdeveloped theoretically in its treatment of human responsiveness to proximal social situational contexts, particularly those associated with social constraint (Graziano et al., 1993). An etching from the medieval

Breviaire d' Amour (Chenevix-Trench, 1970) probably illustrates women's need to behave adaptively in the presence of aggressive males, not women's attraction to dominant men (see Fig. 6.1). Briseis's affiliation with Achilles probably did not reflect an authentic choice, nor is it likely that it displayed an attractive woman's preference for a dominant man.

For the moment, however, let us acknowledge the kernel of truth in Darwin's theorizing about female choice. Trivers (1972) refined Darwin's notion by suggesting that one driving force in intersexual selection was different investment in offspring by males and females. According to Trivers, in the vast majority of species, females make much larger contributions to the survival of their offspring than do males. Consequently, females adopt a reproductive strategy in which they confine sexual contact to males who would be most likely to give an advantage to any offspring (Small, 1992). Trivers notes that females may selectively mate with dominant males because the female can ally her genes with "a male who, by his ability to dominate other males, has demonstrated his reproductive capacity" (p. 170).

As noted previously, Trivers' conjectures are based largely on the behavior of nonhuman animals, mostly birds. Some empirical research based on human behavior has begun (Buss, 1989; Buss & Barnes, 1986), but nearly all of this work is nonexperimental, correlational, and based on self-reports or are responses to vignettes. Almost no research allows females to observe patterns of male behavior, from which they could draw their own conclusions and express their preferences.

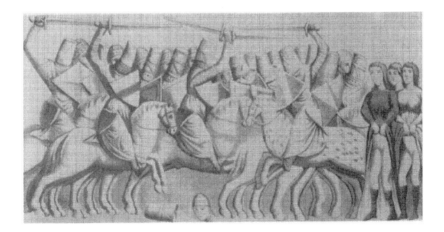

FIG. 6.1. The rewards of the victor were not limited to the accoutrements of the vanquished, but included carnal prizes here illustrated as being (sometimes) on display. From *A History of Horsemanship*, (p. 83), by C. Chenevix–Trench, 1970, Garden City, NY: Doubleday. Copyright 1970 by Doubleday. Reprinted with permission.

In a program of four studies, Sadalla, et al. (1987) explored the hypothesis that behavioral expressions of dominance in males increased males' sexual attractiveness to females. Dominant behavior in males was found to increase females' sexual attraction to them; findings did not link dominant behavior in females to sexual attraction in males. Male dominance enhanced only sexual attractiveness; it did not enhance general likability. Sadalla et al. noted that there is nothing explicit in sociocultural theories of attraction that would predict these results, whereas sociobiological theories would. Specifically, Sadalla et al. asserted that "male dominance is an attribute whose genetic mechanism spread because it conferred a reproductive advantage to its carriers" (p. 737).

Sadalla et al. focused on competition and dominance as key attributes in males' attractiveness to females. But as theorists have noted, dominance is a complex construct, and its behavioral expression will have different consequences depending on the social context (Bernstein, 1980; Buss, 1986; Dunning & McElwee, 1995; Graziano, Brothen, & Berscheid, 1978; Hinde, 1978). There are alternatives to the Sadalla et al. approach, that are not necessarily incompatible with evolutionary accounts that refocus on cooperation and altruism (Gould, 1988; Hinde, 1978). In an early critique of Spencerian interpretations of evolution, Kropotkin (1914) noted that the success of many species was attributable not to "survival of the fittest," but to an ability to organize into groups and engage in mutual aid (see Wilson, chapter 13, this volume). Trivers himself (1972) acknowledged that certain forms of cooperation (e.g., ability to contribute to the care of offspring) may influence the choices of some females.

At a general theoretical level, cooperation and altruism can be understood as social resources, or in Trivers' terms, "investments." That is, males who are disposed to cooperate with their partners, or who show nurturance and altruism, may be selectively preferred to other males. This preference may reflect not only the female's attribution that such males would be rewarding to her personally, but also that such males will be disposed to invest more heavily in their offspring (Ellis, 1992; Feingold, 1992; Graziano & Eisenberg, 1997; Simpson & Gangestad, 1992).

There are potential problems, however, in reconciling our theoretical conjectures with the outcomes of the Sadalla et al. studies, which appear to demonstrate that sexual attraction is unrelated to general attraction or to such positive male behaviors as cooperation and altruism. Worse still, it is possible that females may derogate a cooperative, altruistic male for being a nonmasculine weakling.

A potential reconciliation can be found in research by Godfrey and Lowe (1975). These authors did not conduct research on interpersonal attraction per se but on women's belief in a "just world" and their corresponding reactions to persons who incur costs in helping others. Godfrey and Lowe found that females made negative attributions to persons who incur costs but only when they were driven to incurring the costs by external circumstances. When costs were incurred as a result of intrinsic choice, attributions were positive. Extrapolating from these results, we can refine our prediction: Males whose cooperation and altruism seem intrinsically motivated will be selectively preferred by females, whereas males whose cooperation appears to be externally motivated will not (cf., Hinde, 1978).

This line of reasoning may explain the Sadalla et al. results. Male dominance may be seen as an expression of agentic behavior, in which the male actively makes choices and is effective in dealing with others (Wiggins, 1991). Without additional, qualifying information, male agentic behavior may be generally attractive to females. The agency–tion relation may be moderated, however, by additional information about the specific content of the agency. That is, men whose agency is linked to cooperative, altruistic tendencies should be preferred; men whose agency is linked to competitive, selfish tendencies should not. Male dominance may signal to females the ability to channel resources to her and her offspring, whereas altruism may signal the willingness to do so. A man who has resources but is unwilling to share them is probably not an attractive mate, at least not a long-term relationship (Buss, 1995; Buss & Schmitt, 1993).

Given that human interpersonal attraction is a complex multifaceted process (Berscheid & Walster, 1978), we need a research strategy that recognizes different aspects of attraction, such as sexual attractiveness, dating desirability, physical attractiveness, and overall social desirability. For example, a prosocial man may be seen as socially desirable, but not more physically attractive or desirable as a date. The conceptualization would require a multivariate measurement methodology. Such an approach is especially important in probing hypotheses from evolutionary psychology in that dominance is presumed to affect some aspects of attraction (sexual attractiveness), but not necessarily others (general social desirability).

Male Dominance Is Not Necessarily Incompatible With Prosocial Tendencies

It is not necessary to assume that dominance is a bipolar opposite to altruism, or even that these two aspects of male behavior are incompatible, either conceptually or empirically. That is, two orthogonal (or at least separable) dimensions of dominance and prosocial tendencies may be present in many aspects of male social behavior. If these two aspects of males (and people overall) are not mutually exclusive, then it is possible that high dominance and high prosocial tendencies could coexist in the same individual. More generally, we could construct a conceptual 2 X 2 matrix for male behavior, crossing factorially dominance (high vs. low) with prosocial tendencies (high vs. low). Such a conceptualization would allow us to see four distinct male personae, or configurations, that bear potential relevance to evolutionary theorizing about attraction. Conceptually, such a scheme would also allow us to partition the separate influence of male dominance and prosocial tendencies on female attraction. More important, however, is the possibility that the two separate dimensions of dominance and prosocial tendencies could *interact* in affecting female attraction.

Data exist regarding the (non)bipolarity of dominance and prosocial tendencies in men. First, evidence from natural language descriptions across diverse language and age groups suggests that dominant behaviors are seen as relatively independent of positive, prosocial behaviors (Goldberg, 1981; John, 1990; Wiggins, 1991). Patterned regularities in natural language descriptions tell us about the structures of social behavior that are important enough to warrant a semantic representation. It is possible to generate words that represent reality poorly (e.g., "unicorn"), but patterned regularities in language descriptions probably inform us about the dimensions of social relationships that natural selection has retained in language, and to which, no doubt, it is useful to be attentive.

The premier illustration of this linguistic approach comes from Goldberg's (1981) work on the five-factor semantic structure underlying personality descriptions. In the five-factor model, surgency/dominance and agreeableness are two very important, but independent, dimensions of interpersonal evaluation (see also McCrae & John, 1992; Wiggins, 1991). These two personality dimensions are potentially related to the dominance and altruistic tendencies we discussed previously (see Graziano & Eisenberg, 1997).

To facilitate empirical work on the five-factor model, Goldberg (1992) constructed standard sets of verbal markers for these factors. Persons are typically asked to rate themselves on a Likert-type continuum (i.e., 1 = *strongly disagree* -to 5 = *strongly agree*) trait words like *kind, cooperative, bold,* and *assertive.* From this simple, direct format, self-reports can be combined to produce dimensional scores (i.e., a score for surgency/dominance and a score for agreeableness). Goldberg constructed his measures so that the dimensions would be orthogonal, and presented data showing that factor 1 (surgency) and factor 2 (agreeableness) formed from his adjectives in fact were not significantly correlated.

With these tools, it is possible to reconsider the 2 × 2 matrix we discussed earlier, with surgency/dominance on one dimension and agreeableness on the other. Conceptually, we could use Goldberg's terms to generate four male personae: a high-dominant/high-agreeable male, a high-dominant/low-agreeable male, a low-dominant/high-agreeable male, and a low-dominant/low-agreeable male. If our conjectures are valid, then dominance and agreeableness will interact in influencing female attraction to these four classes of males.

Jensen-Campbell, et al. (1995) experimentally manipulated behavioral expression of male dominance and prosocial orientation, and female subjects evaluated the males. Additional variables were manipulated in the studies (e.g., male physical attractiveness, male or female targets of male dominance) to explore the reliability of effects across variations in social context.

Jensen-Campbell and colleagues predicted first that women's subjective ratings of men's physical attractiveness would be affected not only by dominance, but also by the men's prosocial orientation. Second, men's prosocial orientation would increase his desirability as a date, his general social desirability, and his sexual attractiveness to women. This would be true, however, only if his prosocial orientation was seen as an individually initiated decision (i.e., emerging from dominance, not from submission to the demands of others).

Women students enrolled at Texas A&M University were randomly assigned to the cells of a 2 (dominance) × 2 (agreeableness) between-subjects factorial design. Video scripts were constructed to manipulate dominance and agreeableness. Each script involved three confederates. One confederate portrayed an experimenter and the other two portrayed research participants. The subject's task was to evaluate one of the two target men who posed as research participants.

The Jensen-Campbell team closely followed the dominance manipulations employed by Sadalla et al. (1987). Dominance was manipulated both with nonverbal cues and in overt actions. Within the constraints of credibility, "normal" variations in agreeableness were also manipulated. We used Goldberg's (1992) marker adjectives to build an agreeableness script for the confederates to play. In the high-agreeableness conditions, the male solicited the opinions of his partner, was sympathetic to the perspectives of the partner, and was warm. In the low-agreeableness conditions, without being hostile or antisocial, the male criticized the opinions of his partner, was insensitive to his or her perspective, and was not especially warm. The substantive content of all the conversations was virtually identical, within the constraints of the specific manipulations.

The subject saw one of four conversations in which the male target person displayed either high or low levels of dominant and of agreeable behaviors toward a male or female partner participant. She then rated him on several composite measures (i.e., overall social desirability, dating desirability, physical attractiveness, and sexual attractiveness) based on summed Likert-type scales.

The experimental manipulations were very successful. Manipulation checks on the women's perception of the men showed that the high-dominant confederate was seen as more dominant than the low-dominant confederate. Similarly, the high-agreeable confederate was seen as more agreeable than the low-agreeable confederate. Thus, our manipulations were effective.

We now turn to our hypotheses linking prosocial tendencies and dominance to attraction. Results for each of the four attraction-dependent variables (rated physical attractiveness, sexual attractiveness, social desirability, and dating desirability) corroborated the predictions and showed very similar patterns. In each case, low-agreeable men were not attractive (neither sexually nor physically, nor were they desirable as dating partners); adding or subtracting dominance did not alter that outcome. For men who were high in agreeableness, however, dominance enhanced their attractiveness significantly. Of the four male personae generated by the 2 × 2 agreeableness-dominance matrix, the high-agreeable, high-dominant man was the most attractive to women. This pattern in women's evaluations of the man held when the target man interacted with another man or with a woman partner.

The Phenomenology of Women's Attraction to Men

We used a social–psychological paradigm so that we could probe the subjective reactions of our women subjects (at least as they were reflected in ratings). As we noted previously, our manipulations of agreeableness and dominance were very successful in affecting the subject's phenomenology. One way to consider the outcome is in terms of the relative impact of each variable on women's attraction. Informally, we could compare the F-values for our independent variables of agreeableness and dominance. The multivariate F-values associated with the attraction variables were 19.11, 6.04, and 6.20 for agreeableness, for dominance, and for the agreeableness × dominance interaction, respectively. When considered as a direct main effect on attraction, agreeableness had more than three times the impact of dominance in predicting women's attraction.

Such informal analyses of multivariate main effects are potentially misleading, however. To be even minimally valid, such comparisons require that the independent variables be orthogonal, not only in their manipulation, but also in their phenomenological interpretation by women subjects. Using main-effect F-values as "impact indicators" is complicated further by the significant multivariate agreeableness × dominance interaction.

We had randomly assigned participants to the conditions, so we had reasonable confidence that our manipulation of agreeableness and dominance as main effects were independent of each other. The manipulation checks, however, told us that the women's interpretations of our manipulations were more complicated than we had anticipated. There was some apparent cross-influence of the manipulations on their respective checks. When we manipulated dominance, we obtained the predicted effect on the dominance manipulation check, but we also found an unexpected effect of dominance on the agreeableness check. The high-agreeable confederates were seen as less dominant than were the low-agreeable confederates. Besides the significant main effect for agreeableness, there was also a significant dominance × agreeableness interaction on the agreeableness manipulation check.

Mediational analyses using structural equations. Cross influence on manipulation checks raises potential problems in the interpretation of the independent variables in randomized experiments. If the manipulation of variable A effects the checks on both A and B, then how can we partition the effects of A from those of B? To address this potential problem, we conducted a series of structural equation analyses. The structural equation analyses allowed us to partition the relative impact of correlated predictors on the criterion. We treated the manipulation checks as mediating variables between the manipulations and the dependent variables. (The sex of a participant did not influence any manipulation check or dependent variable, hence it was dropped from all analyses as a moderator. We do, however, report structural equations separately by sex of participants for comparison purposes.)

Reflecting the suggestions of Baron and Kenny (1986), our first structural model jointly examined the effects of the manipulations of dominance, altruism, attractiveness, and the dominance × altruism interaction on the three manipulation checks (see West, Aiken, & Todd, 1993). For each of the three manipulation checks, the largest amount of predictive variance was attributable to the appropriate manipulation, with no impact or smaller impact from other manipulations. For the dominance-manipulation check, the largest portion of variance was attributable to the dominance manipulation, with a smaller portion attributable to the agreeableness manipulation (see Fig. 6.2). For the agreeableness-manipulation check, the largest portion of the variance was attributable to the agreeableness manipulation, with a smaller contribution from the agreeableness × dominance cross product.

We then used a series of structural equation models for each outcome measure to test our hypotheses about mediation. We formed contrast terms corresponding to each of the three manipulations, the three two-way and one three-way interactions. An initial no-mediation (common cause) model specified paths from the contrast terms representing the three manipulations and the agreeableness × dominance interaction to the three mediators and from all seven contrast terms to the outcome measure under consideration. The full-mediation model additionally specified paths from each of the three mediators to the outcome variable. (None of the no-mediation models had an adequate fit to the data, whereas each of the full mediational models fit the data.)

TAKING A LONGER PERSPECTIVE ON ATTRACTION

Our complex statistical analyses may be telling us more about the phenomenology of women's immediate, situational attraction to men than about the planned deliberations that occur when women make relationship choices with long-term implications. These longer-term relationships may be more relevant to evolutionary psychology theory (cf. Small, 1992). Buss and Schmitt (1993), as well as Kenrick, Sadalla, Groth, and Trost (1990), argued that candidates for long-term relationships (e.g., marriage) probably require attributes different from candidates for short-term commitments (e.g., date or sex). It is plausible that prosocial tendencies may increase in importance as the anticipated length of the relationship with a man increases. For the present chapter, we collected a length-of-perspective measure of attraction to probe the possibility that agreeableness in a prospective partner may become more valuable as the length

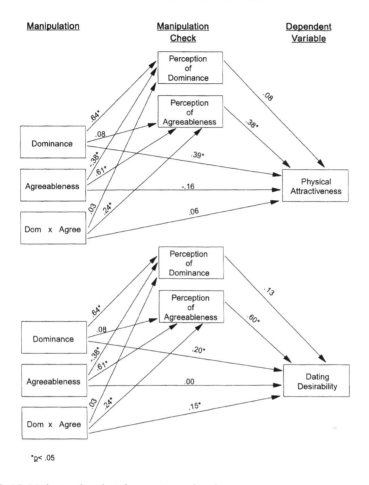

FIG. 6.2. Mediational analysis for experimental study.

of the anticipated relationship increases. In addition, we solicited the opinions of both men and women.

To probe these hypotheses, and to replicate the findings reported by Jensen-Campbell and colleagues, we collected additional data from 372 Texas A&M University students (males = 153; females = 219). The participants were randomly assigned to the cells of a 2 (dominance) × 2 (agreeableness) × 2 (sex of participant) factorial design. To frame the manipulations, we again selected words from Goldberg's (1992) marker adjectives from factors 1 (surgency) and 2 (agreeableness) of the five-factor model of personality.

We created four distinct targets with a 2 (dominance)× 2 (agreeableness) factorial. The dominant and nondominant targets were framed in terms of surgency (factor 1). The dominant individual was described as active, assertive, bold, talkative, and verbal. The nondominant confederate was described as introverted, quiet, reserved, timid, and untalkative. The agreeable and disagreeable targets were framed in terms of agreeableness (factor 2). The agreeable person was described as considerate, cooperative, generous, kind, and sympathetic. The disagreeable confederate was described as rude, selfish, uncooperative, unkind, and unsympathetic.

The participants were randomly assigned to one of the four conditions as they arrived. When all participants were in the room, they were told that people form impressions from very limited information. They were directed to look at a list of words describing one of the four target persons, then try to form an image of the person. Finally, the experimenter then told everyone to turn the page over and rate the person. They were not permitted to look back at the individual words once they began giving us their impressions.

Using a single-item scale from 1 (*not at all dominant*) to 9 (*very dominant*), the high-dominant target persons were seen as more dominant than the low-dominant target persons. Using a single-item scale from 1 (*not at all agreeable*) to 9 (*very agreeable*), the high-agreeable target persons were seen as more agreeable than the low-agreeable target persons. The low-agreeable targets were also seen as more dominant than the high-agreeable targets, suggesting cross-influence of our manipulations.

We first analyzed data for overall patterns common to the three attraction-dependent measures. We found overall multivariate main effects for agreeableness and for dominance on attraction. (The multivariate agreeableness effect was approximately six times larger than the dominance effect.) These multivariate main effects were qualified by a smaller multivariate dominance × agreeableness interaction.

If we look at the attraction variables one at a time, we see that both men and women raters reported that the high-agreeable targets were significantly more physically attractive than were the low-agreeable targets. However, we also found much smaller main effects for dominance, and a dominance × sex interaction. There was a tendency for men and women to rate the high-dominant target as more physically attractive than the low-dominant target, but this effect was larger for women rating men than for men rating women.

For men, high-agreeable women were significantly more desirable for dating than were low-agreeable women. For women's ratings of men, however, we found a small dominance × agreeableness interaction. Dominance enhanced their dating desirability. There was no evidence that dominance affected dating desirability for low-agreeable men targets. These results replicate the findings of Jensen-Campbell et al. (1995).

Regarding desirability for a long-term relationship, we found a significant dominance × agreeableness interaction. The results again replicated the findings of Jensen-Campbell et al. (1995). Dominance enhanced their long-term

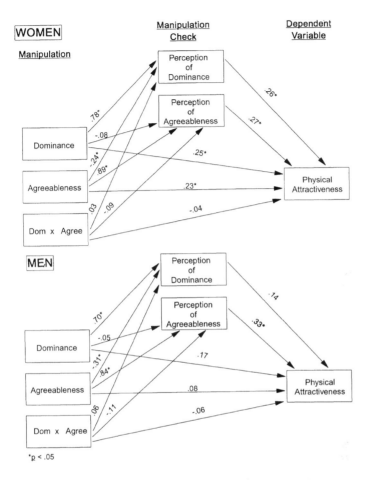

FIG. 6.3. Mediational analysis for physical attractiveness by women and men raters.

desirability high-agreeable men targets; there was no evidence that dominance had any enhancing effect in the evaluation of low-agreeable men. In men's ratings of women targets, agreeable women were rated as much better long-term partners by men than were low-agreeable women targets. Dominant women also were seen as significantly better partners that nondominant women, but this effect was approximately 3% of the size of the agreeableness effect.

As in the Jensen-Campbell et al. study, there were some apparent cross-influences of the manipulations. Agreeableness also affected the dominance-manipulation check. The low-agreeable confederate was seen as more dominant than the high-agreeable confederate.

The first structural model examined jointly the effects of the manipulations of dominance, agreeableness, and the dominance × agreeableness interaction on the two manipulation checks (Aiken & West, 1991). Again, for each manipulation check, the largest predictive variance, by far, was attributable to the appropriate manipulation (see Figures 6.3 to 6.5).

A series of structural equation models was then estimated separately for each outcome measure to test the hypotheses that the effects of the manipulations on each of our dependent measures were mediated by the manipulation checks. Contrast terms were formed corresponding to each of the two manipulations and their interaction. The initial no-mediation (common cause) model speci-

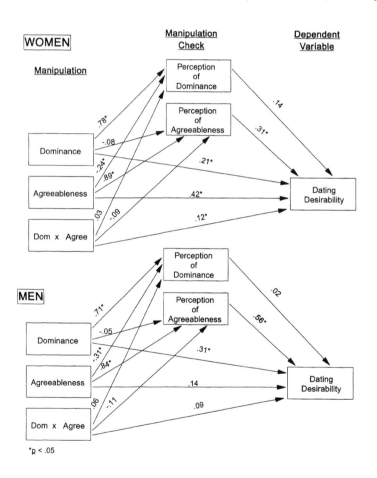

FIG. 6.4. Mediational analysis for dating desirability for women and men raters.

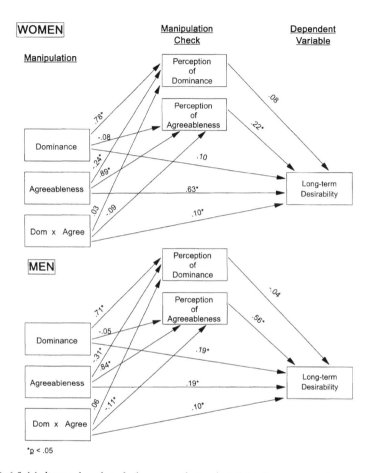

FIG. 6.5. Mediational analysis for long-term dating desirability by women and men raters.

fied paths from the contrast terms representing the two manipulations and their interaction to the two mediators and from the contrast terms to the outcome measure under consideration. The full-mediation model additionally specified paths from each of the two mediators to the outcome variable. As in the Jensen-Campbell et al. studies, each of the no mediation models failed to provide an adequate fit to the data; each of the full mediational models was just identified, and, by definition, fit the data perfectly For women rating men, as well as for and men rating women, tests showed that significant overall mediation occurred for physical attractiveness, dating desirability, and long-term commitment.

For both men and women, the path from the agreeableness manipulation check to the outcome variable was significant in each model (see Fig. 6.3 to 6.5). For women, the path from the dominance–manipulation check to the outcome variable was significant only for physical attractiveness. For men rating women, there was no evidence that the path from the dominance-manipulation check to the outcome variables was significant.

The dominance–manipulation also had a small, direct (nonmediated) effects on outcomes variables. For women rating men, dominance had a direct effect on dating desirability and physical attractiveness. In addition, the Dominance × Agreeableness interaction also had a direct effect on long-term commitment and dating desirability. For men rating women, dominance had a small direct effect on dating desirability, physical attractiveness, and long-term commitment.

In summary, the substantive message here is that perception of agreeableness in a partner is a major contributor to interpersonal attraction, for both men and women. Agreeableness is a large star, around which rotates the much smaller, dark moon of dominance. Dominance effects are elusive, and probably occur only when other aspects of persons are also present. The mediational–structural equation analyses helped us to see how the variables fit together in influencing attraction. More technically, they showed that we can experimentally manipulate variables relevant to evolutionary psychology, and that the largest predictive variance was attributable to the appropriate manipulation. Across all of the outcome variables, the mediational model provided a better fit to the data than the no mediation model. The potent variable here is agreeableness. The manipulation check for agreeableness appeared to serve as a mediator of the effects of the manipulations for all outcome variables. In contrast, the effects of the dominance-manipulation check were smaller, and when they did occur, were more consistent with a direct, nonmediated model for our dependent variables.

CONCLUSIONS

In the concluding paragraph of his epic work *The Origin of Species*, Darwin (1859) contemplated the life of interdependent organisms embedded in a tangled bank. In this tangled bank, complex behaviors of the various organisms have all been produced by evolutionary laws. From the perspective of current evolutionary psychology, the laws that Darwin described are expressed through a set of evolved mechanisms. Our task is to identify those mechanisms and show how they are linked to behavior and cognition.

The precise nature of evolved mechanisms in evolutionary psychology remains unclear, but theorists have suggested that these mechanisms be regarded as kinds of "mental organs," dedicated to solving specific tasks. Tooby and Cosmides (1992) suggested that evolved mechanisms relevant to psychology are internal representations "most closely allied with the cognitive level of explanation than with any other level of proximate causation. This is because the cognitive level seeks to specify a psychological mechanism's function, and natural selection is a theory of function" (p. 284).

If this line of thinking is applied to human mate selection and interpersonal attraction, we can see complications. It is plausible that more than one mechanism (or mental organ) is involved in a process as complex as interpersonal attraction (Cunningham et al. 1990; Cunningham et al., chapter 5, this volume). Selecting a mate is probably a more multifaceted task than detoxifying the blood. Even if we restrict ourselves to cognitive mechanisms, there is still a problem of specifying what is meant by "cognition." To be useful for evolutionary psychology, cognitive mechanisms cannot be restricted to conscious, accessible, easily retrieved processes. In this regard, our structural equation analyses suggest that some attraction processes may be easily accessed through conscious cognition (e.g., those associated with the agreeableness of another person), but others (e.g., those associated with dominance) may not.

Our mediational analyses are complex and depend on technical assumptions about measurement and validity (West et al., 1993). We do not pretend to offer the final word on cognitive mechanisms underlying interpersonal attraction. Our analyses suggest, however, that the cross-influence of independently constructed manipulations are telling us something potentially important about women's attraction to men. Prosocial tendencies such as agreeableness have a potent impact on women's evaluation of men. The agreeableness-manipulation check appeared to serve as a mediator of the effects of the manipulations for all outcome variables. In contrast, the effects of the dominance manipulation were more elusive, and when they occurred, were smaller than the effects of the agreeableness manipulation. Although caution must be used in interpreting these results, they suggest that subjects' cognitive awareness may play a lesser role in the influence of dominance (when it occurs) than in the influence of agreeableness on measures of attraction.

These results also have implications for conceptualizations of cognitive mechanisms discussed by Buss (1995) and by Tooby and Cosmides (1992). First, women's attraction to men is probably affected by more than one cognitive mechanism. Cunningham et al. (1990) have suggested this hypothesis with respect to facial attractiveness, but their conclusion may apply more generally to interpersonal attraction (Cunningham et al., chapter 5, this volume). Single variable or single-mechanism explanations of attraction ("Women are sexually attracted to dominant men") are almost certainly gross oversimplifications. Second, if there are multiple mechanisms underlying interpersonal attraction, then different variables may be linked to different mechanisms, and these may differ in their pattern of impact, their levels of accessibility to perceivers, and the directness with which they affect interpersonal attraction. If this is true, then it is possible that in some circumstances the mechanisms may be working at cross-purposes.

Consider, for example, the case of women's judgments of men's physical attractiveness. (See Fig. 6.3). Manipulation of men's agreeableness has a significant negative impact on women's judgments of men's dominance (manipulation check), but judgments of dominance were positively related to judgments of physical attractiveness. At the same time, the manipulation of men's agreeableness had a direct (unmediated) positive impact on judgments of men's physical attractiveness. The preponderant impact of agreeableness on judgments of men's physical attractiveness is positive, but there is a smaller, negative impact as well. If these speculations are valid, then ambivalence and even "paradoxical attraction" (Graziano, Brothen, & Berscheid, 1980) may be more common in interpersonal attraction than we currently recognize (Perper & Weis, 1987).

From an evolutionary perspective, why should agreeableness influence attraction through conscious cognition, and dominance influence attraction through another route? It is important to recognize that in modern scientific psychology, the boundary between conscious and nonconscious cognition is highly permeable. Current cognitive psychology probably prefers terms such as *accessible* or *available* over the term *conscious*. Nonetheless, we offer some preliminary speculations, knowing that the brickbats of critics cannot be far away. Agreeableness is an abstract summary for individual differences in prosocial behavioral tendencies. Apparently, agreeableness is a salient aspect of persons, generally appearing as the first and largest factor underlying natural language descriptions of persons (Graziano & Eisenberg, 1997). The dimension may be salient in social cognition because it is related to the kinds of outcomes people receive, or expect to receive, from social interaction. During social interactions, agreeableness is related to perceptions of trust. In addition, there is now empirical evidence that agreeableness is related to patterns of interpersonal conflict tactics and conflict resolution (Graziano et al., 1996). The high agreeable person appears to be well socialized, to minimize conflicts with others, and is probably likely to generate more positive outcomes for partners and members of his or her group than will low-agreeable persons.

There is no evidence yet to support this claim, but we might turn Trivers' Parental Investment Theory around to suggest that high-agreeable persons will be more sympathetic, nurturant parents than will low-agreeable persons. At a general theoretical level, cooperation and altruism can be understood as social resources, or in Trivers' terms, "investments." That is, males who are disposed to cooperate with their partners, or who show nurturance and altruism, may be selectively preferred to other males. This preference may reflect, not only the female's attribution that such males would be rewarding to her personally, but also that such males would be disposed to invest more heavily in their offspring (e.g., Ellis, 1992; Feingold, 1992; Graziano & Eisenberg, 1997; Simpson &

Gangestad, 1992). Unless there is some important countervailing reason to select a low-agreeable person (e.g., special expertise for a needed project; uncommon wealth, family connections, or good looks), why would any rational person not prefer a high-agreeable partner? Given the potentially large differences in interpersonal payoffs, it is probably not surprising that preferences for high-agreeable partners over low-agreeable partners could reach the level of conscious cognition.

There is more to interpersonal attraction, however, than rational choices. Some aspects of persons may speak to us through some less conscious mechanisms (Cunningham et al., chapter 5, this volume; Murphy, Monahan, & Zajonc, 1995). There may be more primitive forms of attraction, such as the dominance–sexual attraction link noted by Freud and reported by Sadalla et al., that have their origins in our remote evolutionary past. These mechanisms may have been in place long before humans evolved more sophisticated mechanisms that allowed them to adapt to the increased complexity of social life. Like an out-of-date-but-not-deleted computer routine that lies dormant during normal operation, and that no longer shows up on the new tree diagram, the darker forms of attraction may be less open to conscious inspection. Like the old computer routine, the old mechanism still can be activated with suitable input. Our guess is that a person who experiences the activation of such a mechanism may be as surprised as the computer operator looking at the output that the old routine generates (Murphy et al., 1995).

Here may reside a new line of inquiry about human interpersonal attraction. First, current discussions in evolutionary psychology of dedicated cognitive mechanisms pay no attention to the way such mechanisms moderate, or interact, with each other. (For a parallel analysis of problems in cross-module communication in socialization, see Costanzo, 1991). There are probably multiple motives and mechanisms underlying human interpersonal attraction, and these reside at different altitudes within the psychological structure. Choices and preferences probably reflect compromises among these motive–mechanisms. Furthermore, the compromises must be ecology-sensitive. In harsh, hostile environments, women's usually strong preference for an agreeable partner may be overcome by the need for a dominant partner. In more hospitable ecologies, the pattern may be reversed. In both cases, the choices have implications for survival and reproduction. More generally, it would be important to understand how people arrive—consciously or less consciously—at these compromises, how ecological factors influence compromises, and how specific mechanisms combine to influence the final compromises that persons reach. In this sense, Darwin's tangled bank may be an apt metaphor for *intrapsychic* processes as they go about the complicated business of making social choices.

Our descriptions of evolutionary approaches to interpersonal attraction have been painted with a broad brush, and are very speculative. This is a consequence of the limited empirical data base dedicated to evolutionary psychology. As more work is conducted, however, we will gain a clearer picture of the processes underlying attraction, and a better understanding of the dynamics of life in the interpersonal and intrapersonal tangled bank that is human attraction.

ACKNOWLEDGMENTS

The authors are grateful for the help and advice of Stephen G. West on the structural equation mediation analyses. They are also grateful to Martha Cornog, Philip Costanzo, Michael R. Cunningham, David M. Buss, Shaun D. Campbell, Stephen Gangestad, Clyde Hendrick, Timothy Perper, and Michele Tomarelli for their comments, advice, and help on various aspects of the work. Both editors also provided extensive comments. That we did not use all of their suggestions reflects limited space, not limited appreciation for their insights. This work was supported in part by an NIMH grant to W. Graziano.

REFERENCES

Aiken, L. S., & West, S .G. (1991). Multiple regression: Testing and interpreting interactions. Newbury Park, CA: Sage.

Baron, R. M., & Kenny, D. A. (1986). The moderator–mediator variable distinction in social psychological research: Conceptual, strategic, and statistical considerations. Journal of Personality and Social Psychology, 51, 1173–1182.

Bernstein, I. S. (1980). Dominance: A theoretical perspective for ethologists. In D. R. Omark, F. F. Strayer, & D. G. Freedman (Eds.) Dominance relations (pp. 71–84). New York: Garland STMP Press.

Berscheid, E., & Graziano, W. G. (1979) The initiation of social relationships and interpersonal attraction. In R. L. Burgess & T. L. Huston (Eds.) Social exchange in developing relationships (pp. 31–60). New York: Academic Press.

Berscheid, E., & Walster, E. (1978). Interpersonal attraction. (2nd ed.). Reading, MA: Addison-Wesley.

Brewer, M. B., & Caporael, L. R. (1990). Selfish genes versus selfish people: Sociobiology as origin myth. Motivation & Emotion, 14, 237–243.

Burbank, V. K. (1995). Gender hierarchy and adolescent sexuality: The control of female reproduction in an Australian aboriginal community. Ethos: Journal of the Society for Psychological Anthropology, 23, 33–46.

Buss, A. (1986). Social behavior and personality. Hillsdale, NJ: Lawrence Erlbaum Associates.

Buss, D. M. (1989). Sex differences in human mate preferences: Evolutionary hypotheses tested in 37 cultures. Behavioral and Brain Sciences, 12, 1–49.

Buss, D. M. (1990). The evolution of anxiety and social exclusion. Journal of Social and Clinical Psychology, 9, 196–201.

Buss, D. M. (1992). Mate preference mechanisms: Consequence for partner choice and intrasexual competition. In J. H. Barkow, L. Cosmides, & J. Tooby (Eds.), The adapted mind: Evolutionary psychology and the generation of culture (pp. 249–266). New York: Oxford University Press.

Buss, D. M. (1995). Evolutionary psychology: A new paradigm for psychological sciences. Psychological Inquiry, 6, 1–30.

Buss, D. M ., & Barnes, M. (1986). Preferences in human mate selection. Journal of Personality and Social Psychology, 50, 559–570.

Buss, D. M., Larsen, R., Westen, D., & Semmelroth, J. (1992). Sex differences in jealousy: Evolution, physiology, and psychology. Psychological Science, 3, 251–255.

Buss, D. M., & Schmitt, D. P. (1993). Sexual strategies theory: A contextual evolutionary analysis of human mating. Psychological Review, 100, 204–232.

Cantor, N. (1990). Social psychology and sociobiology: What can we leave to evolution? Motivation and Emotion, 14, 245–254.

Chenevix-Trench, C. (1970). A history of horsemanship. Garden City, NY: Doubleday.

Cosmides, L., & Tooby, J. (1987). From evolution to behavior: Evolutionary psychology as the missing link. In J. Dupré (Ed.), The latest on the best: Essays on evolution and optimality (pp. 277–306). Cambridge, MA: MIT Press.

Costanzo, P. R. (1991). Morals, mothers, and memories: The social context of developing social cognition. In R. Cohen & R. Siegel (Eds.) Context and development, (pp. 91–132). Hillsdale, NJ: Lawrence Erlbaum Associates.

Cunningham, M. R., Barbee, A. P., & Pike, C. L. (1990). What do women want? Facialmetric assessment of multiple motives in the perception of male facial physical attractiveness. Journal of Personality and Social Psychology, 59, 61–72.

Dahlstrom, W. G. (1991). Psychology as a historical science: Meehl's effort to disentangle Science B from Science A. In D. Cicchetti & W. M. Grove (Eds.), Thinking clearly about psychology. Volume 1: Matters of Public Interes. (pp. 40–60) Minneapolis, MN: University of Minnesota Press.

Darwin, C. (1859). On the origins of species by means of natural selection, or, Preservation of favoured races in the struggle for life. London: Murray.

Darwin, C. (1871). The descent of man, and selection in relation to sex. London: Murray.

Dunning, D., & McElwee, R.O. (1995). Idiosyncratic trait definitions: Implications for self-description and social judgment. Journal of Personality and Social Psychology, 68, 936–946.

Ellis, B. J. (1992). The evolution of sexual attraction: Evaluative mechanisms in women. In J. H. Barkow, L. Cosmides, J. Tooby (Eds., The adapted mind: Evolutionary psychology and the generation of culture (pp. 267–288). New York: Oxford University Press.

Feingold, A. (1992). Gender differences in mate selection preferences: A test of the parental investment model. Psychological Bulletin, 112, 125–139.

Godfrey, B. W., & Lowe, C. A. (1975). Devaluation of innocent victims: An attribution analysis within a just world paradigm. Journal of Personality and Social Psychology, 31, 944–951.

Goldberg, L. R. (1981). Language and individual differences: The search for universals in personality lexicons. In L. Wheeler (Ed.), Review of Personality & Social Psychology (Vol. 2, pp. 141–165), Beverly Hills, CA: Sage.

Goldberg, L. R. (1992). The development of markers for the Big-Five factor structure. Psychological Assessment, 4, 26–42.

Gould, S. J. (1986). Evolution and the triumph of homology, or why history matters. American Scientist, 74, 60–69.

Gould, S. J. (1988). Kropotkin was no crackpot: Understanding the spell of his homeland, a peace-loving Russian anarchist argued cogently against a narrow Darwinian view of evolution. *Natural History, 97,* 12–21.

Graziano, W. G., Brothen, T., & Berscheid, E. (1978). Height and attraction: Do men and women see eye-to-eye? *Journal of Personality, 46,* 128–145.

Graziano, W. G., Brothen, T., & Berscheid, E. (1980) Attention, attraction, and individual differences in reaction to criticism. *Journal of Personality & Social Psychology, 38,* 193–202.

Graziano, W. G., & Eisenberg, N. (1997). Agreeableness: A dimension of personality. In R. Hogan, J. Johnson, & S. Briggs (Eds.), *Handbook of Personality Psychology.* San Diego: Academic Press.

Graziano, W. G., Jensen-Campbell, L. A., & Hair, E. C. (1996) Perceiving interpersonal conflict and reacting to it: The case for agreeableness. *Journal of personality & social psychology, 70,* 820–835.

Graziano, W. G., Jensen-Campbell, L. A., Shebilske, L., & Lundgren, S. R. (1993). Social influence, sex differences, and judgments of beauty: Putting the "Interpersonal" back in interpersonal attraction. *Journal of Personality and Social Psychology, 65,* 522–531.

Hendrick, C. (1995). Evolutionary psychology and models of explanation. *Psychological Inquiry, 6,* 47–49.

Hinde, R. A. (1978). Dominance and role—two concepts with dual meanings. *Journal of Social and Biological Structures, 1,* 27–38.

Jackson, L. A. (1992). *Physical appearance and gender: Sociobiological and sociocultural perspectives.* Albany, NY: State University of New York Press.

Jensen-Campbell, L. A., Graziano, W. G., & West, S. (1995). Dominance, prosocial orientation, and female preferences: Do nice guys really finish last? *Journal of Personality and Social Psychology, 68,* 427–440.

John, O. P. (1990). The "Big Five" factor taxonomy: Dimensions of personality in the natural language and in questionnaires. In L. A. Pervin (Ed.), *Handbook of personality: theory and research* (pp. 67–100). New York: Guilford.

Kelley, H. H. (1980). The causes of behavior: Their perception & regulation. In L. Festinger (Ed.), *Retrospections on social psychology* (pp. 78–108). New York: Oxford.

Kenrick, D. T., Groth, G. E., Trost, M. R., & Sadalla, E. K. (1993). Integrating evolutionary and social exchange perspectives on relationships: Effects of gender, self-appraisal, and involvement on level of mate selection. *Journal of Personality and Social Psychology, 64,* 951–969.

Kenrick, D. T., Sadalla, E. K., Groth, G., & Trost, M. R. (1990). Evolution, traits, and the stages of human courtship: Quantifying the parental investment model. *Journal of Personality, 58,* 97–116.

Kropotkin, P. (1914). *Mutual aid: A factor of evolution.* London: William Heinemann.

McCrae, R. R., & John, O. P. (1992). An introduction to the five-factor model and its applications. *Journal of Personality , 60,* 175–219.

Murphy, S. T., Monahan, J. L., & Zajonc, R. B. (1995) Additivity of nonconscious affect: Combined effects of priming and exposure. *Journal of Personality & Social Psychology, 69,* 589–602.

Nisbett, R. E. (1990). Evolutionary psychology, biology, and cultural evolution. *Motivation and Emotion, 14,* 255–263.

Perper, T. (1989) Series of observations on sexual selection and female choice in human beings. *Medical Anthropology, 12*, 59–64.

Perper, T., & Weis, D. L. (1987). Proceptive and rejective strategies of U.S. and Canadian college women. *Journal of Sex Research, 23*, 455–480.

Rusbult, C. E., Verette, J., Whitney, G. A., Slovik, L. F., Lipkus, I. (1991). Accommodation processes in close relationships: Theory and preliminary empirical evidence. *Journal of Personality and Social Psychology, 61*, 641–647.

Sadalla, E.K., Kenrick, D.T., & Vershure, B. (1987). Dominance and heterosexual attraction. *Journal of Personality and Social Psychology, 52*, 730–738.

Simpson, J. A., & Gangestad, S. W. (1992) Sociosexuality and romantic partner choice. *Journal of Personality, 60*, 31–51.

Simpson, J. A., Gangestad, S. W., & Lerma, M. (1990). Perceptions of physical attractiveness: Mechanisms involved in the maintenance of romantic relationships. *Journal of Personality & Social Psychology, 59*, 1192–1201.

Singh, D. (1993). Adaptive significance of waist-to-hip ratio and female attractiveness. *Journal of Personality and Social Psychology, 65*, 293–307.

Slovic, P. (1995). The construction of preference. *American Psychologist, 50*, 364–371.

Small, M. F. (1992). Female choice in mating: The evolutionary significance of female choice depends on why the female chooses her reproductive partner. *American Scientist, 80*, 142–151.

Tooby, J., & Cosmides, L. (1990). On the universality of human nature and the uniqueness of the individual: The role of genetics and adaptation. *Journal of Personality, 58*, 17–68.

Tooby, J., & Cosmides, L. (1992). Psychological foundations of culture. In J. Barkow, L. Cosmides, & J. Tooby (Eds.), *The adapted mind* (pp. 19–136). New York: Oxford University Press.

Trivers, R. L. (1972). Parental investment and sexual selection. In B. Campbell (Ed.), *Sexual selection and the descent of man 1871–1971* (pp. 136–179). Chicago: Aldine.

West, S. G., Aiken, L. S., & Todd, M. (1993). Probing the effects of individual components in multiple component prevention programs. *American Journal of Community Psychology, 21*, 571–605.

Wiggins, J. S. (1991). Agency and communion as conceptual coordinates for the understanding and measurement of interpersonal behavior. In D. Cicchetti & W. Grove (Eds.), *Thinking clearly about psychology: Essays in honor of Paul E. Meehl* (pp 89–113). New York: Cambridge University Press.

Zajonc, R. B., & Markus, H. (1984). Affect and cognition: The hard interface. In C. E. Izard, J. Kagan, & R. B. Zajonc (Eds.), *Emotions, cognition, and behavior* (pp. 73–102). Cambridge: Cambridge University Press.

IV

Pair Bonding and
Mating Strategies

7

Human Sexual Selection and Developmental Stability

Steven W. Gangestad
Randy Thornhill
University of New Mexico

Mating in sexually reproducing species is rarely an equal-opportunity event. Some individuals, relative to others, have access to more mates or mates of higher quality (here, a term strictly reserved for advantages having to do solely with reproduction). One reason for variation in mating success is that members of the other sex prefer mates who possess traits not possessed by all. A second reason is that individuals vary in their abilities to intimidate competitors of their own sex. Selection that results from differences in traits affecting access to number and quality of mates is sexual selection.

Darwin introduced the concept of sexual selection to account for certain design features of organisms that he thought could not be due to natural selection. Why do peacocks have brightly colored, elaborately designed, and extraordinarily developed tails? Why do male widowbirds carry tails well over the length of their bodies? Why are various species of fish "painted" with iridescent splashes of color? Why are many male insects adorned with elegant ornaments? Finally, why do the males of many species engage in vigorous and flagrant displays that draw attention to themselves and their ornamentation? Darwin could not imagine that these features evolved because they enhance their owners' survival. He thus concluded that they evolved because they benefit their owners' mating success. In particular, females in these species prefer males who display these characters. Although it may seem that the processes responsible for peacocks' tails and the iridescent coloration of tropical fish are hardly relevant to the evolved design features of *Homo sapiens*, Darwin thought otherwise; indeed, he elaborated the concept of sexual selection in a volume largely devoted to human evolution (Darwin, 1871).

In this chapter we discuss a line of research designed to explore the nature of historic sexual selection pressures on humans, the design features of humans that those selection pressures have produced, and the implications of these design features for an understanding of human sexual relationships. In particular, this line of research concerns the role of developmental stability—the ability of an individual to precisely express developmental design in the face of environmental and genetic perturbations—in sexual selection. We first discuss major sexual selection theories. Then we turn to consider the role of purported honest advertisements of phenotypic and genotypic mate quality relevant to sexual selection. Different sexual selection models make different predictions about the associations of markers of developmental quality with mating success and sexually selected traits. We discuss these predictions and briefly review research testing them in nonhuman species. Finally, we discuss research that examines these predictions in humans.

SEXUAL SELECTION THEORY
AND RELEVANT RESEARCH

Soon after its publication, Darwin's theory of sexual selection encountered resistance and, ever since, few topics within evolutionary biology have generated so much controversy. Darwin's claim that certain males are advantaged because they can effectively intimidate other males (sexual selection due to intrasexual competition) has never been seriously challenged, but not so the emphasis he gave to female choice. Darwin's codiscoverer of natural selection, Alfred Russel Wallace, was a strident opponent, and few architects of the modern synthesis of Darwinism and Mendelism endorsed Darwin's version of sexual selection. A notable exception was Sir Ronald Fisher (1930), who vindicated Darwin with a preliminary analysis suggesting that sexual selection could indeed work; O'Donald (1962) later developed this analysis into a formal model. (See Cronin, 1991, for an excellent review of this controversy.)

Overall, little progress on sexual selection was made during the century following the publication of Darwin's book. With a paper appearing in a volume dedicated to the centennial of Darwin's work, Trivers (1972) invigorated interest in the topic. Characters, purportedly due to sexual selection, generally are sexually dimorphic: These usually exist (or are exaggerated) in males rather than females. Darwin never provided a convincing rationale for why, in many species, female choice apparently exerts sexual selection pressures on males and not vice versa. Trivers explicated such a rationale: The sex that devotes greater amounts of its total reproductive effort to parental investment is a limited reproductive resource for members of the other sex. The noninvesting sex (relatively speaking) competes for access to members of the investing sex, who are selected for using good sense in choosing mates. Moreover, the investing sex's good sense produces sexual selection pressures on the noninvesting sex to satisfy it. In most species, the sex that devotes the most effort to parental investment is the female; the sex whose features undergo sexual selection, then, is the male. Exceptions exist, however, and the fact that these exceptions show sex-reversed patterns of sexual selection (with females having well-developed sexually selected characters) provides convincing evidence for Trivers' notions. Owing to the considerable parental investment incurred by females during internal gestation and lactation, in nearly all mammals male features are under stronger sexual selection pressures than those of females.

What, however, is the basis for females' good sense? The last two decades of the 20th century witnessed a proliferation of theory and research seeking to address this question.

Models of Sexual Selection

Three major models of sexual selection currently exist (Andersson, 1994; Cronin, 1991; Kirkpatrick & Ryan, 1991), one assuming that sexually selected features are nonfunctional and two assuming that these features are functional.

The Arbitrary Model.   Darwin himself claimed that female choice exerts selection pressures independent of natural selection pressures. That is, he thought that the criteria of female choice are not related to males' own viability or health. In this sense, he thought these criteria to be arbitrary or nonfunctional. According to Darwin's notions and their explication by Fisher (1930), certain females might initially favor certain males over others, perhaps owing to some slight attentional advantage (e.g., because of their somewhat brighter color or larger tails). On the average, males favored by these females have greater mating success than, and thus outreproduce, their competitors. As a result, the favored males' daughters—and, hence, females with the preference for these males—increase in frequency in the next generation, leading to an even greater mating advantage of those males who are somewhat brighter or long-tailed than others. Over a number of generations, the runaway coevolution of female preferences and favored male traits can lead to extreme levels of the male trait (e.g., a peacock's tail) and female aesthetic preferences for these extreme levels (e.g., Fisher, 1930). The preferred male trait provides no natural selection advantage to his female mate's offspring (and, indeed, can even confer natural selection disadvantages to the offspring). Offspring are nonetheless advantaged by females' use of the preference because they are genetically disposed to be advantaged by mating: Sons have more mates, and daughters have sons who have more mates.

The Good-Genes Model.   Contrary to Darwin, Wallace (1891) believed that female choice criteria *are* functional—that females prefer males advantaged by natural selection. Thus, Wallace argued that favored male traits such as bright color or vigorous display behavior evidence health and well-being (Cronin, 1991). One reason why female preferences for more viable males might arise is that if viability is partly heritable, males who have greater viability provide genetic benefits to the offspring of females who mate with them. The good-genes model of sexual selection proposes that female preferences for male features advertising the presence of genes that enhance offspring viability evolve and that these preferences lead to sexual selection for those male features.

Good-genes sexual selection was vigorously debated over the last two decades of the 20th century. One problem that this model encounters has to do with the assumption just mentioned: Good-genes sexual selection can operate only if viability is heritable. According to classical population genetics, however, selection removes heritable variation in viability such that its heritability should be near lyzero. If viability then is not at all heritable, good-genes sexual selection clearly cannot operate; the viability of a potential father would in no way predict the genetic benefits bestowed upon his offspring (Kirkpatrick, 1982, 1985; Lande, 1981).

In recent years, assumptions underlying the expectation that viability has near-zero heritability have been questioned. One assumption is that new genetic variants entering the population have a negligible effect. A process through which new variants are introduced, of course, is mutation (approximately 1 in 5 humans are thought to have a mutation no more than a single generation old; Dobzhansky, 1962). Until recently, it was thought that new, slightly deleterious alleles introduced through mutation would be removed by selection quickly and have a negligible effect on the heritability of viability. In fact, however, mutation-selection balance can leave viability considerably heritable (Rice, 1988).

A second assumption is that the selective environment is constant. Recently, population biologists have noted that at least certain aspects of the environment likely are not constant. Perhaps most notably, pathogens and the immune defenses of their hosts coevolve in response to one another, such that the selection pressures they impose on each other continually change across generations (Anderson & May, 1982). Host–parasite coevolution has been one of the most important concepts to be introduced into evolutionary genetics in the last several decades of the 20th century, having been applied to our understanding of the immense genetic diversity in the natural world, sexual recombination, and the evolution of all other aspects of sexual reproduction—major issues in evolutionary biology (Hamilton, 1980, 1982; Tooby, 1982).

These developments in population genetics have led researchers to suggest two different sources of good-genes sexual selection: (a) Because mutations are a source of genetic differences in viability, individuals should evolve to choose mates who possess indicators of a relative lack of mildly deleterious mutant alleles (Pomiankowski, Iwasa, & Nee, 1991), and (b) because host–parasite coevolution maintains heritable variation in pathogen resistance, host individuals should evolve to choose mates who possess indicators of pathogen resistance (Hamilton & Zuk, 1982).

A second problem encountered by good-genes sexual selection concerns the ways by which "good genes" may be advertised. We discuss this problem later in the section "Honest Advertisements of Mate 'Quality.'"

The Good-Provider Model.   Mates can benefit offspring not only by providing adaptive genes, but also by providing material benefits to the offspring or mate. These benefits may be resources such as food (e.g., male scorpionflies give their mates a food piece that the female consumes during mating; Thornhill, 1976; many male migratory songbirds defend foraging territories; Orians, 1969) or protection (e.g., some female primates may gain protection from others through mating; Smuts, 1985). The good-provider model claims that traits favored by sexual selection advertise the bearer's ability to provide material benefits to a mate or offspring (Hoelzer, 1989).

Avoidance of contagious disease as a direct benefit of mate choice relates to the good-provider model. Phenotypic health is a desirable characteristic that advantages offspring, relative to a diseased mate, independent of material benefits given to the offspring. Hence, selection can favor preferences for healthy mates. In this case, heritable disease resistance does not drive sexual selection. Instead, mate choice increases choosers' survival or that of offspring or both by reducing the probability of their contracting infectious disease (Borgia, 1986).

Which Model Is Correct?   Which sexual selection model is correct? This question is probably not the right one to ask. All models, to the best of our current knowledge, can account for sexual selection, and thus might describe processes that operate in nature. Of course, one or another model might most commonly account for sexually selected traits. (In the presence of high costs of mate choice, for instance, nonfunctional sexual selection may be unlikely to account for preferences; Pomiankowski et al., 1991.) Moreover, with regard to a specific species, one or another model might best describe sexual selection, although multiple sexual selection processes clearly can operate on the same species (Andersson, 1994; Moller & Hoglund, 1991). At the level of particular sexually selected traits in specific species, we should probably expect that one or another process (largely) accounts for sexual selection. Appropriate questions to ask, then, concern whether some specific sexual selection process has operated on a specific species and, if so, what traits and preferences it has produced.

Honest Advertisements of Mate "Quality"

Until recently, the arbitrary model has been perhaps most popular, partly by default. The good-provider model can only account for sexual selection in species in which males provide nongenetic benefits to females, and, as we have seen, the good-genes model has encountered problems. We have already addressed one issue—processes that can maintain the heritability of viability, of which there now appear to be at least two. However, a second issue—the honest advertisement of "quality"—must also be addressed, which brings us to another major aspect of current sexual selection theory.

For females to make functional mate choices, preferred males must possess traits that distinguish them from other males. But if preferred males have greater mating success by virtue of those traits, what prevents other males from cheating by also displaying favored traits? For functional sexual selection to take place, certain traits must honestly advertise good genes or good provisioning—traits that cannot be feigned. Recent theory identifies two sorts of these traits.

Side Effects of Nonadaptive Traits.   Certain honestly advertised traits are merely by-products of the conditions that threaten survival. For instance, by-products of disease (e.g., open sores due to parasitic infection) may indicate lack of mate quality. Similarly, the effects of mutations, such as unusualness, may indicate lack of mate quality.[1]

Handicaps.   Wallace (1891) suggested that many sexually selected traits result from the sorts of side effects of nonviability just discussed. Thus, he suggested that, in the absence of disease, the plumage of many male birds is naturally colorful and full, merely due to the physiological side effects of plumage development. Only diseased male birds have drab and sparse plumage. According to this view, then, traits such as the color of the peacock's tail are not evolved through direct selection at all; rather, these traits are natural, unselected for by-products of the absence of disease that simply do not develop when disease is present, such that females can detect disease by their absence. Darwin criticized this position for good reason: A striking feature of many traits, purportedly evolved through sexual selection, is their energetic costliness. A modern evolutionary view gravely doubts that organisms would evolve to incur these costs were they not offset by considerable benefits (Andersson, 1994; Cronin, 1991).

Recent theory argues that costly traits may evolve through sexual selection precisely because they honestly advertise quality (Zahavi, 1975). The idea is that individuals who have adaptive phenotypes can better afford to develop costly traits than individuals possessing less adaptive phenotypes. Thus, these costly traits function as honest advertisements of quality. In such cases, female preferences for these costly traits get selected through functional sexual selec-

---

[1]Selection should favor organisms who do not show the effects of parasitic infection, all else being equal. Either not all else is equal (e.g., disguise of these effects is too costly to offset its benefits) or adaptive variations have not yet arisen, however, and thus these effects do reliably occur.

tion (e.g., these females acquire mates with better genes), which feed back into selection for honest advertisement in the traits themselves. According to this idea, then, many sexually selected male traits should be veridical indicators of phenotypic and genotypic quality, and many sexually selected female preferences should be for such traits.

We can think about this selection process in terms of a cost–benefit analysis. A costly trait has benefits to its owner, including, for males, sexual selection benefits. The costs of the trait are owing to the fact that the resources going into the development of the trait could have been put into other productive activity (and are scaled in terms of lost reproductive opportunities by investing in costly traits). At some point along a continuum of investment in the trait, the marginal gains of investment will diminish; individuals should be selected because they do not invest in the trait beyond that point. The sexual selection benefits of the trait can be maintained if males of different quality maximize their benefits-to-costs ratio at different levels of the trait. It simply shouldn't pay males of lower quality to develop the trait as extensively as males of higher quality.

Why, however, should males of different quality maximize their benefits/costs at different levels of the costly trait? Two models can be proposed, both not incidentally drawing upon models relevant to why heritable variation in viability may be maintained. First, mutations affect the metabolic efficiency, and thereby increase the costs of a trait (Pomiankowski et al., 1991). Second, pathogens may similarly affect metabolic efficiency. (For quantitative models, see Grafen, 1990; Heywood, 1989; Iwasa, Pomiankowski, & Nee, 1991; Pomiankowski, 1987; Pomiankowski et al., 1991.)

In many avian species, sexually selected handicaps may include large, elaborate head and tail ornaments and, in certain species of fish, the handicaps likewise often involve gaudy displays. In mammalian species, however, costly, sexually selected traits that may function as honest advertisements in males often do not involve elaborate ornamentation; nonetheless, they exist. Large body size, enhanced musculature and robustness, and expenditures of energy demonstrating vigor all are common sexually dimorphic traits in polygynously mating mammals, including primates (Andersson, 1994).

Intrasexual Competition.   Costly handicapping traits may most effectively advertise quality when they affect the outcomes of male–male competitions (Borgia, 1979). Lek species provide prototypic examples. Males gather in a single mating ground and fight for small territories. Females preferentially mate with those males who are able to defend particular territories. Outside of lek species as well, the sexual selection processes imposed by female choice typically are not independent from those that involve intrasexual competition; males best able to intimidate other males are, in many species, preferred by females as well because outcomes of competitions depend on costly physical traits. Processes of intrasexual competition are not entirely redundant with those of female choice (to the extent that markers of male quality do not all involve male–male competition), however, and thus intrasexual competition is an important form of sexual selection above and beyond mate choice.

Of course, in species in which males invest in offspring, traits that advantage males in intrasexual competition may also advantage them in their efforts to garner resources. Thus, it is often not easy to pull apart empirically a costly trait good-genes model, a good-provider model, and an intrasexual competition model of sexual selection. (See Cronin, 1991, for an extended discussion of these difficulties, as well as some attempts to do so.)

Honest Advertisements in Females?   When males invest, their investments become a limiting resource for females. Thus, sexual selection can, in principle, also operate on female traits. Males receive benefits in such species by investing particularly in offspring produced with females with "good genes." Why, then, do females usually show less development of the sorts of costly traits sexually selected in males? One problem has to do with the costs and potential benefits of those handicaps. Productive activities that compete with investment in costly traits with sexual selection benefits include somatic maintenance and parental investment. Males in many species, however, can successfully reproduce even with very little parental investment. Becuase of internal gestation and lactation, however, mammalian females simply cannot reproduce without considerable investment in offspring. Thus, the opportunity costs of heavy investment in costly traits that serve mating are much greater for females than for males. Add to these differential costs the fact that variation in female mating success (owing largely to differences in mate quality and investment) is less than variation in male mating success (in many species due largely to differences in mate number), and we expect female investment in costly honest advertisements of mate quality rarely to reach male levels of investment in these advertisements. Nonetheless, females are expected to evolve honest signals of phenotypic quality when males significantly invest in females and their offspring. Also, males may evolve to prefer females who lack natural by-products of factors that lower viability, such as mutations and pathogens.

Sexual Selection and Developmental Stability

Although costly traits may be the major features that are sexually selected, their existence does not convincingly demonstrate that they honestly discriminate among males of different viability. In certain species, they may be explained by other forms of sexual selection (e.g., the arbitrary model, intrasexual competition). To test models based on differential viability, then, researchers have searched for markers of viability in a variety of species that may be less susceptible to alternate explanation. Perhaps the most fruitful of these attempts concerns fluctuating asymmetry.

Fluctuating asymmetry (FA) is deviation from symmetry in bilateral morpho-
logical traits for which the signed differences have a mean near zero and are
near-normally distributed within a population (i.e., are, on the average, symmetrical; Van Valen, 1962). In light of the fact that the morphologies of many
organisms are built around a bilaterally symmetrical plan, FA is, to a large
extent, simply asymmetry itself. (Exceptions are asymmetries that result from
design and hence are species typical; for instance, human hearts are displaced
away from the center of the body.) Because the same genes control the
development of the two sides of the body on such traits, their asymmetry is
thought to result from *developmental instability*, the imprecise expression of
developmental design owing to disruptions during development. Thus, FA
increases with exposure to disruptive environmental factors such as parasites
(Bailit, Workman, Niswander, & MacLean, 1970; Møller, 1992) and toxins
(Parsons, 1990), as well as genetic disruptions such as mutations and inbreeding
(Parsons, 1990). Within-population differences in FA reflect variation in the
extent to which individuals have had robust, well-canalized development, either
because of variation in resistance to specific environmental disruptions (e.g.,
pathogen resistance; Møller & Pomiankowski, 1993b), or variation in exposure
to disruptions (e.g., mildly deleterious mutations). In a range of species, individuals' FA negatively predicts their fecundity, growth rate, and survival (Mitton &
Grant, 1984; Palmer & Strobeck, 1986; Watson & Thornhill, 1994).

If, as it appears, FA is a (negative) viability marker that results from
developmental instability (particularly that owing to mutations and pathogens),
two predictions follow from functional sexual selection models: Males who
demonstrate developmental stability through lack of FA should experience
greater mating success than their less symmetrical counterparts, and these same
males should be particularly likely to possess highly developed expressions of
costly, sexually selected handicapping traits. These predictions have now been
tested in a wide range of species, and they have often been supported (for
reviews, see Møller & Pomiankowski, 1993; Watson & Thornhill, 1994).

Thus, for example, male barn swallows have a pair of long tail feathers that
are used in sexual displays (Møller, 1988, 1990). Experimental manipulations
of the symmetry of these feathers reveal that males with symmetrical tails enjoy
greater reproductive success than males with asymmetrical tails (Møller,
1992a), partly because favored males obtain mates earlier in the season (Møller,
1990) and partly because they obtain greater numbers of extrapair matings
(Møller, 1988). Experimental work similarly shows that symmetrical male
scorpionflies have greater reproductive success than less symmetrical males
(Thornhill, 1992a, 1992b). In fact, female scorpionflies prefer symmetrical
males on the basis of pheromones alone under conditions in which males are
not visible (Thornhill, 1992a). Developmentally stable males in a variety of
species also have more developed sexually selected characters (e.g., sexually
dimorphic teeth in primates; Manning & Chamberlain, 1993).

These studies support functional sexual selection models in many species ranging from insects to mammals. In no way can they be explained by the nonfunctional model; that model simply does ont expect sexually selected characters to be associated with viability. Interestingly, FA often possesses nonzero heritability (in a metaanalysis across 26 studies of 13 species, including *Homo sapiens*, the overall effect size was .16 and highly signficant; for humans, it was about .3; Møller & Thornhill, in press). Thus, these results are also consistent with good-genes sexual selection in many species. (Indeed, so long as FA is a marker of viability and has nonzero heritability, good-genes sexual selection will occur.) Nevertheless, because symmetrical males may better provide resources in many of these species, good-provider sexual selection cannot be ruled out. (One test that examined this issue in birds supported good-genes sexual selection; more symmetrical male barn swallows provided fewer provisions than less symmetrical males; de Lope & Møller, 1993). Moreover, FA does not relate to mating success or sexually selected characters in some species (Watson & Thornhill, 1994); sexual selection in different species may be best explained by different models.

HUMAN SEXUAL SELECTION
AND FLUCTUATING ASYMMETRY

Over the past several years, we have attempted to assess the relevance of developmental stability to sexual selection in humans by examining the association between male fluctuating asymmetry and measures of male mating success, and associations between fluctuating asymmetry and potentially sexually selected characters. We also have explored how these sexual selection processes might affect human sexual relationships more generally.

The measure of fluctuating asymmetry that we typically have used was developed by Livshits and Kobyliansky (1991) and involves measurement with steel calipers of 7 bilateral features: foot width, ankle width, hand width, wrist width, elbow width, ear width, and ear length. In one recent study, we also assessed index finger length and pinky finger length. For each characteristic, we took the absolute difference between the left and right measurements and, to standardize relative to total characteristic size, divided by the average of the left and right measurements (Palmer & Strobeck, 1986). To create a composite index of FA for each individual, we totalled the seven individual FA measures. The average absolute differences of these measures were small (on the average, less than 1.5 mm) and thus reliability is a potential concern. Two assessments of reliability across two independent measurers (with sample sizes of 49 and 80) revealed reliability satisfactory for correlational research: .63 and .62 (Simpson, Gangestad, & Christensen, unpublished data; Thornhill, Gangestad, & Comer, in press). It should be emphasized, however, that the FA we assess is subtle and probably not observable without precise measurement. Thus, we have assessed FA merely as a marker of developmental stability; the skeletal FA we measured probably does not function as an observable cue that people assess in others. Relations between FA and phenotypes relevant to sexual selection (e.g., sexual behavior) are informative nonetheless because, just as with non-human species, they advance our theoretical understanding of sexual selection processes in humans.

FA and Sexual Behavior

In contemporary human societies, the number of offspring is not an appropriate measure of mating success. Effective contraceptive techniques unlink sexual intercourse and risk of pregnancy. Although surely not an infallible measure of potential male mating success, the number of sexual partners is one measure that should nevertheless partly reflect potential male mating success (see also Pérusse, 1993). To assess the prediction that men's potential mating success relates to their developmental stability, in a sample of 59 college men we correlated the self-reported number of sexual partners (during the lifetime to date) with our FA measure, partialling out age and age squared. As expected, FA significantly and negatively predicted number of partners: $r = -.32$ (Thornhill & Gangestad, 1994). Partialling out additional potential confounding variables (height, minor physical anomalies, ethnicity, marital status) increased the correlation to -.39. In an independent sample of 203 college men, we estimated through LISREL modeling a correlation of -.29 (Gangestad & Thornhill, 1995a; see below under "FA and Costly Traits"). (In our replication study, we specifically asked respondents not to include sex exchanged for payment or forced sex.)

We did not predict that we would find a comparable correlation for women; sexual selection theory provides no explicit rationale to explain why mammalian females in general should convert their own mate quality into partner number. Surprisingly, we did find in our initial sample of 62 women a correlation of -.36, $p < .01$; symmetrical women had more partners (Thornhill & Gangestad, 1994). In a larger sample of 200 college women, however, this effect did not replicate; LISREL-estimated r was -.04 (Gangestad & Thornhill, 1995a).

In summary, college men's FA clearly relates to their number of sexual partners. Women's FA appears to relate to their number of partners more weakly. This pattern of results is consistent with functional theories of sexual selection. It cannot be explained by a nonfunctional model of sexual selection.

FA and Extrapair Partners

Might men who exhibit little fluctuating asymmetry have more sexual partners simply because they are less able to maintain their romantic relationships? That is, do these men have more partners because they ultimately are judged as less desirable mates and hence have a higher turnover of partners? Good-genes sexual selection theory does, in fact, expect men of differing FA to behave differently in relationships, as we consider later. Nevertheless, sexual selection theory does not suggest that low FA men will have more partners simply because they can not keep their mates.

One way to assess men's sexual attractiveness independent of how they are judged in relationships is to examine their extrapair sexual behavior—sex with someone other than a current mate. We asked both members of 104 romantically involved dating couples whether they had had extrapair sex, either during their current relationship or during any relationship. In a probit analysis controlling for relationship duration, partner's extrapair sex, and both partners' physical attractiveness, we found that men's FA significantly and negatively predicted their probability of having had extrapair sex during the current relationship, $\beta = -.39$ (Gangestad & Thornhill, 1995b). When we controlled for additional variables (both partners' self-reported socio economic status (SES), both partners' self- and partner-predicted future salary, and both partners' self-reported romantic attachment styles; Simpson, 1990), this effect increased, $\beta = -.67$. Men's FA similarly predicted their probability of extrapair sex in any relationship, not just the current one. Moreover, low FA men were not merely in relationships in which more extrapair sex took place; men's FA did not significantly predict their partners' extrapair sex (and the effect observed in the sample was actually in the opposite direction; $\beta = .15$). In general, women's FA did not predict either their own or their partners' extrapair sex.

Another measure of men's sexual attractiveness independent of how they are judged in a current relationship is the number of sexual partners who have chosen them as extrapair partners, whether they themselves were in a romantic relationship at the time or not. As predicted, college men's FA appears to correlate with this variable as well. In a sample of 99 men, men's FA negatively related to the self-reported number of partners who were involved with another man at the time of sex, LISREL-estimated $r = -.35$ (Thornhill & Gangestad, unpublished data).

FA and Facial Attractiveness

Why do men who evidence developmental stability have relatively many sex partners? What features of these men might women find more sexually attractive? One possibility is that these men are more physically attractive. To assess this possibility, we examined the association between FA and facial attractiveness, as judged from photographs by multiple raters. In two studies, men's FA negatively predicted their attractiveness (controlling for potential confounds such as age), $rs = -.33$ (Gangestad, Thornhill, & Yeo, 1994) and $-.28$ (Thornhill & Gangestad, 1994). Women's FA did not, $rs = -.17$ and $.00$. In a third study, however, neither sex's FA predicted their facial attractiveness, LISREL-estimated $rs = -.02$ and $-.01$ for men and women, respectively (Gangestad & Thornhill, 1995a). Sampling variability may account for these variations, such that men's FA weakly but negatively relates to their attractiveness.[2]

The face itself is constructed along a symmetrical bilateral plan, and facial asymmetry may affect facial attractiveness (Thornhill & Gangestad, 1993). Grammer and Thornhill (1994) found that facial symmetry predicts both men's and women's attractiveness. Thornhill and Gangestad (1995) found similar but weaker relations (moreover, using a different method of defining symmetry, Swaddle and Cuthill [1995] found an opposite relation). Interestingly, facial symmetry appears to relate to certain sexually dimorphic characters of the face. Women have shorter lower faces relative to total face length than do men, and their symmetry predicts short lower face length. By contrast, men's symmetry predicts longer lower face length (Thornhill & Gangestad, 1995). Thus, more symmetrical faces appear to be more sexually dimorphic; perhaps developmental instability or its underlying causes disrupts full expression of sexual dimorphic traits. Part of this effect may be owing to the fact that the sex hormones partially mediating the expression of these traits are costly (in particular, androgens). Indeed, these hormones not only are energetically costly (Bardin & Catterall, 1981); they also suppress the immune system and thus may advertise effective pathogen resistance (Folstad & Karter, 1992; Thornhill & Gangestad, 1993; Wedekind, 1992).

An issue still being explored concerns the relationship between body asymmetry and facial asymmetry. In one study we found substantial correlations between our body and facial asymmetry indices, $rs = .52$ and $.66$ for men and women, respectively (Thornhill & Gangestad, 1995). Finally, physical attractiveness not restricted to facial attractiveness (see Beck, Ward-Hull, & Lear, 1976; Lavrakas, 1975) may relate to FA. We are currently exploring this issue.

FA and Handicapping Traits

Although physical attractiveness may negatively relate to men's FA and thus partially mediate the association between their FA and sexual history, it cannot fully account for that association. We have thus explored the role of other potential mediators (Gangestad & Thornhill, 1995a). Specifically, based on notions about the nature of sexually selected costly handicaps in mammals and their involvement in intrasexual competition, we examined relations of FA with (a) body mass: Humans show moderate sexual dimorphism in body size, consistent with their purported ancestral polygyny (Alexander, Hoogland,

---

[2]We have found few relationships between FA and women's attractiveness, despite the importance of attractiveness to men as a mate preference. Women's attractiveness may primarily signal qualities of a mate other than developmental health (e.g., fertility, current health status; see Buss, 1994).

Howard, Noonan, & Sherman, 1979), and this sexual dimorphism is partly due to sexual selection on men's size; moreover, two studies had already shown an association between FA and men's weight (Manning, 1995; Thornhill et al., 1996); (b) "Physicality": We developed a measure of men's "physicality," which included ratings of muscularity, robustness, and vigor; (c) Social dominance: Because sexually selected characters may play some role in male–male competitive interactions (Andersson, 1994), we included a measure of men's social dominance based on the California Adult Q-Sort (Block, 1961).

The sample for this study was 203 dating or married couples (see Gangestad & Thornhill, 1995a, for details). By using couples, we could obtain an observer's report as well as a self-report of each person's physicality and social dominance. In addition to costly handicapping traits, we examined the mediating role of several other variables: facial attractiveness, expected resource potential (indexed by self- and partner estimates of future salary), self-reported heterosexual assertiveness (persistence in "hitting on" members of the opposite sex), and narcissism (the Narcissism Personality Inventory; Raskin & Terry, 1988).

LISREL estimates based on a full model applied to men's data are presented in Fig. 7.1 (see Gangestad & Thornhill, 1995c, for further details). As can be seen, men's FA (or the developmental instability it reflects) was estimated to predict their number of sexual partners, $r = -.29$. All three purported costly traits potentially involved in intrasexual competition significantly related to men's FA, $\beta = -.31$, $-.39$, and $-.39$ for body mass, physicality, and social dominance, respectively, and thus potentially mediate this effect. Although only one variable (body mass) significantly related to the number of sexual partners (partially owing to sample size restrictions and multicollinearlity), a composite of the mediators weighted by their associations with FA significantly predicted men's number of sexual partners, $r = .35$. Moreover, indirect effects mediated by the three costly traits alone could account for over 70% of the total effect of FA on the number of sexual partners. Very little of the effect of FA on partner number in our model was direct, $\beta = -.05$.

We also applied this same model to women's data. The results greatly contrasted with those for men (see Fig. 7.1). Women's FA did not significantly relate to their number of sexual partners (as discussed previously). Neither, however, did it relate to any of the potential mediators, including the three costly traits purportedly involved in intrasexual competition: body mass, physicality, and social dominance (average $\beta = -.06$). Women's social dominance did positively predict their number of sexual partners (see also Gangestad & Simpson, 1990), but this effect did not mediate any relationship between FA and number of sex partners.

In summary, men's FA appears to predict costly traits involved in the sort of intrasexual competition that could have been important ancestrally. In addition to affecting the number of sexual partners through intrasexual competition itself, these traits may collectively function as cues that women use to evaluate

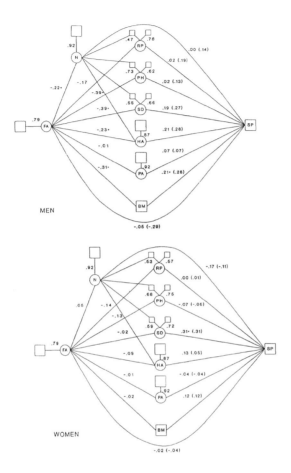

FIG. 7.1. LISREL analyses examining potential mediators of the relationship between FA and number of sexual partners. Top model: Men. Bottom model: Women. $^*p<$ .05; $^+p<$ .10; FA = fluctuating asymmetry; RP = resource potential; PH = physicality; SD = social dominance; HA = heterosexual assertiveness; PA = facial physical attractiveness; BM = body mass; N = narcissism; SP = number of sex partners. For RP, PH, and SD, both self-reports and partner reports were markers; for each, the box on the left represents the self-report and the box on the right represents the partner report. Numbers outside the parentheses are standardized beta weights; numbers in parentheses are correlations. Significance tests applied only to beta weights in the causal model. χ^2 of fit = 41.25 (df = 41), p = .46 for men, 51.39 (df = 44), p = .21 for women. From S. W. Gangestad and R. Thornhill, 1995a, *An Evolutionary Psychological Analysis of Human Sexual Selection: Developmental Stability, Male Sexual Behavior, and Mediating Features,* Manuscript submitted for publication.

partners and thereby mediate the association between men's FA and sexual history. The negative relationships between men's FA and each of the two physical measures are consistent with the pattern predicted of sexually selected honest signals of quality that has now been found in the secondary sexual traits of a vast diversity of taxa of animals (Møller, 1990). Interestingly, it appears that men's FA correlates with their metabolic level at rest, such that low FA men have lower metabolism and perhaps more efficient metabolic processes (Manning, Koukourakis, & Brodie, 1995). The fact that women's FA does not relate to these same traits is anticipated by precisely the same sexual selection theories that predict associations for men: In the absence of substantial sexual selection pressures, we know of no reason to expect women's developmental stability to relate to costly traits involved in physical forms of intrasexual competition. It is worth noting that these relationsips predicted by sexual selection notions are not entirely psychological in nature; they include relationships between physical measures (FA with weight, physicality). It is thus difficult to account for the entire pattern in terms of culture-specific norms about the standards of evaluating men and women as potential sexual partners.

Men's FA and Female Orgasm

Thus far, we have not addressed the question of which functional model—good-genes or good-provider—might best account for the role of FA in human sexual selection. The measure of FA we use is moderately heritable (narrow h^2 [additive genetic variance] $= .3$; Livshits & Kobyliansky, 1989), consistent with good-genes sexual selection. Moreover, in our study of mediators, men's resource potential related to neither men's FA nor their number of sexual partners (Gangestad & Thornhill, 1995a). Nonetheless, the traits that do relate to FA—men's physicality, social dominance, and body mass—may have related to resourcefulness in ancestral environments and, thus, good-provider sexual selection cannot be ruled out by these findings.

Although no data that we have yet gathered can conclusively rule out good-provider sexual selection, a couple of additional findings may be relevant. One finding concerns men's FA and their partners' orgasms. The function of female orgasm has long been a topic of interest to evolutionary scientists. Theories range from the idea that orgasm functions to cement the pair bond (Morris, 1967) to the notion that it is merely an unselected-for by-product of the orgasmic physiology selected for in men (Symons, 1979). Recent evidence supports a theory posed by Baker and Bellis (1993) that female orgasm functions to retain sperm, particularly when it occurs simultaneously with male orgasm (or somewhat after). In light of the fact that conception clearly does not require female orgasm, the likely context in which sperm retention was selected is multiple mating, including mating with extrapair partners. A sperm retention mechanism could allow women to selectively retain the sperm of certain men and not others. Generally, women would have no need to retain the sperm of their in-pair partner; frequent sex could ensure conception in the absence of competition from an extrapair partner's sperm. Extrapair partners may provide sperm and no investment, however, and hence might be particularly chosen because they advertise good genes. Selective retention of extrapair partners' sperm through orgasm could provide an advantage to sperm likely to carry good-genes (see also Smith, 1984). Consistent with these notions, women are particularly likely to have sex with extrapair partners during the most fertile time of their ovulatory cycle; they do not show this pattern with their in-pair partners (Bellis & Baker, 1990). Moreover, women's pattern of orgasm suggests that retention of sperm from extrapair matings is greater than that from in-pair sex (Baker & Bellis, 1993).

As we have already noted, men who evidence developmental stability are particularly likely to be chosen as extrapair partners. Because they possess characteristics that signal good genes, however, do their partners also have more sperm-retaining orgasms, even during in-pair sex? To test this hypothesis, Thornhill et al. (1996) regressed women's relative frequency of orgasm during sexual intercourse (based on reports made by both the women themselves and their partners) on their mate's FA, while controlling for a variety of other partner features, such as SES, expected future salary, age, and physical attractiveness. Men's FA negatively predicted frequency of orgasms, $\beta = -.31$. Men's facial attractiveness and body mass also significantly predicted women's orgasms, but FA had a significant effect independent of these features. (Of course, certain features associated with FA probably mediate this effect, but we do not yet know what they are.) Interestingly, women's orgasms roughly categorized by Baker and Bellis (1993) as high sperm-retention orgasms (based on time proximity to male ejaculation) were especially associated with men's low FA. Men's resource potential, as indexed by their expected future salary, did not predict the frequency of their partners' orgasms, $\beta = .08$.

Although Baker and Bellis' ideas about the function of women's orgasms remain speculative, these findings, together with associations between FA and extrapair sex, are clearly consistent with good-genes sexual selection.

FA and Relationship Investment

How might men's developmental stability affect their relationships? One notion derives from good-genes sexual selection, which brings us to a second finding potentially relevant to the issue of which functional model accounts for the role of FA in sexual selection. In many ancestral conditions, men who could successfully mate without substantial investment could benefit by putting effort into doing so (Buss & Schmitt, 1993; Symons, 1979; Trivers, 1972). In absence of investment benefits (e.g., for extrapair sex), women should be most disposed to mate with those men who evidence genetic benefits for their offspring. Given that many investment activities detract from mating effort (i.e., there exists a trade-off between parental investment or substantial investment in a single relationship and mating effort), men who evidence good genes should put less effort into maintaining a single relationship. Thus, we expect women to be faced with a trade-off in their relationships between the indicators of their mate's good genes and their mate's investment in the relationship (Gangestad, 1993; Gangestad & Simpson, 1990; Simpson & Gangestad, 1991). If the association between men's developmental stability and sexual success is owing to good-genes sexual selection, then, men who evidence developmental stability should put fewer efforts into the maintenance of a single relationship.

In fact, however, not all activities that benefit a mating partner entail equal trade-offs with mating effort outside the relationship. Some activities may particularly take away from such mating effort. Others may not take away from it at all and, in fact, may effectively function as mating effort as well. Ellis (1995) systematically developed a measure of relationship-specific investment, defined as activities that serve to maintain a relationship. Through nominations made by men and women, he identified 10 different content domains of acts that do so: *expressive and nurturing, committed, giving of time, sexually proceptive, monetarily investing, honest, physically protective, attentive in social contexts, good relationship with partner's family,* and *not sexualizing of others.* Several of these activities appear to particularly entail trade-offs with mating effort outside the relationship. Clearly, such mating effort entails sexualizing other potential partners, time not spent with the partner, and dishonesty about what is done with time not spent with the partner. Thus, *giving of time, honest,* and *not sexualizing of Others* reflect activities that clearly entail trade-offs with extrapair mating effort. One of these content domains of activities may be achieved without compromising this mating effort: *physically protective.* Indeed, if the primary kind of physical protection a man provides a mate is protection from other men, and if men protect their mates against other men primarily by being intimidating to other men, then men can benefit their mates in this way by doing precisely what they can do to demonstrate good genes: win intrasexual competitions (or show that they could win if such competitions were to take place).

If men who evidence good genes are least likely to benefit their mates in ways that take away from extrapair pursuits and most likely to benefit them in ways that complement these pursuits, then we expect their mates to most compromise their sexualizing of other women, the time they spend with them, and their honesty, while being compensated with increased physical protection. We might expect moderate compromises with regard to *expressiveness and nurturance, commitment, monetary investment,* and *attentiveness in social contexts.* Of course, if the association between men's developmental stability and sexual success is due to good-genes sexual selection, then these predictions should hold for the partners of men who evidence developmental stability.

Thus, the good-genes model of the role of developmental stability in sexual selection offers two predictions: First, all else being equal, FA should positively predict overall levels of investment in single romantic relationships; second, all else being equal, FA should predict a more precise pattern of investment in romantic relationships. To test these hypotheses, we had 203 romantically involved couples fill out Ellis's (1995) 63-item Relationship-Specific Investment Inventory, both with regard to what they do for their partner and with regard to what their partner does for them. This inventory has 10 scales, each corresponding to one of the identified act content domains. For each scale, we regressed the report of what the male did on the report of what the female did, and thereby derived residual scores of male investment adjusted for female investment. After standardizing all scale scores, we then calculated two aggregates: First, one that merely summed across the 8 scales listed to arrive at an overall investment score; second, one that weighted *giving of time, honest,* and *not sexualizing of others* -2 (to reflect the fact that mates of "good-genes men" should be particularly underbenefitted in these regards, relative to mates of other men), *physically protective* +2 (to reflect the fact that mates of "good-genes men" should be overbenefited in this regard, relative to mates of other men), and the remaining scales -1 (to reflect the fact that mates of "good-genes men" should be somewhat underbenefited in these regards). Thus, the first aggregate could be used to test the hypothesis that men's developmental stability relates to the overall level of investment their mates receive. The second aggregate could be used to test the more specific hypothesis that men's developmental stability relates to the pattern of investment their mates receive.[3]

LISREL analyses testing these two hypotheses are presented in Fig. 7.2 (see Gangestad & Thornhill, 1995c). The model we fit to the data also examined the impact of men's resource potential, as assessed by self- and partner reports of expected future salary, as well as women's physical attractiveness based on ratings from facial photographs (included because evidence suggests that these characteristics are particularly desired by women and men, respectively; Buss, 1989; Kenrick, Sadalla, Groth, & Trost, 1990). In support of the first hypothesis, men's FA did affect the overall level of investment in their relationships, $\beta = .26$. Even more powerfully supported, however, was the more specific hypothesis about the effect of men's FA on the pattern of investment their mates receive, $\beta = -.41$.

---

[3]We did not include two of Ellis's (1995) content domains. *Good relations with partner's family* concerns good relationships with a partner's relatives. Because many students are not from the immediate vicinity, however, many do not know those relatives well. *Sexual proceptivity* concerns efforts to satisfy the partner sexually. It is not clear to us that there ought to be any trade-off between efforts to make sex pleasureable to one's partner and mating effort outside the relationship. Thus, we could make no prediction regarding this scale.

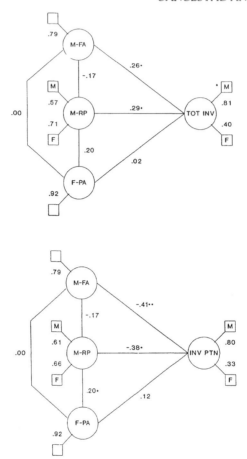

FIG. 7.2. LISREL analyses examining the impact of men's FA (M-FA), men's resource potential (M-RP), and women's attractiveness (F-PA) on either total investment (TOT INV) or a predicted investment pattern (INV PTN; see text).
**$p < .01$
*$p < .05$
+$p < .10$
From S. W. Gangestad and R. Thornhill, 1995, *Sexual Selection and Relationship Dynamics: Trade-offs Between Partner Investment and Fluctuating Asymmetry*, Unpublished manuscript.

To examine the effect of men's FA on individual types of investment, we performed LISREL analyses using each of the individual investment measures as the outcome variable. Results of major interest are reported in Table 7.1. As can be seen, FA significantly or marginally significantly predicted 5 of the 8 scales, including all tapping aspects of relationship investment that particularly

TABLE 7.1

Effects of Male Fluctuating Asymmetry, Male Resource Potential,
and Female Attractiveness on Male Relationship-Specific Investment

	Beta Weight		
RSI Scale	Male FA	Male Resource Potential	Female Attractiveness
Expressive/nurturing	.15	.31[a]	-.05
Commitment	.12	.08	.08
Giving of time	.24[a]	.26[b]	-.11
Monetarily investing	.04	-.06	.08
Honest	.30[a]	.33[a]	.08
Physically protective	-.17[b]	-.03	.25[a]
Attentive in social contexts	.20[b]	.31[b]	-.09
Not sexualing of others	.32[a]	.24[a]	.00

Note: RSI = Relationship-Specific Investment Inventory (Ellis, 1995).

[a] $p < .05$

[b] $p < .10$

Beta weights estimated from LISREL analyses (see text).

From S. W. Gangestad & R. Thornhill (1995). *Sexual Selection and Relationship Dynamics: Trade-offs Between Partner Investment and Fluctuating Asymmetry,* Manuscript submitted for publication.

compromise extrapair mating effort. Men's FA predicted all investment domains in the same direction, except for one: physically protective.[4]

Interestingly, men's resource potential positively predicted a number of investment domains, none of which was *monetarily investing.* In particular, men who were expected to earn more money, relative to others, were more expressive and nurturing, honest, attentive, and giving of time, and tended to not sexualize others. Thus, these men actually appear to sacrifice mating effort outside of their relationships to invest in their current relationships. Buss (1989, 1994) and Buss and Schmitt (1993) argued that men's resource potential importantly enhances their mate value and should thus give them a wide array of mating options. Based on these notions, then, we might expect men of relatively great resource potential to exert mating effort outside of their relationships. This does

---

[4]It might be thought that, rather than the *physically protective* measure reflecting men's protection of their mates when they are with them, this measure might reflect men's willingness to spend time with their partners in potentially dangerous situations. In the latter case, one would have little reason to suspect that low-FA men would score higher. In fact, the measure appears to tap both. Two of the three items on the measure have to do with concern for a partner's safety and willingness to spend time protecting her—"I make sure my partner doesn't have to go out alone at night" and "I escort my partner in potentially dangerous situations (such as walking her home at night"). The third item has to do with the partner feeling safe when she is with her partner ("When my partner is with me, she feels physically safe"). LISREL analyses substituting the individual items for the full scale revealed that the scale effect is carried by the last item. For that item, the a was -.48, p< .05; for the former items, betas were .03 and -.16, respectively, ns. These results suggest that the full scale effect has more to do with men's physical presence than with their willingness to invest time in protecting a mate.

not seem so of college men with resource potential. One possible reason is that these men seek to enhance partner quality rather than partner number. Men's resource potential did correlate somewhat (although not significantly) with their partners' attractiveness: $r = .20$. The analysis of the estimated effect of men's resource potential on their investment in these analyses, however, controlled for their partners' physical attractiveness. Another possible explanation is that men's perceived resource potential is affected by how hard they work at their studies. Hard work at studies may be a long-term investment that men who currently do not have a great many mating opportunities are most likely to make—men who thus invest heavily in their current relationships as well. More generally, the long-term investment that hard studies represent may be one most likely to be made by men who are predisposed toward parenting effort (for any reason, including experiential ones; Belsky, Steinberg, & Draper, 1991). Thus, investment in relationships and investment in a money-making career may each reflect a male strategy associated with high parental investment.

Although our findings are clearly consistent with a good-genes model, they do not rule out good-provider sexual selection. The physical protection that low-FA men provide their mates may have been a highly important benefit in ancestral conditions, such that, even though these women do receive less investment in other ways, they would have experienced an overall net benefit in ancestral conditions. Also, the same characteristics that lead men to be able to provide physical protection could have led them to outcompete with other men for resources in ancestral conditions (e.g., through hunting).

SUMMARY AND CONCLUSIONS

What does research on FA and the underlying developmental instability it taps reveal about sexual selection processes in humans and design features that those processes have produced? First, it provides further physical evidence for sexually dimorphic historical sexual selection pressures (Buss, 1994; Symons, 1979; Trivers, 1972). Relationships between men's FA and phenotypic characteristics that are energetically costly do exist, as anticipated by sexual seleciton notions. Relationships between women's FA and these same characteristics appear minimal, as also anticipated by these same sexual selection notions.

Second, this work provides clear evidence that important historical sexual selection processes include functional sexual selection: selection for good genes or good provisioning. The reliable associations between men's FA, their sexual histories, and their costly phenotypic traits simply cannot be explained by nonfunctional sexual selection. Although intrasexual competition may play some role in generating these associations, it seems likely that the outcomes of intrasexual competitions also advertise ancestrally relevant fitness components, ancestrally relevant provisioning abilities, or both, and partly account for women's preferences for men's athletic phenotypes (Beck et al., 1976; Lavrakas, 1975) and social dominance (Sadalla, Kenrick, & Vershure, 1987).

Third, although no findings to date pinpoint the nature of benefits that ancestral women received through their mate preferences resulting in sexual selection on men's phenotypic features associated with FA, results do constrain reasonable interpretations. Clearly, those women could have received good genes; FA is moderatedly heritable, and we have no reason to believe that is was not heritable during ancestral times. If those women also received material benefits, those benefits are most likely to have been physical protection from other males, resources that could be provisioned through athleticism (e.g., perhaps hunting), or both. It also seems likely, however, that women traded off benefits associated with direct time investment, and thus it remains unclear whether men's sexually selected traits evolved because they resulted in any increased net material benefits to women. Most likely, women mated with sexually successful men and gained good genes and certain material benefits (e.g., protection) while compromising benefits of time-consuming paternal investment, such that they suffered a net loss in benefits resulting from men's investment (Gangestad, 1993; see also de Lope & Møller, 1993).

We find no evidence that qualities associated with men's resource potential account for the sexual success of low FA men. Indeed, we find that men's expected resource potential behaves in certain ways inconsistent with that expected of a sexually selected trait (e.g., it is associated with greater attentiveness to and investment in a single romantic relationship). This finding may have to do with the fact that these qualities (e.g., diligent studying) have little to do with men's competitiveness in ancestral environments. It may also have to do with the restricted range of expected income in this population.

A great many questions remain unresolved and, indeed, are raised by this research. For instance, if good-genes sexual selection is partly responsible for our results, what specific population genetic model accounts for the heritability of viability? We find that low-FA men are perceived to be more muscular, robust, and vigorous than other men, but do actual performance measures covary with men's FA? Is FA associated with metabolic efficiency (see Manning et al., 1995, for one study that suggests that it is)? Despite not being associated with costly traits, might women's developmental stability be associated with certain phenotypic markers that serve as honest signals and are preferred by men? Might women's FA relate to their reproductive capabilities? (Evidence suggests that women's breast size asymmetry correlates negatively with fertility; Møller, Soler, & Thornhill, 1995) If women trade off certain material benefits (e.g., physical protection) for others (e.g., time investment), then what factors influence their relative preferences for these benefits? For instance, which women will take physical protection over time investment, and vice versa? Despite the progress toward an understanding of human sexual selection processes and their implications achieved thus far, work in this area has just begun.

REFERENCES

Alexander, R. D., Hoogland, J. L., Howard, R. D., Noonan, K. M., & Sherman, P. W. (1979). Sexual dimorphisms and breeding systems in pinnipeds, ungulates, primates, and humans. In N. A. Chagnon & W. Irons (eds.), *Evolutionary biology and human social behavior: An anthroplogical perspective*. Belmont, CA: Duxbury.

Anderson, R. M., & May, R. M. (1982). Coevolution of hosts and parasites. *Parasitology, 85*, 411–426.

Andersson, M. (1994). *Sexual selection*. Princeton NJ: Princeton University Press.

Bailit, H. L., Workman, P. L., Niswander, J. D., & Maclean, J. C. (1970). Dental asymmetry as an indicator of genetic and environmental conditions in human populations. *Human Biology, 42*, 626–638.

Baker, R. R., & Bellis, M. A. (1993). Human sperm competition: Ejaculate manipulation by females and a function for the female orgasm. *Animal Behaviour, 46*, 887–909.

Bardin, C. W., & Catterall, J. F. (1981). Testosterone: A major determinant of extragenital sexual dimorphism. *Science, 211*, 1285–1294.

Beck, S. P., Ward-Hull, C. I., & Lear, P. M. (1976). Variables related to women's somatic preferences of the male and female body. *Journal of Personality and Social Psychology, 23*, 1200–1210.

Bellis, M. A., & Baker, R. R. (1990). Do females promote sperm competition? Data for humans. *Animal Behaviour, 40*, 997–999.

Belsky, J., Steinberg, L., & Draper, P. (1991). Childhood experience, interpersonal development, and reproductive strategy: An evolutionary theory of socialization. *Child Development, 62*, 647–670.

Block, J. (1961). *The Q-sort method in personality assessment and psychiatric research*. Springfield, IL: Charles C. Thomas.

Borgia, G. (1979). Sexual selection and the evolution of mating systems. In M. S. Blum & N. A. Blum (Eds.), *Sexual selection and reproductive competition in insects*. New York: Academic Press.

Borgia, G. (1986). Satin bowerbird parasites: A test of the bright male hypothesis. *Behavioral Ecology and Sociobiology, 19*, 355–358.

Buss, D. M. (1989). Sex differences in human mate preferences: Evolutionary hypotheses tested in 37 cultures. *Behavioral and Brain Sciences, 12*, 1–14.

Buss, D. M. (1994). *The evolution of desire*. New York: Basic Books.

Buss, D. M., & Schmitt, D. P. (1993). Sexual Strategies Theory: A contextual evolutionary analysis of human mating. *Psychological Review, 100*, 204–232.

Cronin, H. (1991). *The ant and the peacock*. Cambridge: Cambridge University Press.

Darwin, C. (1871). *The descent of man, and selection in relation to sex*. New York: Appleton.

de Lope, F., & Møller, A. P. (1993). Female reproductive effort depends upon the degree of ornamentation of their mates. *Evolution, 47*, 1152–1160.

Dobzhansky, T. (1962). *Mankind evolving: Evolution of the human species*. New Haven, CT: Yale University Press.

Ellis, B. J. (1995). *Investment in dating relationships*. Manuscript submitted for publication.

Fisher, R. A. (1930). *The genetical theory of natural selection*. Oxford: Clarendon.

Folstad, I., & Karter, A. J. (1992). Parasites, bright males, and the immunocompetence handicap. *American Naturalist, 139*, 603–622.

Gangestad, S. W. (1993). Sexual selection and physical attractiveness: Implications for mating dynamics. *Human Nature, 4,* 205–235.

Gangestad, S. W., & Simpson, J. A. (1990). Toward an evolutionary history of female sociosexual variation. *Journal of Personality, 58,* 69–96.

Gangestad, S. W., & Thornhill, R. (1995a). *An evolutionary psychological analysis of human sexual selection: Developmental stability, male sexual behavior, and mediating features.* Manuscript submitted for publication.

Gangestad, S. W., & Thornhill, R. (1995b). *The evolutionary psychology of extrapair sex: The role of fluctuating asymmetry.* Manuscript submitted for publication.

Gangestad, S. W., & Thornhill, R. (1995c). *Sexual selection and relationship dynamics: Trade-offs between partner investment and fluctuating asymmetry.* Unpublished manuscript.

Gangestad, S. W., Thornhill, R., & Yeo, R. A. (1994). Facial attractiveness, developmental stability, and fluctuating asymmetry. *Ethology and Sociobiology, 15,* 73–85.

Grafen, A. (1990). Biological signals as handicaps. *Journal of Theoretical Biology, 144,* 517–546.

Grammer, K., & Thornhill, R. (1994). Human facial attractiveness and sexual selection: The roles of averageness and symmetry. *Journal of Comparative Psychology, 108,* 233–242.

Hamilton, W. D. (1980). Sex versus non-sex versus parasite. *Oikos, 35,* 282–290.

Hamilton, W. D. (1982). Pathogens as causes of genetic diversity in their host populations. In R. M. Anderson & R. M. May (Eds.), *Population biology of infectious diseases* (pp. 269–296). New York: Springer-Verlag.

Hamilton, W. D., & Zuk, M. (1982). Heritable true fitness and bright birds: A role for parasites. *Science, 218,* 384–387.

Heywood, J. S. (1989). Sexual selection by the handicap principle. *Evolution, 43,* 1387–1397.

Hoelzer, G. A. (1989). The good parent process of sexual selection. *Animal Behaviour, 38,* 1067–1078.

Iwasa, Y., Pomiankowski, A., & Nee, S. (1991). The evolution of costly mate preferences. II. The "handicap" principle. *Evolution, 45,* 1431–1442.

Kenrick, D. T., Sadalla, E. K., Groth, G., & Trost, M. R. (1990). Gender and trait requirements in a mate: An evolutionary bridge between personality and social psychology. *Journal of Personality, 58,* 97–116.

Kirkpatrick, M. (1982). Sexual selection and the evolution of female choice. *Evolution, 36,* 1–12.

Kirkpatrick, M. (1985). Evolution of female choice and male parental investment in polygynous species: The demise of the "sexy son." *American Naturalist, 125,* 788–810.

Kirkpatrick, M., & Ryan, M. J. (1991). The evolution of mating preferences and the paradox of the lek. *Nature, 350,* 33–38.

Lande, R. (1981). Models of speciation by sexual selection on polygenic traits. *Proceedings of the National Academy of Science USA, 78,* 3721–3725.

Lavrakas, P. J. (1975). Female preferences for male physiques. *Journal of Research in Personality, 9,* 324–334.

Livshits, G., & Kobyliansky, E. (1989). Study of genetic variance in the fluctuating asymmetry of anthropometrical traits. *Annals of Human Biology, 116,* 121–129.

Livshits, G., & Kobyliansky, E. (1991). Fluctuating asymmetry as a possible measure of developmental homeostasis in humans: A review. *Human Biology, 63,* 441–466.

Manning, J. T. (1995). Fluctuating asymmetry and body weight in men and women: Implications for sexual selection. *Ethology and Sociobiology, 16,* 145–153.

Manning, J. T., & Chamberlain, A. T. (1993). Fluctuating asymmetry, sexual selection, and canine teeth in primates. *Proceedings of the Royal Society of London B, 251,* 83–87.

Manning, J. T., Koukourakis, K., & Brodie, D. A. (1995). *Fluctuating asymmetry, resting metabolic rate, and sexual selection in human males.* Submitted for publication.

Mitton, J. B., & Grant, M. C. (1984). Associations among protein heterozygosity, growth rate, and developmental homeostatis. *Annual Review of Ecology and Systematics, 15,* 479–499.

Møller, A. P. (1988). Female choice selects for male sexual tail ornaments in the monogamous swallow. *Nature, 332,* 640–642.

Møller, A. P. (1990). Male tail length and female mate choice in the monogamous swallow *Hirundo rustica. Animal Behaviour, 39,* 458–465.

Møller, A. P. (1992a). Female swallow preference for symmetrical male sexual ornaments? *Nature, 357,* 238–240.

Møller, A. P. (1992b). Parasites differentially increase the degree of fluctuating asymmetry in secondary sexual characters. *Journal of Evolutionary Biology, 5,* 691–700.

Møller, A. P., & Hoglund, J. (1991). Patterns of fluctuating asymmetry in avian feather ornaments: Implications for modelling sexual selection. *Proceedings of the Royal Society of London B, 245,* 1–5.

Møller, A. P., & Pomiankowski, A. (1993). Fluctuating asymmetry and sexual selection. *Genetica, 89,* 267–279.

Møller, A. P., Soler, M., & Thornhill, R. (1995). Breast asymmetry, sexual selection, and human reproductive success. *Ethology and Sociobiology, 16,* 207–219.

Møller, A. P., & Thornhill, R. (in press). A meta-analysis of the heritability of developmental stability. *Journal of Evolutionary Biology.*

Morris, D. (1967). *The naked ape.* New York: McGraw-Hill.

O'Donald, P. (1962). The theory of sexual selection. *Heredity, 17,* 541–552.

Orians, G. H. (1969). On the evolution of mating systems in birds and mammals. *American Naturalist, 103,* 589–603.

Palmer, A. R., & Strobeck, C. (1986). Fluctuating asymmetry: Measurement, analysis, patterns. *Annual Review in Ecology and Systematics, 17,* 392–421.

Parsons, P. A. (1990). Fluctuating asymmetry: An epigenetic measure of stress. *Biological Review, 65,* 131–145.

Perusse, D. (1993). Cultural and reproductive success in industrial societies: Testing the relationship at proximate and ultimate levels. *Behavioral and Brain Sciences, 16,* 267–322.

Pomiankowski, A. (1987). Sexual selection: The handicap principle does work—sometimes. *Proceedings of the Royal Society of London B, 231,* 123–145.

Pomiankowski, A., Iwasa, Y., & Nee, S. (1991). The evolution of costly mate preferences, I.: Fisher and biased mutation. *Evolution, 45,* 1422–1430.

Raskin, R., & Terry, H. (1988). A principal-components analysis of the Narcissistic Personality Inventory and some further evidence of its construct validity. *Journal of Personality and Social Psychology, 54,* 890–902.

Rice, W. R. (1988). Heritable variation in fitness as a prerequisite for adaptive female choice: The effect of mutation-selection balance. *Evolution, 42*, 817–820.

Sadalla, E. K., Kenrick, D. T., & Vershure, B. (1987). Dominance and heterosexual attraction. *Journal of Personality and Social Psychology, 52*, 730–738.

Simpson, J. A. (1990). Influence of attachment styles on romantic relationships. *Journal of Personality and Social Psychology, 59*, 971–980.

Simpson, J. A., & Gangestad, S. W. (1991). Personality and sexuality: Empirical relations and an integrative theoretical model. In K. McKinney & S. Sprecher (Eds.), *Sexuality in close relationships*, (pp. 121–146). Hillsdale, NJ: Lawrence Erlbaum Associates.

Smith, R. L. (1984). Human sperm competition. In R. L. Smith (Ed.), *Sperm competition and the evolution of human mating systems* (pp. 601–660). London: Academic Press.

Smuts, B. B. (1985). *Sex and friendship in baboons.* New York: Aldine de Gruyter.

Swaddle, J. P., & Cuthill, I. C. (1995). Asymmetry and human facial attractiveness; Symmetry may not always be attractive. *Proceedings of the Royal Society of London B, 261*, 111–116.

Symons, D. (1979). *The evolution of human sexuality.* Oxford: Oxford University Press.

Thornhill, R. (1976). Sexual selection and nuptial feeding in *Bittacus apicalis* (Insecta: Mecoptera). *American Naturalist, 110*, 529–548.

Thornhill, R. (1992a). Female preference for the pheromone of males with low fluctuating asymmetry in the Japanese scorpionfly (*Panorpa japonica: Mecoptera*), *Behavioral Ecology, 3*, 277–283.

Thornhill, R. (1992b). Fluctuating asymmetry and the mating system of the Japanese scorpionfly (*Panorpa japonica*). *Animal Behaviour, 44*, 867–879.

Thornhill, R., & Gangestad, S. W. (1993). Human facial beauty: Averageness, symmetry, and parasite resistance. *Human Nature, 4*, 237–269.

Thornhill, R., & Gangestad, S. W. (1994). Fluctuating asymmetry and human sexual behavior. *Psychological Science, 5*, 297–302.

Thornhill, R., & Gangestad, S. W. (1995). *Sexual selection and features of facial attractiveness: Toward an evaluation of models.* Unpublished manuscript.

Thornhill, R., Gangestad, S. W., & Comer, R. (1996). Human female orgasm and mate fluctuating asymmetry. *Animal Behaviour, 50*, 1601–1615.

Tooby, J. (1982). Pathogens, polymorphism, and the evolution of sex. *Journal of Theoretical Biology, 97*, 557–576.

Trivers, R. (1972). Parental investment and sexual selection. In B. Campbell (Ed.), *Sexual selection and the descent of man, 1871–1971* (pp. 136–179). Chicago: Aldine de Gruyter.

Van Valen, L. (1962). A study of fluctuating asymmetry. *Evolution, 16*, 125–142.

Wallace, A. R. (1891). *Natural selection and tropical nature: Essays on descriptive and theoretical biology.* London: Macmillan.

Wedekind, C. (1992). Detailed information about parasites as revealed by sexual ornamentation. *Proceedings of the Royal Society of London B, 247*, 169–174.

Watson, P. J., & Thornhill, R. (1994). Fluctuating asymmetry and sexual selection. *Trends in Ecology and Evolution, 9*, 21–25.

Zahavi, A. (1975). Mate selection—a selection for a handicap. *Journal of Theoretical Biology, 53*, 205–214.

8

On the Dynamics of Human Bonding and Reproductive Success: Seeking Windows on the Adapted-for Human-Environmental Interface

Lynn Carol Miller
Stephanie Allison Fishkin
University of Southern California

INTRODUCTION

Are close, relatively enduring relationships fundamental to human beings? In this work, biological and psychological evidence is presented that suggests the intriguing possibility that our current biological design—rooted in our Pleistocene gatherer–hunter roots—strongly favors relatively enduring relationships and few sex differences in mating strategies. Furthermore, this biological design appears to dovetail with the attachment system (Bowlby, 1969/1982), a system with roots in our primate past. The adaptations of humans to their physical environments, it will be argued, interfaced with adaptations to their social environments (Caporael, 1994) to enhance both maternal and paternal infant–caregiver emotional bonding and adult pair-bonding. Because human infants were exceptionally dependent primates (Fisher, 1987, 1989), the involvement of paternal as well as maternal caregivers was critical for offspring survival. As they are today, high levels of paternal involvement would be expected to be associated with close, relatively enduring pair-bonds (Draper & Harpending, 1988). Design features supporting these systems (e.g., caregiving, attachment, pair-bonding) and their interfaces are consistent with Bowlby's (1969/1982) evolutionary ethological position. Consistent with Bowlby, we would argue that *all humans born today share this evolutionary heritage*, as well as the latent design features that support caregiving, attachment, and adult pair-bonding. Thus, we argue for universal design mechanisms influencing mating strategies.

However, such an argument may seem at odds with an extensive literature examining cross-cultural, within-culture, and gender differences in sexual behavior. For example, a variety of researchers have reported marked and consistent sex differences. These include sex differences in minimum standards or criteria for short-term mates (Kenrick, Groth, Trost, & Sadalla, 1993; Townsend, 1993), in seeking a short-term mate (Buss & Schmitt, 1993), in differential willingness to respond affirmatively to requests for sex from attractive opposite-sex partners (Clark & Hatfield, 1989), and in ideal number of partners desired over various future intervals (Buss & Schmitt, 1993), as well as differences between men and women in their desire to have a more unrestricted mating strategy (Simpson & Gangestad, 1992). Others can easily point to the prevalence of divorce and infidelity (Buss & Schmitt, 1993), the diversity of fertility patterns across cultures, or the variability in marital forms (e.g., monogamy, polygamy), and the acceptance of polygyny across most cultures—if not the prevalence of its practice within those cultures (Ford & Beach, 1951; Daly & Wilson, 1983).[1]

Others, noting the tremendous variability in childrearing practices, gender differentiated behavior, and paternal involvement in various cultures, have attempted to cluster this diversity in intriguing ways. For example, Draper and Harpending (1988) and others have suggested a dichotomy between human cultures differing in male reproductive strategies in which fathers are absent and those in which fathers are present. Father-present societies are those with close pair-bonds, and high levels of provisioning directed toward a mate and their male and female offspring. Those offspring, in turn, seek more enduring pair-bonds as their predominant mating strategy. Father-absent societies are those with more transient bonds, higher prevalence of polygyny, and lower

---

[1]Actually, there is often what appears to be conceptual confusion in the literature. The nature of the marital arrangement (e.g., monogamy, polygamy) and the duration of the relationship (e.g., short versus long term) are somewhat distinct issue; and both are distinct again from whether these relationships involve a close pair-bond. Although monogamy and polygamy are likely to involve relatively enduring relationships, they need not. Furthermore, simply knowing the form of marital arrangement or its durability does not tell us very much about the level of intimacy between mates (Draper & Harpending, 1988). *Monogamy* is sometimes used to refer to an institutional form of marriage (e.g., usually in anthropolgy as, "I have one wife") and at other times, *monogamy* is used to refer to whether an individual is having sex with only one other partner (called *behavioral monogamy*). Therefore, we have chosen to use the term *pair-bond* rather than any other to avoid some of this conceptual slippage and confusion. The reader should note that we do not mean to imply that *pair-bond* is synonymous with *monogamy*, and, sexual relationships, whether short- or long-term can either be sequential (behaviorally monogamous) or involve simultaneous multiple partners (behaviorally nonmonogamous). In using the term *pair-bond* we are referring then to an emotionally close, relatively enduring sexual relationship.

levels of parental investment, especially paternal investment, in male and female offspring. Those offspring engage in sex at an earlier age, forging less stable relationships. There are more marked sex differences in behavior and more negative perceptions about the opposite sex. There are also within-culture differences in the extent to which fathers are present or emotionally absent. With such behavioral variability across cultures, within cultures, and between men and women, how can humans share universal design features?

There may be a variety of ways of clustering individuals and explaining such differences using a variety of different evolutionarily based accounts. For example, there are evolutionarily based accounts of sex differences in relative focus on short- and long-term mating strategies (Buss & Schmitt, 1993). Moreover, there are evolutionary-based frameworks to explain within- and between-culture differences based on human early sensitivity to one or another type of social environment that may trigger alternative adaptive mechanisms involving human reproductive strategies (Draper & Harpending, 1988). However, current behavioral variability, including differences between men and women, may well be the result of relatively modern differences in the social environments encountered by humans that were not present in the Pleistocene era; such recent changes are unlikely to be the result of adaptations (Tooby & Cosmides, 1992). We examine this argument in greater detail.

In contrast to environments encountered by some humans today, that involve relatively low levels of paternal investment (e.g., father-absent societies), the stable pattern for humans over hundreds of thousands of years is likely to have been one that included maternal and paternal involvement with offspring and emotionally intimate relations between mates (Draper & Harpending, 1988). For ancient hominids (e.g., during the Pleistocene era) and for most contemporary hunter–gatherer groups, there was high maternal and paternal involvement with offspring. As Drapter and Harpending (1988) noted, "A prominent part of the environment for parents and dependent offspring must have been each other. . . . Because of the small size of the groups, the nomadism, the lack of alternatives to mother's milk for nourishing young, the mother and her mate would have been the primary target of succorance requests by the child. This kind of child played a major and active role in sustaining a high level of parental investment. Even though other group members could assist the parents, it would be unlikely that a system of substantial surrogate caretaking would develop." (p. 358). Because Pleistocene mothers were apt to have spent much of their time gathering (to provide the primary food source for the family), fathers—perhaps even more so than today—were apt to have been heavily involved with their offspring for a large percentage of time each day.

After the end of the Pleistocene era, and perhaps less than 10,000 years ago, with the advent of agricultural societies and the differential accumulation of individual wealth, close paternal caregiving of offspring—apt to have been so critical to the survival of highly dependent human offspring—was no longer essential. For example, as agricultural societies supported larger and larger human bands, kin and others could more easily have assumed or been assigned the role of caregiver. Procuring food, shelter, and the protection of offspring would have required less and less sustained vigilance and daily maternal and paternal involvement. During this period, increased variability in the importance of paternal caregiving for offspring survival may have emerged. Cross-cultural work as well as work within cultures suggests that such differences in paternal involvement—creating differences in the encountered social environments of human offspring—may have had a dramatic impact on subsequent male and female sexual strategies and behaviors (Draper & Harpending, 1988).

When social environments became more divergent—producing marked variability in sexual practices, mating strategies, and parental relationships—is critical. As Tooby and Cosmides elaborated (1992, p. 5) the "evolved structure of the human mind is adapted to the way of life of Pleistocene hunter-gatherers, and not necessarily to our modern (e.g., post-Pleistocene) circumstances." This possibility suggests that we must be cautious about making inferences about *evolved* human mating strategies based on modern (post-Pleistocene) behavioral patterns.

> What we think of as all of human history—from, say, the rise of the Shang, Minoan, Egyptian, Indian, and Sumerian civilizations—and everything we take for granted as normal parts of life—agriculture, pastoralism, governments, police, sanitation, medical care, education, armies, transportation, and so on—are all the novel products of the last few thousand years. In contrast to this, our ancestors spent the last two million years as Pleistocene hunter-gatherers, and, of course, several hundred million years before that as one kind of forager or another. These relative spans are important because they establish which sets of environments and conditions defined the adaptive problems the mind was shaped to cope with: Pleistocene conditions, rather than modern conditions. This conclusion stems from the fact that the evolution of complex design is a slow process when contrasted with historical time. Complex, functionally integrated designs like the vertebrate eye are built up slowly, change by change, subject to the constraint that each new design feature must solve a problem that affects reproduction better than the previous design. The few thousand years since the scattered appearance of agriculture is only a small stretch in evolutionary terms, less than 1% of the two million years our ancestors spent as Pleistocene hunter-gatherers. For this reason, it is unlikely that new complex designs—ones requiring the coordinated assembly of many novel, functionally integrated features—could evolve in so few generations (Tooby & Cosmides, 1990a, 1990b). Therefore, it is improbable that our species evolved complex adaptations even to agriculture, let alone to postindustrial society. Moreover, the available evidence strongly supports this view of a single, universal panhuman design, stemming from our long-enduring existence as hunter-gatherers. If selection had constructed complex new adaptations rapidly over historical time, then populations that have been agricultural for several thousand years would differ sharply in their evolved architecture from populations that until recently practiced hunting and gathering. They do not. (p. 5)

If most of our design changes took place in the Pleistocene era, and if our goal is to understand the evolved sexual strategies and behaviors of humans, we need to consider what human behaviors emerge given environments most similar to those of our Pleistocene ancestors. If we assume important universal designs relevant to mating, where might we find adults today who experienced a social environment most similar to that encountered by our ancient ancestors? These individuals are apt to provide a clearer window on evolutionarily based adaptations regarding human mating strategies.

One possibility is to identify individuals who report having had responsive maternal and paternal caregivers. Hazan and Shaver (1986), in fact, developed a measure to identify such individuals and found that individuals with more responsive caregivers are more likely to feel comfortable with closeness and to trust others in romantic relationships, just as Bowlby (1969/1982) had predicted. Another possibility is to identify adults who are particularly capable of forming pair-bonds. Secure individuals seem particularly likely candidates (Hazan & Shaver, 1987): That is, they feel comfortable with closeness and can trust others in close relationships. Secures, as adults, are more likely to report having had caregivers who were responsive and warm (Collins & Read, 1990; Hazan & Shaver, 1987). Both those individuals who directly report having had responsive caregivers and those who report being secure in relationships may provide particularly useful windows for examining the role of our evolutionary heritage on adult sexual and relationship practices.

Furthermore, we present evidence to suggest that a variety of design features, including biological and chemical features, argue for the adaptive advantage of both caregiving and having close, relatively enduring relationships, in which emotional bonding and sexuality are apt to be interwoven (Miller, Fishkin, Gonzales-Tumey, & Rothspan, 1996). Still, the same underlying universal system, given environmental variability, can account for differences among humans in their ability to develop and maintain these bonds, and in their resultant mating choices involving a variety of sexual behaviors, preferences, and relationships (Miller & Fishkin, 1996). Then, given space constraints, we briefly contrast Attachment–Fertility Theory (Fishkin & Miller, 1994; Miller & Fishkin, 1996) of human sexual behavior with Sexual Strategies Theory (Buss & Schmitt, 1993), a well-known psychological theory regarding the evolutionary basis of human mating strategies.

MAKING EVOLUTIONARY ARGUMENTS

In making an evolutionary argument, one is on stronger ground to the extent that one identifies underlying universal design features or systems that are apt to date back hundreds of thousands of years and were adaptive for long periods of time across cultures—at least in that distal social and physical environment

(Tooby & Cosmides, 1992). Design features or mechanisms are functionally organized evolved structures and processes that overcome environmentally posed obstacles. A number of design features may support one another as part of an evolved adaptive system. These systems may be intraindividual so that one design feature of an individual supports another in overcoming an environmental obstacle. These systems may also be interpersonal such that design features (e.g., affecting the manifestation of infant behavior) interface with design features of relevant social environments (e.g., affecting the manifestation of adult caregiving behavior) to support interpersonal systems of adaptive responses (e.g., the attachment system). *Universal design does not, however, necessarily translate into universal manifestation of behavior.* Manifestation of behavior is perhaps best viewed as a function of universal design, the interface of a given individual with his or her social and physical environments (those similar to, as well as divergent from, the adapted-for environments), and within-species biological variability.[2] Although acknowledging the potential role of within-species biological variability (see for example, Simpson & Gangestad, 1992), our focus in the current work is on how the first two factors may contribute to emergent differences in human behavior.

As Tooby and Cosmides (1992) noted, for humans, design mechanisms "have been born into one cultural environment or another hundreds of billions of times" (p. 89); only when there was communality across cultures and habitats in environmental features and obstacles encountered would we observe "long-term cumulatively directional effects of selection on human design." (p. 89) For humans, some of our design features and the systems of which they are part may have developed early in our primate history; we would expect these features to be shared with other related primates. Other design features may have resulted as a response to social and physical conditions more unique to humans, including the environments encountered in the Savanna and the resultant move to bipedalism (Barkow, Cosmides, & Tooby, 1992); the systems of which these design features were a part may have evolved to solve somewhat different problems. However, systems of design features that support one another are more likely to sustain a more enduring effect of selection.

Universal designs and behavioral systems developed in response to environmental conditions encountered by humans in our evolutionary past. We would expect these designs to interact with relevant current environmental features to produce emergent behavior across cultures. If current environmental features

---

[2] It is important to emphasize that, unlike other approaches (Gangestad & Simpson, 1990), we do not argue that individual differences in mating strategies depend on genetically based individual difference propensities to favor one set of strategies (e.g., unrestricted) over another (e.g., restricted). Rather, we argue that our universal design is to be pair-bonders: Variability is primarily an emergent product of universal design and its interaction with differing caregiving and varied social environments. Genetic variability (e.g., in impulse control mechanisms) may, however, exacerbate these differences.

are similar to those encountered by our ancient ancestors, the behaviors that unfold might provide a clearer view of those evolved designs and the systems to which they were tied. Similarly, any model of evolutionary processes must explain, not only typical responses but also individual variability in behavior. One source of individual variability in behavior is that which would result from interactions between universal design features and departures from the adapted-for environmental conditions (e.g., differences in caregiving environments).

In the discussion that follows, we begin our quest for evolutionary connections regarding human mating and intimacy by asking three questions. First, what enduring historical conditions, both shared with other primates and uniquely human, would have produced evolved universal latent mechanisms among humans that are made manifest when today's humans interface with the adapted-for environment? Second, which individuals are apt to manifest these patterns and therefore provide the best window on our evolutionary heritage? And, finally, what does an examination of current human sexual behaviors through this "window" suggest regarding our evolutionary heritage as well as current patterns of individual differences in behavior that depart from that heritage?

IN SEARCH OF EVOLUTIONARY MECHANISMS

Enduring Historical Conditions

A search for evolutionary mechanisms must take into account the historically enduring and common conditions under which humans evolved (Tooby & Cosmides, 1992). Recently, it has been suggested (Caporael, 1994) that humans—and other primates—adapted not only to physical environments (such as the Savanna) but also to social environments (e.g., the dyad, various size groups, etc.). For example, even as physical and cultural environments changed, the dyadic unit of mother and infant was constant for humans and other primates (Bowlby, 1969/1982; Goodall, 1968, as cited in McGrew & Feistner, 1992). Infant primates are born dependent on their mothers for survival; given that, the attachment system, Bowlby argued, was highly adaptive. Among other things, it ensured that very dependent infants came equipped with mechanisms for maintaining proximity to the caregiver.

While mother–infant dyadic bonds are universal across primates, father–infant bonds and another type of dyadic bond—pair-bonding—show greater between- and within-species variability. Nevertheless, pair-bonding and paternal involvement with offspring seem to go together across and within human cultures (Draper & Harpending, 1988). If humans were pair-bonding creatures, and if humans were adapted to develop and maintain strong paternal–offspring bonds, what was it about their adaptive heritage that would have produced that outcome? The interaction of early humans' physical and social environments may have had an impact on the development of unique human design features. Anthropologists (Fisher, 1987, 1989; Tanner, 1981; Tanner & Zihlman, 1976; Zihlman, 1978) have argued that the transition from forest to Savanna lifestyles resulted in bipedal behavior and tool use. These changes may have set the stage for two dramatic changes in social patterns: greater shared parental childrearing and pair-bonding.

Bipedalism, which had taken place by about 3½ million years ago, narrowed the birth canal (Abitbol, 1993; Lovejoy, Hieple, & Burstein, 1973; Reynolds, 1931, as cited in Abitbol, 1993; Tague, 1992). Moreover, an expanding human brain about 2 million years ago made it increasingly more difficult for the infant's head to pass through the narrower channel (Fisher, 1987, 1989; Tague, 1992). Given a larger human head and a pelvis that could expand only so far, there would be an increasing "space crunch"—what Fisher (1987) referred to as an "obstetric crisis"—in which numerous infants and their mothers likely perished in labor because the mothers' pelvises placed a constraint on the size of their infants' heads.

Traces of this crisis may be seen even in more modern times. Patterns of high maternal mortality were prevalent in premodern Europe (Dobbie, 1982; Eccles, 1982; Shorter, 1982). Even in the early 1900s, where the standard of medical care for the day was high, the second leading cause of death among women of childbearing age (e.g., 15–44 years) was mortality associated with childbirth (Megus, 1917, as cited in Tague, 1992). During the same time, the rate of maternal mortality among Pueblo Indians was 7 times greater than that for other women in the United States (Aberle, 1934, as cited in Tague, 1992). In the early 1960s, in the least-developed nations of the maternal world, mortality was still high. With a rate of maternal death per birth at .06 and with those women who lived through their childbearing years having an average of 6 children (UNICEF, 1994), the cumulative lifetime risk of death in pregnancy—even in modern times—is indeed striking! Historically, in addition to infection and toxemia, a leading cause of death in delivery was obstruction during labor because of small pelvic size (Aberle, 1934, as cited in Tague, 1992; Dobbie, 1982; Eccles, 1982; Shorter, 1982). In short, although humans may have had 2 million years to adapt to the constraints of an expanding infant head in a narrow channel, the ripples of an obstetrical crisis can still be felt.

This problem would have been much more acute in our ancient past. In contrast to other related primates, pregnancy for human females has historically been a much more dangerous proposition causing difficulties that are unheard of in other primate species (Abitbol, 1993). Curiously, some recent anthropological work suggests a significant covariation between female age of death among early humans and pelvis size such that females with wider pelvises lived longer (Tague, 1994).

During the early phases of this evolutionary transition for bipedal humans these difficulties are apt to have been even more pronounced. Up to several hundred thousand years ago, Homo Sapiens and Neanderthals—two descendants of Homo Erectus—co-existed (Yoon, 1995). Neanderthals—our now extinct evolutionary "cousins"—had a range of pelvic size within the range of variation of modern females; however, Neanderthal newborns were larger in size than those of modern humans (Tague, 1992). Our ancestral humans may have found, via natural selection, a more adaptive path; giving birth to more dependent, less mature offspring.

Might such a mechanism, such as one that shortened human gestation periods—and thereby produced more dependent offspring—have been plausible? Perhaps. As it is, human fetuses today run up against a considerable "space crunch" during pregnancy. In contrast to other primates, human fetuses develop abdominally—because there is insufficient space in the pelvic region for the expanding human head (Abitbol, 1993). When humans concurrently conceive more than one fetus, however, the space constraints for humans—much more so than for other primates—become severe. For today's humans, the adaptive solution to an unusual added "space crunch" (e.g., concurrent multiple fetuses) is a dramatically shorter gestation period, one that also increased the odds of infant mortality. Perhaps, a shorter gestation period was part of an earlier adaptive solution to the spatial constraints of a single fetus—one with an expanding cranium.

Those infants who did squeeze through the birth canal and survived childbirth were apt to have been much smaller and less well developed (Gould, 1977; Montague, 1961; Washburn, 1950; all as cited in Fisher, 1987). In fact, among primates today, humans are the most dependent (Fisher, 1987, 1989). Infant and child mortality rates in the least-developed countries in the world, perhaps our best estimate of conditions in our evolutionary past, reveal the fragility of early human life: Even in the early 1960s, child mortality rates (UNICEF, 1994) were high (17% not surviving the first year, 28% not surviving their first five years).

In our very distant ancestral hominid past, one could easily imagine that infant mortality rates might have been at least as high as those in related species, such as chimpanzees,who need not endure the trauma of a smaller maternal canal with an expanding infant cranium. Those infant mortality rates are exceptionally high: It has been estimated that as many as one half of infant chimpanzees do not survive their first year of life in the wild (C. Stanford, personal communication, September 28, 1995). Probably to increase their survival rates, most primate females typically have only one dependent infant at a time (Tanner & Zihlman, 1976). This is not a serious problem, given that these infants are not dependent for very many years; but, human infants are. Waiting until human infants matured sufficiently to be independent would have drastically reduced the species' ability to maintain its population. For humans to produce more offspring who were simultaneously dependent on adults, an adaptive solution would have been necessary: the typical support of other adults to ensure child survival.

If early hominids experienced high infant and maternal mortality rates, there might have been three noticeable consequences: (a) If maternal mortality rates were high enough for an evolutionary time period, they may have resulted in a sustained sex ratio imbalance with important ramifications for pair bonding; (b) maternal mortality rates would have meant that many mothers died with surviving offspring that were too young to care for themselves; paternal caregivers and familial units would have greatly increased their offspring's chances for survival; and (c) given infant vulnerability and dependency, male caregiving that assisted the female also would have increased the number of surviving offspring who would live to produce their own offspring.

Enduring Sex Ratio Imbalances: Impact on Pair-Bonding

Early maternal morbidity would have created a pronounced sex ratio imbalance with fewer females bearing live offspring (and surviving) for the available males. If many women died in childbirth, younger women and girls might have been sought as mates to grapple with a sex ratio imbalance. Unfortunately, this might only have compounded the problem. Menarche was likely to have been relatively late in prehistoric human groupings, with first pregnancies for most women by about 18 to 20 years of age (Tague, 1994). Women were fertile from menarche until they died, typically, in their late 30s or early 40s (C. Stanford, personal communication, September 28, 1995). Girls, younger than 18 would not only have been less fertile than older women; their pelvises might not have had sufficient time to expand (Tague, 1994). Thus, even those who became pregnant might have been at particularly high risk for maternal and infant death. One intriguing adaptive solution to this problem may have been a preference by males for mates with larger pelvises, a preference that seems consistent with recent work by Singh (1993).

Sex ratio imbalances in the short term are known to be associated with behavioral changes within subpopulations in relationship commitment, patterns of promiscuity, and sexual behavior (Guttentag & Secord, 1983). Examining sex ratio imbalances and their impact across cultures and historical times, Guttentag and Secord found that when there were fewer females compared to males, monogamy and an emphasis on commitment in pair-bonding tended to be stressed.[3] Such a sustained sex ratio imbalance among those able to pass along their genes could have resulted, not only in a sustained period of pair-bonding, but in a number of adaptations that might have favored men and women who were more likely to maintain those bonds, be effective caregivers, or both.

---

[3] This relationship is more likely when men do not have high or absolute structural power or control over societally based resources.

Design Features that Enhanced Pair-Bonding

Early humans encountered a number of serious obstacles to survival; perhaps among the most serious problems were infant vulnerability and maternal death in childbirth. We argue that pair-bonding enhanced the species' survival. With increased levels of infant vulnerability, females probably chose men who would stay around to help care for offspring. Those fathers who played a role in caring for and pair-bonding with their mates, and who cared for their mates' offspring, even if they died, increased their offspring's chances of survival, thereby enhancing the probability that future generations would share characteristics that fostered pair-bonding and parenting propensities.

With a shortage of females, pair-bonding probably also provided the most viable solution for men to have available sexual partners. If those men and women who formed emotional bonds had enjoyable, and therefore more frequent, sex, a desire to form and maintain romantic attachments might well have been selected for. We next discuss this connection between bonding and sexuality as well as chemical and biological design elements that enhanced the bonding, the fertility, or both of those couples who formed more enduring relationships.

Sexuality and Emotional Bonding

What factors would increase the likelihood that couples would engage in sexual behavior? It seems likely that chief among these would be sexual satisfaction (Sprecher & McKinney, 1993). Curiously, one strong correlate of sexual satisfaction is emotional bonding (Miller et al., 1996). At first glance, one might argue that the reinforcing properties of sex may enhance the likelihood that the couple will form and maintain strong emotional bonds. However, the role of sex in affecting later bonding is unclear. For example, among dating couples, having sexual intercourse early in the relationship did not necessarily enhance emotional bonding or relationship longevity (Peplau, Rubin, & Hill, 1977). In recent longitudinal work (Miller et al., 1996), cross-lag part correlations suggest that newlywed couples' early emotional bonding predicted their later sexual enjoyment and not the reverse; this was the case for both husbands and wives.

The links between happiness and sex have been more thoroughly investigated. In a recent survey of sexual practices in the United States, Laumann, Gagnon, Michael, & Michaels (1994) found that men and women who were happiest (or least unhappy) were those who reported having only 1 partner in the past year (compared to those with no partners or 2 or more partners) and sex 2 or 3 times a week (compared to no sex in the past year or sex less than once a month). Furthermore, these researchers found that those partners with only 1 current sexual partner tended to be much more likely to experience both physical pleasure via the sexual relationship and emotional satisfaction in the relationship. For those with more than one partner, both the quality of the sex and the emotional bond were depressed. Frequency of orgasm was also tied to happiness in the relationship for women, but not for men.

Overall, the picture that emerges is one that connects sexual quality and emotional bonding, or as Laumann et al. (1994) noted, "The reality seems to be that the quality of the sex is higher and the skill in achieving satisfaction and pleasure is greater when one's limited capacity to please is focused on one partner in the context of a monogamous, long-term partnership" (p. 365). This extensive survey supports the view that the majority of Americans form pair-bonds that are relatively long-term, and, as Laumann et al. pointed out, it "surely reinforces the view that, in our culture, having a spouse or a cohabitational partner and no other sex partners is far more likely to be associated with positive feelings about one's sex life and, correspondingly, the fewest negative feelings" (p. 367).

Curiously, emotional closeness or its lack (loneliness) seem tied to both psychological health as well as physical health. Recent work suggests significant links between relationship satisfaction, sexual satisfaction, and a variety of health factors (Cutler, 1991; Laumann et al., 1994; Saphier, 1989). Furthermore, as we see in the following discussion, a variety of design features appear to favor closeness in both infant caregiving and adult romantic relationships. We now discuss those design features related to pair-bonding.

Biological and Chemical Design Elements Favoring Pair-Bonding

Sexual Attraction and Attachment.   Various chemical design features may have favored not only attracting male and female humans to one another but maintaining the pair-bond—at least long enough to reproduce, and support newborns through the initial years of dependency as infants (Fisher, 1987; Liebowitz, 1983). Liebowitz (1983) argued that during an initial *attraction phase*, humans experience heightened levels of phenylethylamine (a neurotransmitter that may enhance intense emotional feelings associated with infatuation). Although humans may habituate to this chemical reaction after 2 or 3 years, that time frame is sufficient to increase the probability of pregnancy (Money, 1980). At that point, during what is referred to as an *attachment phase*, Liebowitz (1983) argued that another set of chemicals, the endorphins, may serve to maintain those bonds. These design elements may have enhanced the psychological reward system associated with pair-bonding. Other chemical elements, have been implicated in enhancing the fertility of couples in relatively enduring relationships.

Design Features That Enhanced Our Ancestors' Production of Offspring.
Externally secreted, chemical substances, pheromones, may play an important role in enhancing the fertility of couples who pair-bond, thereby increasing the

chances that those characteristics will be passed along to future generations. As they do in many animal species, pheromones evoke autonomic (Monti-Bloch, Jennings-White, Dolberg, & Berliner, 1994) as well as social (Gower & Ruparelia, 1993) responses, and appear to play an important role in human sexuality.

McClintock's (1971, 1983) classic studies of women living in close proximity to one another found that they tend to synchronize their menstrual cycles. Others have found that female axillary excretions (sweat, pheromones) may play a role in this phenomenon. Women who had these axillary extractions from a female donor applied to their upper lips synchronized their cycles to that of the female donor (Preti, Cutler, Garcia, Huggins, & Lawley, 1986; Russell, Switz, & Thompson, 1980). There is evidence, then, to suggest a link between pheromones and female hormonal patterns.

But do male partners also influence their female partner's hormonal patterns? The available evidence is intriguing. Women who have weekly heterosexual intercourse may have more fertile menstrual cycles than celibate women or women who have sporadic sexual intercourse (Cutler, Garcia, & Krieger, 1979; Cutler, Preti, Huggins, Erickson, & Garcia, 1985; Goldman & Schneider, 1987; Veith, Buck, Getzlaf, Dalfsen, & Slade, 1983). Furthermore, male pheromones are implicated as a causal agent in regulating female cycles (Cutler et al., 1986; Matteo, 1987; Veith et al., 1983). Women who had male axillary extractions from a donor applied to their upper lips had cycles that were more regular and therefore more likely to be fertile (Cutler et al., 1986).

Interestingly, hormonal influences between men and women may not be a one-way street. Men's own hormonal levels may adjust in enduring relationships with those of their female partners, such that they may be most fertile when their partners are also most fertile. Supporting this possibility, testosterone levels of husbands (Persky et al., 1978) and sexual activity in couples tends to be highest in the period leading up to the woman's ovulation (Adams, Gold, & Burt, 1978; Hedricks, Piccinino, Udry, & Chimbira, 1987). Furthermore, in a longitudinal daily "beeper study," males reported feeling closest to their partners at the midfollicular and ovulatory phases of their partner's cycle; this effect was not related to the females' reported levels of closeness (Hedricks, Ghiglieri, Church, Lefevre, & McClintock, 1994). Of course, those couples who are more emotionally close are more likely to have more frequent and enjoyable sex (Lauman et al., 1994; Miller et al., 1996); such couples also appear much less likely to have additional sexual partners (Laumann et al., 1994).

In short, these studies suggest a dynamic interplay between men and women and between human chemistry, biological design, and human sexual behavior. They raise the possibility that more sexually and emotionally responsive couples help to maintain each other's levels of sexual hormones, behavior, and fertility (Cutler, 1991), and therefore the likelihood that these characteristics would be passed to future generations.

Design Elements That Reduced Infidelity.   Although the previous design elements may have increased the probabilities of pair-bonding over extended periods compared to that of most other primates, other evolving design elements, including relative hairlessness, simultaneously may have reduced the opportunities for human infidelity. The relative hairlessness of humans means that scents are not as easily emitted (sweat will evaporate rather than cling to hair). Curiously, the few places where humans still have quantities of hair are the head, the underarm, and the pubic area.

The presence of hair in these regions further supports the idea that humans are designed to operate at close range; their secretions are far less volatile than are the fatty acids of the copulins found in other primates, whose hairs provide a greatly increased surface area that encourages volatilization and dissemination to occur (Stoddart, 1990, p. 232).

Stoddart (1990) argued that with reduced olfactory sensitization of female estrus in males, came the co-occurrence of concealed ovulation and extended female receptivity, all occurring as the cerebral cortex expanded.[4] Thus, males were less likely to detect the scent or visual displays of nonprimary females. These changes, Stoddart argued, all reduced the chances that human males would have sexual relations with a nonprimary female that resulted in the production of offspring.

Similarly, those males who were in daily close contact with a female, because they were emotionally close to her, might still detect her rather subtle ovulation cues, especially during an evolutionary transition phase. Those men who stayed close to females throughout her cycle might have been much more sensitive to within-partner hormonal fluctuations (such as increased sexual desire during peak hormonal phases). Emotionally close mates, might then have been more likely to have had sex frequently with one another and when they were most fertile. The production of offspring with a partner with whom one was emotionally close would have increased the probability that those characteristics related to pair bonding would have been passed to future generations.[5]

---

[4]It has been assumed that ancestral hominids moved from unconcealed to concealed ovulation. However, whether humans' hominid ancestors ever had visual cues to ovulation is an open question. In fact, the vast majority of primates have concealed ovulation (C. Stanford, personal communication, September 28, 1995).

[5] Today emotional closeness is still apt to be highly adaptive. In contrast, for women who are not emotionally close or sexually satisfied (e.g., in terms of frequency, quality) in their primary relationship, the probabilities of being with more than one partner may increase (Laumann et al. (1994), along with the adaptive advantages of engaging in sex with another man. Baker and Bellis (1993) compared the timing of orgasms for women who were in one relationship against the timing of orgasms for women who were in two sexual relationships (e.g., one with a primary partner, one with a secondary partner) for the women's most recent sexual episode. One difficulty in such research, however, is that it is unclear how the frequency of sex (especially before ovulation) for relationships of various types trades off against other mechanisms to result in differentials in fertility. It would also have been useful to examine how emotional closeness relates to the frequency, timing, and summed probability of conception across a cycle for couples of various types.

Other biologically based factors also may have worked against promiscuity as a *primary* sexual strategy. Chief among these were sexually transmitted diseases. The probability of contracting a sexually transmitted disease increases with one's number of sexual interactions and sexual partners (Monahan, Miller, & Rothspan, in press). Because mothers with some sexually transmitted diseases may pass along fatal diseases or life-threatening effects to their offspring, any added advantage of promiscuity in offspring production would have to have been weighed against the costs to offspring survival and subsequent parent and offspring fertility.[6] Furthermore, such infidelities may have played havoc on early social groupings and increased the chance of within- (and between-) group conflict and violence. Although we are not arguing that infidelity did not exist in the Pleistocene era, a number of factors would probably have worked against its prevalence, for humans.[7]

In summary, a variety of design elements and historical adaptations may have come together to increase the adaptive advantage of pair-bonding. First, a variety of chemical transmitters may have increased the likelihood of humans forming and maintaining pair-bonds (at least long enough to ensure the survival of young infants). Second, additional chemical agents may have increased the probability of humans being fertile and producing offspring with mates with whom they were regularly sexually active. Third, an obstetric crisis for extended periods may have enhanced the likelihood that females may have selected men who were more committed to relationships and therefore more likely to take care of their offspring. Finally, a variety of additional factors (including disease and the propensity of within- and between-group conflict) may have reduced the probability of infidelity and promiscuity among humans. In short, changes in the survivability of offspring and in the sexual relationships of adults may have been closely intertwined in our evolutionary past. We now turn to this issue for further discussion.

Pair-Bonding and Infant Survivability

Did human pair-bonding enhance infant survivability? Pair-bonding is rare among mammals. Although more common among primates, it still is found in a minority of primate species, typically emerging under particular environ-

---

[6] Although other primates carry viruses associated with sexually transmitted diseases, including the simian version of human immunodeficiency virus (simian immunodeficiency virus; SIV), these diseases do not kill within-primate species (Craig Stanford, personal communication, September 28, 1995). Had humans been as promiscuous as other primates, would we not also have developed immunities to these diseases? In addition, if women were in short supply, most women would have mates; men who were unfaithful might have simultaneously risked losing a long-term partner and physical harm from the additional partner's mate.

[7] Nevertheless, especially in more recent times (e.g., since the Pleistocene era), infidelity is a very viable strategy for some individuals in achieving their goals (e.g., sex, attention, control, etc.). But certainly, the relative benefits of a sexual strategy for individuals need to be considered against their relative costs.

mental or "niche" conditions. Kleiman, (1977), for example, reported that one type of monogamy is favored in evolution when more than the female is needed to ensure the survival of the young.

Certainly, the conditions encountered by our human ancestors qualify as one such niche. Human infants have a much longer dependency period than most other primates, most likely as a result of the physiological changes in our evolutionary history (e.g., increased bipedalism, decreased body hair, etc.). Because human infants have more difficulty clinging to their mother than other primates, women, no doubt, were likely to encounter great difficulty in trying to simultaneously gather food and protect their infants from potential predators.

Pair-bonding provided a number of advantages for the couple as well as their offspring, in addition to providing opportunities for sex and enhancing infant survivability. For women, pair- bonding provided another adult responsible for the caregiving and protection of the infant, allowing women more time to gather food. For men, pair-bonding may have greatly enhanced men's primary supply of food—vegetation provided by the women. As Tanner (1981) argued, advances in using tools to procure and gather more vegetation, necessary for survival on the Savanna, were probably introduced by women. Current design features involving human saliva production and its role, similar to that of other primates, in vegetation digestion, suggests that vegetables were the bulk of the diet for our early ancestors. Early male hominids, who were able to only occasionally to increase the diet with a successful hunt, benefited from the gathering of the women who were likely to share their food with more sociable and friendly males (e.g., those that did not bare their canine teeth; Tanner, 1981). Offspring produced by these women may have been more likely to stay longer in the familial unit, producing males that were more sociable and sought after as mates by other females (Tanner, 1981). Tanner argued that sexual selection and natural selection processes worked together to reduce the sexual dimorphism of men and women, including differentials in canine tooth size (human male canines are similar in size to those of females in contrast to those of other primates), as well as enhance the propensity of both mates to pass along to their male and female offspring the tendency to be better parental caregivers.

Beyond physical needs (e.g., for food and sex), the needs of humans for social support and emotional closeness probably also date back fairly far into our distant human past. For example, recent work suggests that social support provided by the spouse and others is positively correlated with infant birth weight (within normal ranges), and therefore survivability—of the offspring (Collins, Dunkel-Schetter, Lobel, & Scrimshaw, 1993). Having a spouse around, especially for the first four or five years after the birth of the offspring, also probably greatly enhanced offspring survival. As adults mature, emotional closeness in a relationship may enhance couple health and well-being (Cutler, 1991; Saphier, 1989), which, in turn, would enhance the couple's ability to nurture and protect, not only their own offspring, but, if they were fortunate, their grandchildren as well.

As a result of this codependency, additional infants could be raised before the first was able to survive independently (Zihlman, 1978). Therefore, we argue that what appears to have been selected for was a codependency between men and women, which enhanced not only their own survival, but the survival of future generations of offspring as well.

Parental Caregiving and Infant Survivability
Across Primates

Bowlby (1969/1982) noted that, like humans, other primates have an extended infancy period; without the support, nourishment, and protection of a caregiver, infants are unlikely to survive physically, and if they do, are severely impaired (Harlow & Zimmerman, 1959). Bowlby argued that children have a fundamental human need to feel secure in the world. In search of felt security, Bowlby suggested, children engage in a variety of behaviors to gain the attention and response of their caregivers. For example, behaviors such as crying, are used to bring the caregiver closer and to maintain contact.

In a variety of different species (macaque, baboon, ape, etc.), Bowlby (1969/1982) noted a similarity in the behaviors between the females and their new infants: close physical contact and proximity-seeking for an extended time period. For example, in rhesus monkeys, the infant stays close to its mother for its first three years before exploring its surroundings; baboons are in close maternal contact for at least one year. These similarities in primate behavior led Bowlby to consider the evolutionary significance of attachment. He argued that attachment increased the offspring's protection from predators, thereby enhancing reproductive success. Infants come equipped with a number of design features that would maximize their ability to maintain proximity to the caregiver and form an emotional bond.

Design Elements Enabling Proximity. Some of the design features enabling proximity appear to have "hard wired" concomitants in the various sensory systems. For example, human infants seem particularly responsive to schemas of human faces and learn, early on in their first year of life, to differentiate human caregivers as objects with which to maintain visual and physical proximity.

Interestingly, human infants, both boys and girls, seem responsive to human faces that are more attractive (Langlois et al., 1987). Why might this be? As Langlois found, average faces or composites of many human faces or facial prototypes are judged particularly attractive (Langlois & Roggman, 1990). One intriguing possibility is that a prototype of the human face is hard wired—perhaps to ensure that the infant has an easy time identifying other human faces—human faces upon which she is utterly dependent as a newborn. We judge faces that more closely approximate that prototype as more "attractive." Lending additional credence to this position, cross-cultural work indicates that there is high inter-rater reliability in judgments of attractiveness across a variety of cultures (e.g., Bernstein, Lin, & McClellan, 1982; Cunningham, 1986; Johnson, Dannenbring, Anderson, & Villa, 1983; Maret, 1983; Thakerar & Iwawaki, 1979), suggesting "that ethnically diverse faces possess both distinct and similar structural features; these features seem to be perceived as attractive regardless of the racial and cultural background of the perceiver." (Langlois & Roggman, 1990, p. 115). Thus, preferences for attractiveness, as well as the presence of a variety of curve and shape feature detectors may be by-products of this highly adaptive wiring to ensure that the infant and later, the toddler, identified and maintained proximity with a human caregiver.

No sensory system is more ancient and more apparently tied to species and individual identification (and reproduction) than the olfactory system. Knowledge of the olfactory system's role in mother-infant bonding and in sexual reproduction has increased dramatically in recent years. Research on young human and other primate infants indicates that infants show preference for the odors of their own lactating mother over other lactating females or non-lactating controls (see Bartoshuk & Beauchamp, 1994; and Schaal, 1988 for recent reviews of this research). Thus, odors appear to play an important role in human infant identification of their mothers, and thereby the ability of infants to maintain proximity early in infancy to that particular caregiver. This and additional research suggests that humans may be "hard-wired" to react to pheromones. Although this effect may be more acute in infants (Schaal, 1988), studies of adult responses implicate for pheromones a continuing role in social and sexual responses (Gower & Ruparelia, 1993).

Human Dependency, Caregiver Responses.   The societal environment that the primate child enters is one that recognizes the dependency of the infant and the responsibilities of adults in meeting the infant's needs, including his or her socialization into the broader simian or human culture. In a variety of different nonhuman primate species, in which close physical contact and proximity seeking for an extended time period exists (macaque, baboon, ape, etc.), the period of maternal contact and dependency is much less than that of humans (e.g., under five years of age). It is difficult to imagine human children surviving on their own much before the age of ten. Parental caregiving thus was an essential feature of primate evolutionary design.

A variety of design features may enhance human parental caregiving, both that of the fathers as well as the mothers. Take crying for example. Research on how adults respond to crying infants suggests that "cry perception may be a fundamental, species-general perceptual process that is 'fine tuned' by caregiving experience" (Green, Jones, & Gustafson, 1987, p. 380). Green et al., found that adult females and males in general, whether parents or not, perceived a cry similarly both physiologically and behaviorally. Other researchers comparing how parents and nonparents responded to normal and aversive infant cries found that physiologically the subjects, both women *and* men, showed the same patterns of responses to particular cries. But across all types of cries, parents rated cries as less aversive than did nonparents, regardless of type of cry (Murray, 1979; Zeskind & Lester, 1978).

Similarly, adult humans have a preference for babyish faces (Mitchell, 1979). Babies have a propensity to smile at an early age, much to the enjoyment of adults. Perhaps, such preferences and emotional responses are hard-wired to ensure that parents and other adults will respond more warmly to infants, especially when they prove particularly difficult (Mann, 1992).

Bowlby noted the importance of not only maternal contact, but also the importance of paternal contact in infant survivability. Most adult male, nonhuman primates remain close to mothers with infants, but will come into little or no physical contact with the infant. As the infant gets bigger, however, interactions with the adult male primates increase, and the infant often becomes attached to one particular male. It has been argued that with human infants the equation shifted such that males spent more time with their offspring (Bowlby, 1969/1982). Human infants were so dependent for so long that parental investment by both parents must have greatly enhanced survival. Parents who pair-bonded could better care for their offspring, giving them an adaptive advantage.

Building Strong Attachments. The development of the attachment system depends on the responses of specific caregivers, as well as the responses of the child. In addition, Bowlby argued, these early-infant attachment interactions are likely to carry over into adulthood, affecting adult relationships, including romantic ones (Bowlby, 1969/1982). In nonhuman primates—and, we would suspect, in humans as well—disruption of early caregiver–infant bonding appears to affect later adult sexual relationships (Harlow, Harlow, & Suomi, 1971). Although this disruption may be rare, when it occurs, nonhuman primates often may not survive into adulthood outside of captivity (e.g., primatology laboratories). Humans may exhibit similar dysfunctional patterns, their severity worsening with the departure from having had a warm and responsive caregiver. Humans and other primates who do not, or are unable to, attach to a caregiver show severely retarded psychological development (Fedigan, 1982). In addition, the lack of a secure attachment can be detrimental to one's health, have overt physiological effects, and affect subsequent childrearing practices (see Hazan & Zeifman, 1994 for a discussion).

It therefore seems reasonable to assume that among primates having a secure attachment to a caregiver could be viewed as critical to later normal functioning. With such a secure base, primate infants are able to learn and explore, a behavior which, in humans, allows for cognitive development. As we have noted, humans, even more than nonhuman primates, are especially dependent on caregivers for their physical survival and psychological development. Thus, a responsive caregiver for long periods of time would seem to have been a likely, adapted-for social environment, a social environment without which humans and other primates, were less likely to have survived.

Although many of the above design features, as Bowlby (1969/1982) argued, are apt to have been deeply rooted in our primate past (and shared therefore with nonhuman primates), some relatively unique human features of the infant–caregiver attachment system evolved. These included design features that likely depended heavily on cognitive changes in human information processing systems and abilities (Tooby & Cosmides, 1992). In the attachment domain, this includes the development of children's mental models. Bowlby argued that the child forms expectations, or mental models, concerning caregiver behaviors and what responses the child deserves from the caregiver. These expectations, in turn, influence the types of relationships a person forms, as well as the behaviors a person exhibits in adult relationships. Such mental models would have further increased—for humans compared to other primates—the links between models of early infant–caregiver interactions and later, adult romantic ones. If the caregivers are warm and responsive, meeting the needs of the infant, the child develops in what Bowlby referred to as an environment of evolutionary adaptedness. From this responsive environment emerges a child with a secure attachment style, including a positive model of self and a model of others as dependable and trustworthy.[8] Research to date suggests that the majority of parent–child interactions fall into this category (Ainsworth, Blehar, Waters, & Wall, 1978).

We would concur with Bowlby (1969/1982), who argued that the emergent attachment system that develops in infants who have consistently warm and loving interactions with caregivers reflects a type of "ideal"—an ideal interface between a set of human design adaptations and a social environment (e.g., a responsive caregiver) for which humans are adapted. It was Bowlby who suggested that those early bonding experiences were likely to increase the probability that individuals would form close subsequent attachments in romantic relationships, a prediction consistent with more recent work relating adults' perceptions of their caregivers and adults' reports of their propensity to form close, trusting relationships with others (Collins & Read, 1990; Hazan & Shaver, 1987).

Variability in Caregiving Environment: Subsequent Impact on Adult Romantic Relationships and Mating Strategies

Unfortunately, not all human caregivers today are responsive, and partly as a result, not all children develop secure attachments. Bowlby argued that the

---

[8] Increasing cognitive capacity of the brain clearly enhanced the ability to develop elaborate mental models of self and others. Similarly, the increasing need to develop such "mental models" may also have enhanced cognitive development (Read & Miller, 1995).

infant–adult attachment system ideally resulted in secure attachments, but when the caregiver is not responsive (e.g., when the child does not encounter the adapted-for environmental interface), individual differences emerge based on different patterns of interaction between caregivers and offspring, and not on genetically based differences among children (Ainsworth et al., 1978; Main, 1991).

What is the connection between these infant–caregiver interactions and subsequent romantic relationships? The links between infant–adult attachments and secure attachments in adult relationships are only now being examined via suitable longitudinal research: Published research to date has not yet focused on an adolescent or older sample (Shulman, Elicker, & Sroufe, 1994). However, work linking adult perceptions of childhood interactions with their caregivers with adult patterns of secure attachments in romantic relationships (Hazan & Shaver, 1986) suggests the plausibility of these links.

Using a variety of methods, ranging from categorical descriptions to larger dimensional measures of attachment (Bartholomew, 1990; Collins & Read, 1990; Hazan & Shaver, 1987; Levy & Davis, 1988; Mikulincer, Florian & Tolmacz, 1990; Simpson, 1990), researchers found proportions of each adult attachment style that are similar to those derived in the developmental literature regarding child–parent attachments (Shaver & Hazan, 1994); across a variety of populations these percentages for adult attachments have been similar (Shaver & Hazan, 1994). Consistent with our contention that secure attachments represent a more typical or normal human interface with our adapted-for social environments, the majority of adults (55% to 60%) report being securely attached. An additional group, 25% to 30%, are avoidant, and the rest are anxious–ambivalent (see Hazan & Shaver, 1987; Shaver & Hazan, 1994 for a description of these attachment styles). The parallels between infant–caregiver interactions and romantic interactions are considerable and lend further credence to the links among these systems (Hazan & Shaver, 1994a, 1994b).

Cross-Cultural and Within-Culture Comparisons: Paternal Involvement Level and Its Correlates

As mentioned earlier, Draper and Harpending (1988) noted that across and within cultures, warm spousal relationships and paternal caregiving styles seem to cluster together. In father-present societies, relationships between women and men are intimate and close with higher levels of bonding, whereas those in father-absent societies either involve mutual avoidance or pronounced hostility and sometimes violence (often directed at women), and are more transient, with a higher prevalence of polygyny and more pronounced sex roles reinforcing male dominance and female subordination. In father-present societies, both male and female offspring, as adolescents, are more careful and reticent in choosing partners and entering into sexual relationships, have good skills for forming and maintaining close relationships, and tend to form a pair-bond with a single mate. Where fathers are emotionally aloof, boys are more likely to engage in competitive and aggressive behaviors; daughters are less coy, begin sexual activity earlier, and form less stable pair-bonds. Unfortunately, greater fertility in the father-absent societies does not necessarily translate into more surviving offspring. In father-absent households, in which fathers frequently terminate their paternal care abruptly, maternal interest in the offspring ends at an early age and children are cared for, instead, by siblings or fostered out to relatives. It is a pattern of high fertility and often high toddler mortality.

We would argue that humans, over vast stretches of human history, were adapted to experience responsive caregivers, both fathers and mothers. When humans do not encounter this social environment as offspring, this lack of fit between what humans were adapted for and what they encounter is likely to have a number of emergent outcomes: Chief among these may be the impaired ability to trust and feel that one can get close to and depend on others. Control for such insecure individuals may instead be achieved through emotional withdrawl or attempts to dominate others. Because of their relative size and strength insecure males and females may seek to achieve such goals in different ways with varying success; but, certainly, early experiences with caregivers, in concordance with the findings of Bowlby and others (e.g., Draper & Harpending, 1988; see also Belsky, Steinberg, & Draper, 1991), are likely to affect subsequent sexual strategies and adult romantic relationships.

SEEKING AN EVOLUTIONARY WINDOW: THOSE INDIVIDUALS MOST LIKELY TO MANIFEST BEHAVIOR CONSISTENT WITH OUR LATENT EVOLUTIONARY HERITAGE

If our goal is to understand evolved human propensities, and if human design mechanisms pertaining to mating strategies are universal, then differences in current human environments are likely to account for much of the variance in behavioral differences. It seems reasonable, then, to seek those humans today who experienced a social environment most similar to that of our Pleistocene ancestors: Such humans would be most likely to manifest behavior that provides a window on our human evolutionary heritage. Two groups of humans seem to be likely candidates: those who had responsive caregivers (both maternal and paternal), and those who are comfortable with emotional closeness and may have a greater propensity to form and be able to maintain pair-bonds (e.g., secure individuals).

Secure individuals, as well as those adults who report having had a close relationship with their caregivers, their fathers as well as their mothers, would be most likely to have experienced social environments similar to those experienced by our Pleistocene ancestors, and therefore would be more likely to manifest behaviors most in line with that of our evolutionary design. These individuals should provide a particularly clear window for understanding more about our underlying evolutionary design regarding sexual relationships.

Looking Through the Window

Attachment Styles, Parental Caregiving, and Human Mating Patterns

We now examine the responses of individuals varying in attachment styles and in the perceived parental caregiving of their mothers and fathers. We have argued that our windows for examining how universal design features interacted with the adapted-for environment are those individuals who experienced warm and responsive maternal and paternal caregivers, and those individuals most likely to trust and feel comfortable with others: secure individuals.

People with secure attachment styles seek close, relatively autonomous relationships with others (Collins & Read, 1994). These individuals tend to have long, stable relationships (Kirkpatrick & Hazan, 1994), characterized by trust and intimacy (Collins & Read, 1990; Hazan & Shaver, 1987), and are more likely to seek emotional support under anxiety-provoking experiences (Simpson, Rholes, & Nelligan, 1992). Therefore, secure individuals might be particularly able to develop and maintain the types of pair-bonds that might have been typical of our Pleistocene ancestors. Patterns for secure individuals are contrasted with those of nonsecure individuals (e.g., those who have a more anxious–ambivalent and avoidant attachment style) to ascertain how variability in sexual strategies today is likely to emerge when humans interact with social environments that are apt to have differed from those more typically encountered in the Pleistocene era.

Another window, perhaps the most direct window on our evolutionary past, is to examine the behaviors and practices of individuals who had a warm and responsive relationship with their caregivers. Given our argument that paternal caregivers were probably critical during most of the Pleistocene era—and were less critical for human survival in more recent human history—the role of paternal caregiving is a particularly interesting one to examine in opening up a window on our evolutionary heritage.

Predicting Sexual Practices and Mating Strategies

Secure individuals and those who enjoyed warm relationships with their caregivers may be especially likely to manifest, in their sexual behaviors and responses to preferences for ideal sexual partners, humans' latent evolutionary heritage relevant to reproductive strategies. Although recent work looking at adult attachment suggests that attachment styles predict satisfaction in relationships (Feeney & Noller, 1990; Hazan & Shaver, 1990), the role of these patterns of adult attachment styles or parental caregiving perceptions in predicting preferences for sexual strategies and behaviors in adult romantic relationships is less clear.

Our theoretical position is that humans are social creatures adapted to pair-bond. Secure individuals, and those who experienced a warm relationship with their caregivers are expected to manifest few sex differences in their desire for sexual partners, and to exhibit more desire for and ability to maintain sexually enjoyable relationships, in part owing to their ability to bond emotionally with their partners.

Although a number of predictions follow from this theory, because of limited space we focus here on only a few. For example, based on Attachment Fertility Theory (AFT), we would predict that *most* men and women would prefer relatively few partners (e.g., one or two) over the expected time frame of life expectancy for humans in the Pleistocene (e.g., over 20 or 30 years from the vantage point of 18 year olds), this would be especially the case for our "windows": (1) secure individuals—those who have a self-reported comfortableness being close and trusting others in close relationships and (2) individuals who had a warm relationship with their maternal and paternal caregivers.

At first glance, this prediction seems to fly in the face of findings by Buss and Schmitt (1993). Those researchers reported large *mean* sex differences in the number of partners men and women would ideally seek over long stretches of time, such as 30 years, into the future. Nevertheless, our experience with such questions led us to suspect that such data would be highly skewed, resulting in means that would not provide adequate measures of central tendency. A descriptive breakdown of the data would provide a better picture of what the majority of men and women would ideally prefer. Therefore, we asked college students (sample sizes of 106 men and 160 women) to fill out measures, using items from Buss and Schmitt (1993), that asked individuals to indicate how many mates they ideally sought over various periods of time.

As expected, these variables were highly skewed resulting in severe violations of the assumption of normality. To illustrate this problem, consider Fig. 8.1 depicting one item: How many partners would you ideally desire in the next 30 years? As indicated, the mean value for men is 64.32, whereas the mean value for women is 2.79. With the removal of two outliers, our mean values become very similar to those reported by Buss and Schmitt (1993). Not surprising, when we examined mean differences using the tests they used (purely for comparison purposes), we replicated their reported sex difference in the mean number of partners sought by men and women for various time periods. Because means are highly sensitive to outliers, medians for this type of data provide a better measure of central tendency (Cliff, 1993; Lehman, 1991). Certainly a median

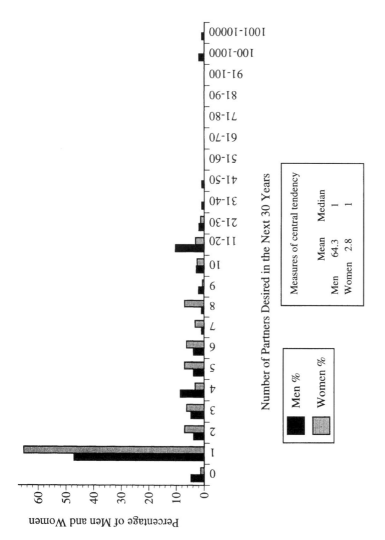

FIG. 8.1. Distributions for men and women for the ideal number of partners desired over 30 years.

Note. To include all of the data, we collapsed across categories further out on the tail of these distributions. If every category represented a single number, it would be more apparent that the tail is very flat, and that distributions are even more skewed than is apparent here.

of 1 here seems to better capture or represent the responses of more of the men and suggest that, in general, most men and women prefer relatively few partners over long stretches of time (e.g., 30 years) into the future. Nevertheless, because more women than men were at or below the median, there were overall median differences between men and women, such that more women than men were at or below a median of 1.

And, what is the pattern for secure versus nonsecure men and women? As indicated in Fig. 8.2, a breakdown of attachment styles and ideal preferences for mates reveals that among women, the median number of partners desired is relatively constant regardless of attachment style. For men there is much more variability; although secure men seek relatively few partners over a 30-year period, anxious–ambivalent and avoidant men show a different pattern. Consistent with AFT, we found that most secure men and women ideally sought relatively few mates over long periods of time (medians of 1 for both men and women up to 30 years into the future) and did not significantly differ from one another, whereas anxious ambivalent and avoidant men and women did.

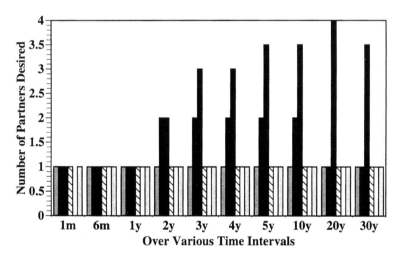

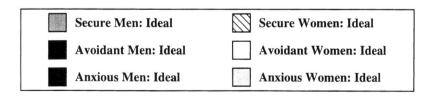

FIG. 8.2. Medians for ideal number of partners desired over 30 years by attachment style for men and women.

We were also interested in how parental caregiving styles predicted the number of partners men and women ideally desired over extended periods of time. To assess parental caregiving, participants were asked to indicate on a scale from (1) extremely uncharacteristic to (9) extremely characteristic, the extent to which they perceived, separately, that their mother and their father were warm, cold, distant, rejecting, responsive, their parent's highest priority, inconsistent, and so forth. These items were drawn from paragraphs developed by Hazan and Shaver (1986). Factor analyses revealed that items for warm and cold fathers loaded separately, as did items for warm and cold mothers. However, warm and cold items were highly negatively correlated with one another (e.g., for the mothers, -.83; for the fathers, -.65). Thus, for simplicity two composite scores were created, one for paternal caregivers ($\alpha = .96$) and one for maternal caregivers ($\alpha = .96$) with higher scores indicating warmer parental caregivers and lower scores indicating more cold, distant, unresponsive, or inconsistent caregivers. These two measures were moderately correlated. For both statistical and theoretical reasons these measures of parental caregiving were kept separate, and not combined.

As mentioned earlier, measures of the number of partners desired over various time periods were very highly skewed. Given the nature of the data, medians instead of means were examined. Because some men and women indicated that they desired 0 partners over the next 30 years (about 5%), these individuals were not included in the data analysis (although this had no effect on the pattern of medians reported here). As indicated in Fig. 8.3, a fascinating pattern for paternal caregiving emerged. Essentially, when men had experienced cold and distant relationships with their paternal caregivers (e.g., were in the lowest quartile of paternal caregiving), they desired more partners in the next 30 years (e.g., 4) than men who had enjoyed warmer relationships with their caregivers. At least 50% of all women per quartile indicated that they would desire only 1 partner in the next 30 years. Maternal caregivers played a less clear linear role in predicting the number of partners that men desired over 30 years, with men in the second quartile desiring the most partners (e.g., 4) and men in the third quartile desiring the fewest (e.g., 1) and the other quartiles (the lowest and the highest) falling between. Nevertheless, men below the median desired more partners than men above the median. For women, maternal caregiving was unrelated to median number of partners desired (e.g., 1). As suggested in Fig. 8.3, for those individuals who perceived that they had warmer mothers and warmer fathers (upper 50%), no sex differences emerged. For those who had colder and more distant relationships with their mothers and fathers, there were statistically significant sex differences in the medians. An especially interesting sex difference here, however, involves the greater variability of the men in their desired number of partners and the apparent greater sensitivity of the men to parental caregiving, with men having especially distant, cold, or rejecting fathers differing most from other men in their patterns of preferences.

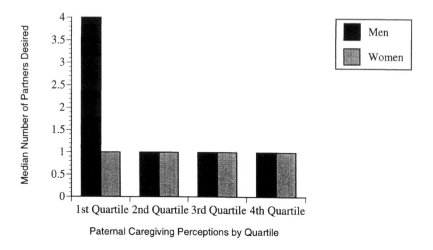

FIG. 8.3. Medians for ideal number of partners desired over 30 years by paternal caregiving by quartile for men and women.

Attachment Fertility Theory, consistent with Sexual Strategies Theory, would argue that the distinction between short- and long-term relationships is a crucial one. Attachment Fertility Theory, however, would also posit that most men and women would more likely be seeking a long-term mate than a short-term one. In fact, we would expect that whereas most men and women would be seeking a long-term mate, their desire to seek a short-term mate would be minimal. To investigate this, we asked 156 men and 232 women a question identical to one used by Buss and Schmitt (1993): We asked them to rate "on 7-point scales (1 = *not at all currently seeking* and 7 = *strongly currently seeking*) the degree to which they were currently seeking a short-term mate (defined as a one-night stand, brief affair, etc.), and, independently, the degree to which they were currently seeking a long-term mate (defined as a marriage partner)" (p. 210). Because these data were only mildly skewed, they could be transformed, allowing the use of parametric tests. As Buss and Schmitt (1993) had found, overall for men and women in our sample, there was a significant difference between men and women in the extent to which they were seeking a short-term relationship, with men more strongly seeking a short-term relationship than women. Similarly, as also reported by Buss and Schmitt (1993) we found no sex differences for the extent to which individuals were strongly seeking a long-term relationship. However, in apparent contrast to Buss and Schmitt's data (1993, p. 210)—although not any of their specific predictions—both men and women were significantly more inclined to be seeking a long term compared to a short-term mate (see Fig. 8.4). Such a pattern suggests that among humans, both women and men are more inclined to seek an enduring relationship compared to a short-term one.

To examine the role of attachment styles, maternal and paternal caregiving, and gender in predicting desire to seek a short-term relationship, we entered all of these variables and their resultant interaction terms into a simultaneous regression analysis. Essentially, we were asking: Controlling for all other variables in the equation, what role did each variable play in predicting how strongly individuals were seeking a short-term relationship? We found that neither gender per se, ($p = .61$) nor any of the interaction terms with gender ($p = .16$ to $p = .96$) were significant unique predictors of how strongly men or women were seeking a short-term relationship. Instead, attachment and parental caregiving variables and their interaction terms were significantly predictive (or marginally predictive), including paternal caregiving ($p < .05$), maternal caregiving ($p = .06$), and secure/avoid attachment scores ($p = .09$).

A second set of predictions from AFT concerns the sexual experiences of secure compared to nonsecure individuals. We would predict that secure compared to nonsecure individuals would be more likely to have enjoyable sex mediated by emotional bonding. Miller et al., (1996) found that, for men and women in marital relationships, more secure wives and less anxious–ambivalent husbands experienced greater sexual enjoyment, as did their spouses: an effect mediated by emotional bonding both at six months into their marriage and one year later. Similar effects were found for both husbands' and wives' paternal and maternal caregiving measures (Miller et al., 1996). For nonmarried women from a variety of ethnic backgrounds, more secure and less anxious–ambivalent women were more likely to have been emotionally bonded in their relationships,

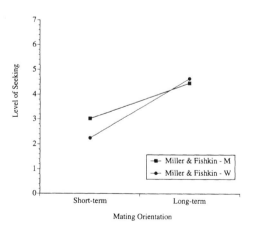

FIG. 8.4. Level of seeking a short- and long-term relationship for men and women on a scale from 1 (*not at all currently seeking*) to 7 (*strongly currently seeking*).

which in turn predicted sexual satisfaction, refusal to be with men who did not turn them on sexually, being able to have sex when they wanted to, and being able to turn a man on sexually. A similar pattern emerged for parental caregiving measures.

The responses of nonsecurely attached individuals may provide a view of how our biological heritage, as it intertwines with less than ideal or adapted-for environmental conditions, results in emergent behavior patterns that are somewhat different from their securely attached counterparts. Indeed, different patterns of partner choice, relational satisfaction, disclosure, and communication seem tied to departures from secure adult attachments (Collins & Read, 1990; Feeney, Noller, & Callan, 1994; Kobak & Hazan, 1991; Miller, Cooke, & Read, 1994; Senchak & Leonard, 1992). Compared to secure individuals, anxious/ambivalent individuals report more jealousy, sexual attraction, and the desire for reciprocation (Hazan & Shaver, 1987). The love-styles of anxious–ambivalents are characterized by high levels of preoccupation, mania, and addictive reliance on partners (Feeney & Noller, 1990), which may reflect their simultaneous fear of rejection and need for closeness (Collins & Read, 1994). Compared to securely attached persons, avoidants appear more likely to desire maintaining distance in relationships (Collins & Read, 1994). Nevertheless, we found that like securely attached persons (of whom 92% had been in a romantic relationship), nearly 85% to 86% of the anxious–ambivalent females, avoidant females, and anxious–ambivalent males reported having been in a romantic relationship, and 71% of the avoidant males reported having been in one. Thus, even among nonsecurely attached persons, being in or wanting to be in a romantic relationship seems common to being human. However, as in our own research, Hazan et al. (1994 as cited in Shaver, 1994) found that it was among the nonsecurely attached that sex differences in sexual behavioral patterns tended to emerge. For example, Hazan et al. found a positive relationship between being anxious–ambivalent and engaging in one-night stands. It is important to emphasize that patterns of emergent behavior for nonsecurely attached persons, as well as for those with more distant caregivers, may well be adaptive, given their difficulties in maintaining enduring relationships. Their behavioral strategies may be their most viable alternatives, given the constraints under which they operate.

Comparison with Sexual Strategies Theory

Attachment fertility theory (AFT) is quite different from sexual strategies theory (SST; Buss & Schmitt, 1993). Although both theories argue that men and women may engage in both types of sexual strategies (e.g., long- and short-term relationships), SST states that behavioral sex differences among humans have been selected for over the course of human evolution, such that human males maximize their reproductive success with multiple, short-term matings, whereas females maximize theirs with fewer, longer-term relationships, and that this pattern is rooted in human selection pressures. Although SST posits that both men and women are apt to seek long-term as well as short-term mates, it is presumed that men, on the average, seek more sexual partners than women and it is this strategy that enhances their reproductive success. It is presumed that women, on the average, seek fewer mates, and it is this strategy that enhances their reproductive success. Because men and women's sexual strategies were unlikely to have differed during the Pleistocene era, attachment fertility theory (AFT) posits that there should be few such sex differences.

These theories make some opposite predictions and can therefore be directly compared, often on exactly the same variables (Miller & Fishkin, 1996). Bacause of limited space, we have focused here on only a few of the predictions deemed "central to Sexual Strategies Theory" (Buss and Schmitt, 1993, p. 210) and relevant to attachment fertility theory.

Although replicating many of Buss and Schmitt's findings using means (e.g., men desire more ideal partners over various time frames; men are more strongly seeking a short-term mate), we found that many of these variables were very highly skewed, such that means were not the best measure of central tendency. When medians were employed, the picture changed considerably. For example, for both men and women, the median number of partners desired in the next 30 years was 1. Employing analyses appropriate given the nature of the data (e.g., parametric or not) and consistent with our predictions, our evolutionary windows (e.g., men and women who had enjoyed more warm and responsive paternal caregivers, men and women who were more secure) evinced few sex differences (e.g., in men or women's ideal number of sexual partners sought over various time intervals). Furthermore, both men and women were more interested in seeking a long-term than a short-term mate. We also found that, controlling for parental caregiving styles and attachment measures, gender was no longer a significant predictor of how strongly men and women were seeking a short-term mate. One of the most intriguing sex differences that emerged in our analyses was the tendency to observe much greater variability in men's responses (e.g., more men desiring 0 partners in the next 30 years and more male outliers ideally desiring more partners). Nevertheless, the majority of men desired only one partner over long stretches of time into the future (e.g., 30 years). Such variability would seem to make it difficult to argue that males, per se, have evolved a strategy such that the typical man seeks more sexual partners than women, and that it is this strategy that enhances their reproductive success. Taken together, these findings provide strong initial support for AFT, and suggest that differences between men and women become apparent when humans encounter social environments that are dissimilar from those of our Pleistocene ancestors (e.g., departures from responsive paternal and maternal caregivers). As such, these initial findings suggest that we may wish to take a second look at the assumptions undergirding sexual strategies theory.

In addition, consistent with AFT, in other research we found that attachment styles and parental caregiving measures for husbands and wives significantly predict spouse and own sexual enjoyment in marriage, an effect mediated by emotional bonding. We have replicated these effects for women in a sample of single individuals across several ethnic groups, and, in line with predictions, adults' perceptions of paternal caregiving, consistent with cross-cultural and within-culture analyses (Draper & Harpending, 1988), were particularly important in predicting their sexual practices and strategies. That sexual enjoyment should be tied to bonding certainly has adaptive value for ensuring the survival of pair-bonds and fertility. That those individuals for whom this linkage is stronger are our "evolutionary windows" suggests the fundamental adaptive value of the Pleistocene pattern: involved and responsive paternal and maternal caregivers, pair-bonds, and strong infant–parent attachments. These systems and their interfaces, as Bowlby (1969/1982) suggested, are fundamental to our evolved heritage as humans.

COMMONALTIES AND CONCLUSIONS

We would argue that after the Pleistocene era, when variability in patterns of paternal caregiving began to emerge, some offspring failed to encounter the adapted-for environment (e.g., warm paternal and maternal caregivers). This resulted in naturally occurring emergent outcomes: greater difficulties in trusting and forming positive views of others, greater insecurity (and concerns with control), and greater difficulties in forming and maintaining emotionally close relationships. Because of such outcomes, nonsecure individuals may have tended to develop less positive views of others, especially out-group others (e.g., of the opposite sex), and because they were more insecure, they may have been more likely to compete with and dominate others (e.g., to gain some measure of perceived control). They likely would have had fewer social skills that enhanced cooperation and would have formed less stable pair-bonds. Because of this, they might have spent a greater proportion of their time seeking less enduring relationships that met at least some of their needs (e.g., short-term relationships). In short, it seems that a propensity to spend more of one's time seeking short-term relationships rather than long-term ones simply may have been "fallout" of a failure to interface with humans' adapted for social environment (e.g., responsive paternal and maternal caregivers). When, for whatever reasons, long-term relationships were not possible or difficult to forge and maintain, humans, both women and men, may have developed a secondary alternative strategy: short-term relationships. Many humans may seek both types of relationships over time, in part, because at some point or chronically they are unable to achieve and maintain (or based on their caregiving histories, less motivated to seek) enduring, emotionally close, long-term relationships.

What was the strategy for which we were adapted? We would argue that although short-term mating strategies may be fall-out from a failure of humans to interface with their adapted for environments, seeking a long-term mate for a close and enduring relationship is based on universal design features, (i.e., part of our evolutionary heritage). We have argued that a number of design features support the view that emotional closeness, pair-bonding, and parental caregiving—including the enhanced role of paternal involvement—were important human adaptations. Consistent with this and his theoretical position regarding the prevalence and centrality of long-term relationships as an evolutionary strategy, Buss et al.'s (1990) cross-cultural findings indicated that mutual attraction and love—as a "state of the relationship, one that signifies mutuality and, perhaps, reciprocity," (p. 18)—is at the top of characteristics desired in mates.[9] Furthermore, in related research (Miller & Fishkin, 1996), we found that mate characteristics such as emotionally warm and kind and understanding, in general, are highly desirable, especially for a long-term relationship.[10] Humans who pair-bonded and provided a supportive environment for offspring were probably more apt to experience enjoyable sexual encounters and more apt to have offspring that survived to adulthood. Perhaps, as Harlow's work might suggest (Harlow & Zimmerman, 1959), in our most intimate relationships, it is not basic needs (e.g., food, sex) alone or even primarily that we seek. Rather, in our own way, we are attempting to achieve the emotional closeness that we, as humans—both with our offspring and with our mates—have evolved to desire.

REFERENCES

Abitbol, M. M. (1993). Growth of the fetus in the abdominal cavity. *American Journal of Physical Anthropology, 91*, 367–378.

Adams, D. B., Gold, A. R., & Burt, A. D. (1978). Rise in female-initiated sexual activity at ovulation and its suppression by oral contraceptives. *New England Journal of Medicine, 299*, 1145–1150.

Ainsworth, M. D. S., Blehar, M. C., Waters, E., & Wall, S. (1978). *Patterns of attachment: A psychological study of the strange situation.* Hillsdale, NJ: Lawrence Erlbaum Associates.

Baker, R. R., & Bellis, M. A. (1993). Human sperm competition: Ejaculate manipulation by females and a function for the female orgasm. *Animal Behaviour, 46*, 887–909.

Barkow, J. H., Cosmides, L., & Tooby, J. (Eds.) (1992). *The adapted mind: Evolutionary psychology and the generation of culture.* New York: Oxford University Press.

Bartholomew, K. (1990). *Attachment styles in young adults: Implications for self concept and interpersonal functioning.* Unpublished doctoral dissertation, Stanford University, CA.

---

[9] Curiously, Buss et al., also argued that these cross-cultural findings suggest that the variance accounted for by culture was probably greater than that accounted for by sex differences.

[10] See also work by Buss as referred to in Buss and Schmitt (1993) who examined these same variables.

Bartoshuk, L. M., & Beauchamp, G. K., (1994). Chemical senses. *Annual Review of Psychology, 45,* 419–449.

Belsky, J., Steinberg, L., & Draper, P. (1991). Childhood experience, interpersonal development, and reproductive strategy: An evolutionary theory of socialization. *Child Development, 62,* 647–670.

Bernstein, I. H., Lin, T., & McClellan, P. (1982). Cross- vs. within-racial judgments of attractiveness. *Perception and Psychophysics, 32,* 495–503.

Bowlby, J. (1969/1982). *Attachment and loss: Volume 1. Attachment.* New York: Basic Books.

Buss, D. M., Abbott, M., Angleitner, A., Asherian, A., Biaggio, A., & Blanco-Villasenor, A. (1990). International preferences in selecting mates: A study of 37 cultures. *Journal of Cross-Cultural Psychology, 21,* 5–47.

Buss, D. M., & Schmitt, D. P. (1993). Sexual strategies theory: An evolutionary perspective on human mating. *Psychological Review, 100,* 204–232.

Caporael, L. (1994, August). Nested hierarchy in biology and psychology. In L. C. Miller (Chair), *Evolutionary dynamics: On the emergence of social behavior.* Symposium presented at the American Psychological Association, Los Angeles, CA.

Clark, R. D., & Hatfield, E. (1989). Gender differences in receptivity to sexual offers. *Journal of Psychology and Human Sexuality, 2,* 39–55.

Cliff, N. (1993). Dominance statistics: Ordinal analyses to answer ordinal questions. *Psychological Bulletin, 114,* 494–509.

Collins, N. L., Dunkel-Schetter, C., Lobel, M., & Scrimshaw, S. C. (1993). Social support in pregnancy: Psychosocial correlates of birth outcomes and postpartum depression. *Journal of Personality and Social Psychology, 65,* 1243–1258.

Collins, N. L., & Read, S. J. (1990). Adult attachment, working models, and relationship quality in dating couples. *Journal of Personality and Social Psychology, 58,* 644–663.

Collins, N. L., & Read, S. J. (1994). Cognitive representations of attachment: The structure and function of working models. In K. Bartholomew & D. Perlman (Eds.), *Advances in personal relationships. Vol. 5: Attachment* (pp. 53–90). London, England: Jessica Kingsley Publishers.

Cunningham, M. R. (1986). Measuring the physical in physical attractiveness: Quasi-experiments on the sociobiology of female facial beauty. *Journal of Personality and Social Psychology, 50,* 925–935.

Cutler, W. B. (1991). *Love Cycles: The Science of Intimacy.* New York: Villard Books.

Cutler, W. B., Garcia, C. R., & Krieger, A. M. (1979). Sexual behavior frequency and menstrual cycle length in mature premenopausal women. *Psychoneuroendocrinology, 4,* 297–309.

Cutler, W. B., Preti, G., Huggins, G. R., Erickson, B., & Garcia, C. R. (1985). Sexual behavior frequency and biphasic ovulatory type menstrual cycles. *Physiology & Behavior, 34,* 805–810.

Cutler, W. B., Preti, G., Krieger, A. M., Huggins, G. R., Garcia, C. R., & Lawley, H. J. (1986). Human axillary secretions influence women's menstrual cycles: The role of donor extract from men. *Hormones and Behavior, 20,* 463–473.

Daly, M., & Wilson, M. (1983). *Sex, evolution, and behavior* (2nd ed.). Belmont, CA: Wadsworth.

Dobbie, B. M. W. (1982). An attempt to estimate the true rate of maternal mortality, sixteenth to eighteenth centuries. *Medical History, 26,* 79–90.

Draper, P., & Harpending, H. (1988). A sociobiological perspective on the development of human reproduction strategies. In K. B. MacDonald (Ed.), *Sociobiological perspectives on human development* (pp. 340–372). New York: Springer-Verlag.

Eccles, A. (1982). *Obstetrics and gynaecology in Tudor and Stuart England*. Kent, OH: Kent State University Press.

Fedigan, L. M. (1982). *Primate Paradigms: Sex roles and social bonds*. Montréal, Canada: Eden Press.

Feeney, J. A., & Noller, P. (1990). Attachment style as a predictor of adult romantic relationships. *Journal of Personality and Social Psychology, 58*, 281–291.

Feeney, J., Noller, P., & Callan, V. J. (1994). Attachment style, communication, and satisfaction in the early years of marriage. In K. Bartholomew and D. Perlman (Eds.), *Advances in personal relationships. Vol. 5: Attachment processes in adulthood*, (pp. 269–308). London: Jessica Kingsley Publishers.

Fisher, H. E. (1987). The four year itch: Do divorce patterns reflect our evolutionary heritage? *Natural History, 96*, 22–33.

Fisher, H. E. (1989). Evolution of human serial pairbonding. *American Journal of Physical Anthropology, 78*, 331–354.

Fishkin, S. A., & Miller, L. C. (1994, August). The attachment system: Dynamics of bonding, fertility, and offspring survival. In L. C. Miller (Chair), *Evolutionary dynamics: On the emergence of social behavior*. Symposium presented at the American Psychological Association, Los Angeles, CA.

Ford, C. S., & Beach, F. A. (1951). *Patterns of sexual behavior*. New York: Harper and Row.

Gangestad, S. W., & Simpson, J. A. (1990). Toward an evolutionary history of female sociosexual variation. *Journal of Personality, 58*, 69–96.

Goldman, S. E., & Schneider, H. G. (1987). Menstrual synchrony: Social and personality factors. *Journal of Social Behavior and Personality, 2*, 243–250.

Gower, D. B., & Ruparelia, B. A. (1993). Olfaction in humans with special reference to odorous 16-androstenes: Their occurrence, perception and possible social, psychological and sexual impact. *Journal of Endocrinology, 137*, 167–187.

Green, J. A., Jones, L. E., & Gustafson, G. E. (1987). Perception of cries by parents and nonparents: Relation to cry acoustics. *Developmental Psychology, 23*, 370–382.

Guttentag, M., & Secord, P. F. (1983). *Too many women? The sex ratio question*. Beverly Hills: Sage.

Harlow, H. F., Harlow, M. K., & Suomi, S. J. (1971). From thought to therapy: Lessons from a primate laboratory. *American Scientist, 59*, 538–549.

Harlow, H. F., & Zimmerman, R. R. (1959). Affectional responses in the infant monkey. *Science, 130*, 421.

Hazan, C., & Shaver, P. (1986). *Parental caregiving style questionnaire*. Unpublished manuscript.

Hazan, C., & Shaver, P. (1987). Romantic love conceptualized as an attachment process. *Journal of Personality and Social Psychology, 52*, 511–524.

Hazan, C., & Shaver, P. (1990). Love and work: An attachment-theoretical perspective. *Journal of Personality and Social Psychology, 59*, 270–280.

Hazan, C., & Shaver, P. R. (1994a). Attachment as an organizational framework for research on close relationships. *Psychological Inquiry, 5*, 1–22.

Hazan, C., & Shaver, P. R. (1994b). Deeper into attachment theory. *Psychological Inquiry*, 5, 68–79.

Hazan, C., & Zeifman, D. (1994). Sex and the psychological tether. In K. Bartholomew and D. Perlman (Eds.), *Advances in personal relationships. Vol. 5: Attachment processes in adulthood* (pp. 151–177). London: Jessica Kingsley Publishers Ltd.

Hedricks, C. A., Ghiglieri, M., Church, R. B., Lefevre, J., & McClintock, M. K. (1994). Hormonal and ecological contributions toward interpersonal intimacy in couples. *Annals of the New York Academy of Sciences*, 18, 207–209.

Hedricks, C. Piccinino, L. J., Udry, J. R., & Chimbira, T. H. K. (1987). Peak coital rate coincides with onset of lutenizing hormone surge. *Fertility & Sterility*, 48, 234–38.

Johnson, R. W., Dannenbring, G. L, Anderson, N. R., & Villa, R. E. (1983). How different cultural and geographic groups perceive the attractiveness and unattractiveness of active and inactive feminists. *The Journal of Social Psychology*, 119, 111–117.

Kenrick, D. T., Groth, G. E., Trost, M. R., & Sadalla, E. K. (1993). Integrating evolutionary and social exchange perspectives on relationships: Effects of gender, self-appraisal, and involvement level on mate selection criteria. *Journal of Personality and Social Psychology*, 64, 951–969.

Kleiman, D. G. (1977). Monogamy in mammals. *Quarterly Review of Biology*, 52, 39–69.

Kirkpatrick, L. A., & Hazan, C. (1994). Attachment styles and close relationships: A four-year prospective study. *Personal Relationships*, 1, 123–142.

Kobak, R. R., & Hazan, C. (1991). Attachment in Marriage: Effects of security and accuracy of working models. *Journal of Personality and Social Psychology*, 60, 861–869.

Langlois, J. H., & Roggman, L. A. (1990). Attractive faces are only average. *Psychological Science*, 1, 115–121.

Langlois, J. H., Roggman, L. A., Casey, R. J, Ritter, J. M., Rieser-Danner, L. A., & Jenkins, V. Y. (1987). Infant preferences for attractive faces: Rudiments of a stereotype? *Developmental Psychology*, 23, 363–369.

Laumann, E. O., Gagnon, J. H., Michael, R. T., & Michaels, S. (1994). *The social organization of sexuality*. Chicago: The University of Chicago Press.

Lehman, R. S. (1991). *Statistics and research design in the behavioral sciences*. Belmont, CA: Wadsworth.

Levy, M. B., & Davis, K. E. (1988). Lovestyles and attachment styles compared: Their relations to each other and to various relationship characteristics. *Journal of Social and Personal Relationships*, 5, 439–471.

Liebowitz, M. R. (1983). *The chemistry of love*. Boston: Little, Brown.

Lovejoy, C. O, Hieple, K. G., & Burstein, A. H. (1973). The gait of Australopithecus. *American Journal of Physical Anthropology*, 38, 757–780.

Mann, J. (1992). Nurturance or negligence: Maternal psychology and behavioral preference among preterm twins. In J. H. Barkow, L. Cosmides, & J. Tooby (Eds.) (1992). *The adapted mind: evolutionary psychology and the generation of culture*. New York: Oxford University Press.

Main, M. (1991). Metacognitive knowledge, metacognitive monitoring, and singular (coherent) vs. multiple (incoherent) model of attachment: Findings and directions for future research. In C. M. Parkes, J. Stevenson-Hinde, & P. Marris (Eds.), *Attachment across the lifecycle* (pp. 127–159). London: Tavistock/Routledge.

Maret, S. M. (1983). Attractiveness ratings of photographs of Blacks by Cruzans and Americans. *The Journal of Psychology, 115*, 113–116.

Matteo, S. (1987). The effect of job stress and job interdependency on menstrual cycle length, regularity, and synchrony. *Psychoneuroendocrinology, 12*, 467–476.

McClintock, M. K. (1971). Menstrual synchrony and suppression. *Nature, 229*, 244–245.

McClintock, M. K. (1983). Pheromonal regulation of the ovarian cycle: Enhancement, suppression, and synchrony. In J. G. Vandenbergh (Ed)., *Pheromones and reproduction in mammals*. New York: Academic Press.

McGrew, W. C., & Feistner, A. T. C. (1992). Two nonhuman primate models for the evolution of human food sharing: Chimpanzees and Callitrichids. In J. H. Barkow, L. Cosmides, & J. Tooby (Eds.), *The adapted mind: Evolutionary psychology and the generation of culture* (pp. 229–243). New York: Oxford University Press.

Mikulincer, M., Florian, V., & Tolmacz, R. (1990). Attachment styles and fear of personal death: A case study of affect regulation. *Journal of Personality and Social Psychology, 58*, 273–280.

Miller, L. C., Cooke, L., & Read, S. J. (1994, May). *Mental models of caregiving styles: Do they predict marital satisfaction and communication?* Paper presented at the International Communication Association, Sydney, Australia.

Miller, L. C., & Fishkin, S. A. (1996). *Is it in our nature that men and women use different sexual strategies? Questioning assumptions and inferences.* Unpublished manuscript.

Miller, L. C., Fishkin, S. A., Gonzales-Tumey, C., & Rothspan, S. E. (1996). *When sex is hot and when its not: For women do attachments matter?.* Unpublished manuscript.

Mitchell, G. (1979). *Behavioral sex differences in nonhuman primates*. New York: Van Nostand Reinhold.

Monahan, J. L., Miller, L. C., & Rothspan, S. (in press). Power and intimacy: On the dynamics of risky sex. *Health Communication*.

Money, J. (1980). *Love and love sickness*. Baltimore: The John Hopkins University Press.

Monti-Bloch, L. Jennings-White, C., Dolberg, D. S., & Berliner, D. L. (1994). The human vomeronasal system. *Psychoneuroendocrinology, 19*, 673–686.

Murray, A. D. (1979). Infant crying as an elicitor of parental behavior: An examination of two models. *Psychological Bulletin, 86*, 191–215.

Peplau, L. A., Rubin, Z., & Hill, C. T. (1977). Sexual intimacy in dating relationships. *Journal of Social Issues, 33*, 86–109.

Persky, H., Leif, H., Strauss, D., Miller, W., & O'Brien, C., (1978). Plasma testosterone level and sexual behavior of couples. *Archives of Sexual Behavior, 7*, 157–173.

Preti, G., Cutler, W. B., Garcia, C. R., Huggins, G. R., & Lawley, H. J. (1986). Human axillary secretions influence women's menstrual cycles: The role of donor extract in females. *Hormones and Behavior, 20*, 474–482.

Read, S. J., & Miller, L. C. (1995). The centrality of stories: For social creatures how could it be otherwise? In R. S. Wyer & T. C. Srull (Eds.), *Advances in social cognition*. Hillsdale, NJ: Lawrence Erlbaum Associates.

Russell, M. J., Switz, G. M., & Thompson, K. (1980). Olfactory influences on the human menstrual cycle. *Pharmacology, Biochemistry, and Behavior, 13*, 737–738.

Saphier, D. (1989). Neurophysiological and endocrine consequences of immune activity. *Psychoneuroendocrinology, 14*, 64–87.

Schaal, B. (1988). Olfaction in infants and children: Developmental functional perspectives. *Chemical Senses, 13*, 145–190.

Senchak, M., & Leonard, K. E. (1992). Attachment styles and marital adjustment among newlywed couples. *Journal of Social and Personal Relationships, 9*, 51–64.

Shaver, P. (1994, August). *Attachment, Caregiving, and Sex in Adult Romantic Relationships.* Paper presented at the American Psychological Association, Los Angeles, CA.

Shaver, P. R., & Hazan, C. (1994). Adult romantic attachment: Theory and evidence. In D. Perlman & W. Jones (Eds.), *Advances in Personal Relationships.* (Vol. 4, pp. 29–70). London: Jessica Kingsley Publishers.

Shorter, E. (1982). *A History of Women's Bodies.* New York: Basic Books.

Shulman, S., Elicker, J., & Sroufe, L. A. (1994). Stages of friendship growth in preadolescence as related to attachment history. *Journal of Social and Personal Relationships, 11*, 341–361.

Simpson, J. A. (1990). Influence of attachment styles on romantic relationships. *Journal of Personality and Social Psychology, 59*, 971–980.

Simpson, J. A., & Gangestad, S. W. (1992). Sociosexuality and romantic partner choice. *Journal of Personality, 60*, 31–51.

Simpson, J. A., Rholes, W. S., & Nelligan, J. S. (1992). Support-seeking and support-giving within couple members in an anxiety provoking situation: The role of attachment styles. *Journal of Personality and Social Psychology, 62*, 434–446.

Singh, D. (1993). Adaptive significance of female physical attractiveness: Role of waist-to-hip ratio. *Journal of Personality and Social Psychology, 65*, 293–307.

Sprecher, S., & McKinney, K. (Eds.). (1993). *Sexuality.* Newbury Park, CA: Sage.

Stoddart, D. M. (1990). *The Scented Ape.* Cambridge, MA: Cambridge University Press.

Tague, R. G. (1992). Sexual dimorphism in the human bony pelvis, with a consideration of the Neandertal pelvis from Kebara Cave, Israel. *American Journal of Physical Anthropology, 88*, 1–21.

Tague, R. G. (1994). Maternal mortality or prolonged growth: Age at death and pelvic size in three prehistoric Amerindian populations. *American Journal of Physical Anthropology, 95*, 27–40.

Tanner, N. M. (1981). *On becoming human.* Cambridge, MA: Cambridge University Press.

Tanner, N. M, & Zihlman, A. L. (1976). Women in evolution, part I: innovation and selection in human origins. *Signs: Journal of Women in Culture and Society, 1*, 585–608.

Thakerar J. N., & Iwawaki, S. (1979). Cross-cultural comparisons in interpersonal attraction of females toward males. *The Journal of Social Psychology, 108*, 121–122.

Tooby, J., & Cosmides, L. (1992). The psychological foundations of culture. In J. H. Barkow, L. Cosmides, & J. Tooby (Eds.), *The adapted mind: evolutionary psychology and the generation of culture* (pp. 119–136). New York: Oxford University Press.

Townsend, J. M. (1993). Sexuality and partner selection: sex differences among college students. *Ethology and Sociobiology, 60*, 31–51.

UNICEF [United Nations Children's Fund] (1994). *The state of the world's children* New York: Oxford University Press.

Veith, J. L., Buck, M., Getzlaf, S., Dalfsen, P. V., & Slade, S. (1983). Exposure to men influences the occurrence of ovulation in women. *Physiology & Behavior, 31*, 313–315.

Yoon, C. K. (1995). New hominid species was bipedal 3.9-4.2 million years ago. *The Journal of NIH Research, 7*, 30–32.

Zeskind, P. S., & Lester, B. M. (1978). Acoustic features and auditory perceptions of the cries of newborns with prenatal and perinatal complications. *Child Development, 49*, 580–589.

Zihlman, A. L. (1978). Women in evolution, Part II: Subsistence and social organization among early hominids. *Signs: Journal of Women in Culture and Society, 4*, 4–20.

9

Attachment: The Bond in Pair-Bonds

Debra Zeifman
Cindy Hazan
Cornell University

A recent issue of the *New York Times Magazine* featured an article about summer love songs. Included was a photograph of the rapper, Method Man, whose song is the current favorite among young lovers; in the photo, he is sitting with his arms wrapped around his long-term girlfriend, Shortie. The caption reports that, while on tour, Method "got lonely" for Shortie, arranged to have her fly cross country to join him, and then wrote the song, "I'll Be There for You," upon her arrival. It seems reasonable to assume that Method had other options for female company available to him. As a physically strong, high-status, amply-re-sourced male (musician), he likely had easy access to a variety of novel and fertile potential sexual partners (fans). But it was Shortie he longed for. Only she would do.

This kind of emotional bond between two individuals that makes each unique in the other's life, that evokes feelings of loneliness during separations, and that manifests itself in intense longing for reunion and physical contact is the essence of attachment and the focus of this chapter. Incorporating attachment phenomena into the new evolutionary social psychology requires no new interpretation of behavior and no major reconceptualization of existing research findings. Attachment theory (Bowlby, 1958, 1969, 1973, 1980) is and always has been well grounded in evolutionary principles. It is a theory about one among many complex and specialized psychological mechanisms that evolved through natural selection to solve specific adaptive problems.

Human nature, like that of other animals, is largely the product of these solutions to a wide range of problems that directly or indirectly enhanced reproductive fitness. To achieve reproductive fitness, individuals must success-fully negotiate at least three major adaptive challenges: survive to reproductive age, mate, and provide adequate care for offspring so that they, too, will survive to reproduce. Evolutionary theory has already been fruitfully applied to the study of mating, and the resulting research has led to profound insights into the solutions to such crucial adaptive challenges as how to select and attract a mate. Attachment theory complements this growing body of work by providing a different evolutionary perspective on human pair-bonds. Its strength lies not so much in explaining strategies of selection and attraction as in providing insights into the nature and functions of the bond that holds two individuals together. It further supplements existing models by highlighting some broad-based similarities between the sexes. In the realm of attachment, the differences of interest are intrasex variations resulting from the effects of environmental input on evolved mechanisms.

Attachment theory began as an attempt to explain the behavior of human infants in relation to their primary caregivers. It ultimately led to the postulation of a behavioral system with important implications for adaptation and functioning, not just in infancy, but across the entire life span. Thus, the theory covers issues of ontogeny as well as phylogeny. The main focus of this chapter is attachment in pair-bond relationships, but we begin at the ontogenetic beginning, with attachment in the first year of life.

INFANT–CAREGIVER ATTACHMENT

Ethological attachment theory (Bowlby, 1958, 1969, 1973, 1980) grew out of the observation that infants and young children manifest profound and lasting distress if separated from their primary caregivers, even when their nutritional and hygienic needs are adequately met by surrogates. Moreover, reactions to such separations take the form of a predictable and universal sequence of stages. The first stage, *protest*, is characterized by obsessive searching for the absent caregiver, inconsolable crying, and resistance to others' offers of comfort. This stage is reliably followed by *despair*, a period of passivity and depressed mood. If the separation is sufficiently prolonged, the eventual response is emotional *detachment*. Bowlby noted that the duration of these reactions as well as the timing of the transition from one stage to the next varied as a function of age and other factors, but their appearance and ordering was invariant across individuals.

He also documented evidence that protracted separations result in long-term deleterious effects. For example, retrospective examination of the life histories of 44 juvenile delinquents revealed that a disproportionate number of them suffered episodes of maternal separation during infancy and early childhood (Bowlby, 1944). The total absence of attachment appeared to be even more costly than disruption. Infants reared in institutions and cared for collectively by nursing staff suffered alarming health decrements and often failed to thrive

(Robertson, 1953; Spitz, 1946). Thus, the establishment and maintenance of an attachment bond with a particular caregiver appeared to be essential for normal physical and psychological development.

An Evolved Psychological Mechanism

The search for an explanation of the reactions to and long-term effects of separation eventually led Bowlby to examine the work of ethologists such as Lorenz (1970) and Harlow (1958), who had demonstrated that the offspring of many altricial species (i.e., those who bear immature young) possess an innate tendency to imprint on (become attached to) a particular caregiver. In such species, the establishment of a strong and enduring bond between infant and caregiver is clearly adaptive. Such a bond greatly enhances the infant's chances for survival by ensuring that it receives the care and protection needed to survive. Bowlby surmised that in our own (altricial) species, there may be a similar innate mechanism with a similar function.

Within this framework, the observed reactions to separation were comprehensible. In earlier human environments, an infant separated from its mother faced the risk of death by starvation or predation. The infant would likely first let out a stream of loud calls to attract the attention of its mother (protest). Loud cries would be adaptive only if the mother were nearby, but continued calling might attract predators, and so passivity would soon set in (despair). Finally, when little hope of the mother's return remains, the youngster would let go of the bond (detachment), and attempt to survive on its own or be open to establishing a new attachment.

Insofar as the work of ethologists was helpful in explaining the obvious importance of attachment bonds to human young and in providing support for the notion that the tendency to form them is a species-typical behavior, it was control systems theory (Miller, Gallanter, & Pribram, 1960; Young, 1964) that proved useful in explaining the dynamics of attachment. All normal human infants exhibit a common pattern of behavior in relation to their primary caregiver: They tend to seek and maintain relatively close proximity to this person, retreat to him or her for comfort and reassurance in the face of perceived or real danger, are distressed by and resist separations and, in the caregiver's presence (and in the absence of threat), engage in exploration and play. From this universally observed dynamic, Bowlby postulated the existence of a behavioral system or psychological mechanism that served to regulate proximity to a protector, just as physiological systems serve to maintain set goals for blood pressure and body temperature. The set goal of the attachment system is a sufficient degree of physical proximity to an attachment figure to ensure safety, a goal that necessarily varies as a function of endogenous (e.g., state, age) and exogenous (e.g., novelty, threat) factors. The emotional triggers for the system are anxiety, which activates it, and calm (or what Sroufe and Waters, 1977, referred to as *felt-security*), which deactivates it.

The form of the behavior suggests the primary function of procuring care and protection, but it also implies a secondary function. Through distress alleviation, the attachment system facilitates exploratory and affiliative behaviors which also ultimately contribute to reproductive fitness. This affect-regulation function of the attachment bond helps to ensure that human young will engage in competence-building activities, but only when it is judged safe to do so.

An Interpersonal Relationship

Whereas the attachment mechanism itself is innate, the target of attachment behavior (i.e., the attachment figure) must be selected. Several cues seem to be particularly influential in the choice. Familiarity appears to be a key factor, as is responsiveness, especially in the context of distress alleviation. The primary caregiver is, by definition, the one who most reliably responds to the infant's signals of distress and, as such, is typically the person to whom an attachment is formed. Bowlby speculated that maturity and wisdom may be additional cues, based on the fact that infants tend to direct their bids for comfort to adults rather than children, that is, to the ones who appear competent and seem to know what to do (Bowlby, 1969). Infants also appear to be sensitive to variations in social interactions that signal differences in the nature or quality of social relationships. For example, compared to interactions with other care providers, interactions with primary caregivers tend to be more intimate, often involving prolonged mutual gazing, nuzzling, cuddling, sucking, extended skin-to-skin and ventral–ventral contact, and the touching of body parts otherwise considered private (Schaffer, 1971). This combination of cues—familiarity, responsiveness, competence, and intimate contact—informs the infant as to whom it should stay close, to whom it should turn for comfort or protection in threatening situations, from whom it should resist being separated, and in whose presence it is safe to explore.

The process of selecting a specific caregiver and forming an attachment bond unfolds gradually over the course of the first year of life. Across all cultures in which it has been examined, and regardless of rearing conditions, the developmental course of attachment formation is virtually identical (Kagan, 1971). At birth, infants are prepared in a variety of ways to engage in social interaction and, when distressed, accept care indiscriminately. At around 3 to 4 months of age, they begin to direct bids for interaction and contact with familiar adults, especially the primary caregiver. Later, typically around the age of 6 or 7 months, a major transformation occurs. Just as infants become capable of self-produced locomotion (and are capable of wandering dangerously far from their protector), they begin to exhibit two new fears: of strangers and separations from the caregiver. In fact, separation protest has traditionally been considered the clear-cut marker of attachment (Ainsworth, Blehar, Waters, & Wall, 1978; Bowlby, 1969; Sroufe & Waters, 1977). Although infants may enjoy interacting with many individuals, and even solicit and accept care from a number of people, they rarely object to or exhibit distress as a result of being separated from anyone other than their primary attachment figure.

Beyond Infancy

It took Bowlby more than two decades to construct his theory of attachment, and more than a thousand pages to lay out the evidence he had amassed in support of it. The results of his efforts left little room for doubt about the central postulates of the theory. An attachment mechanism had evolved to solve an adaptive problem faced by all members of our altricial species. Extreme immaturity at birth necessitates reliable protection and care; without first ensuring our survival through childhood, forces of natural selection operating on strategies for selecting and attracting mates would have been of limited utility. Thus, the first adaptive challenge of reproductive fitness—surviving to reproductive age—was solved by the selection of a mechanism to promote the formation of an enduring emotional bond to a particular individual who could be trusted to provide protection and care, a mechanism that would guarantee resistance to separations and an intense longing for close proximity.

Although the need for attachment during infancy is indisputable, its importance and relevance beyond childhood has not been as convincingly demonstrated. Perhaps the most controversial aspect of attachment theory was Bowlby's (1979, 1988) claim that the attachment system is functional throughout the human life span. He acknowledged that the system would undergo major modifications and even qualitative transformations in the course of normative development. For example, the extent and duration of physical distance from an attachment figure judged to be tolerable would be expected to change, not only from situation to situation, but from one developmental period to the next. Whereas 1-year-old infants are typically upset when their primary caregiver merely leaves the immediate vicinity, most 3-year-olds can comfortably tolerate the day-long separation imposed by nursery school. Bowlby also hypothesized that the target of attachment behavior—the person who is sought out for comfort and from whom separations are distressing—would naturally shift from a parental figure to a peer (Bowlby, 1979, 1988).

It is on this point that his hypotheses become most relevant to the thesis of this chapter. Bowlby viewed pair-bonds as the adult instantiation of attachment, and, implicit in this view is the suggestion that attachment serves a similar affect–regulatory function in adulthood as it does in infancy. In other words, the need for felt security is ageless; during all phases of development, humans will function optimally when they have a trusted figure on whom they can rely for support and reassurance. In essence, Bowlby proposed a very different function of pair-bonds than is generally postulated. Independent of the possible benefits to offspring, attachments satisfy a basic human need "from the cradle to the grave" (Bowlby, 1979). This hypothesis and the possible role of attachment beyond infancy is the focus of what follows.

PAIR-BONDS

It can be and has been argued that the adaptive problems associated with reproductive fitness are dramatically different for males and females, at least in terms of the criteria used to select a mate and the strategies effective for attracting one (see Gangestad & Thornhill, chap. 7, this volume; also, Buss & Schmitt, 1993; Daly & Wilson, 1983; Symons, 1979). The accepted explanation of differences in mate preferences and selection strategies is that they are the result of differential parental investment (Trivers, 1972). Differential investment begins before conception: Males have an abundant supply of small sperm cells produced at a rate of approximately 500 million per day (Zimmerman, Maude, & Moldawar, 1965), whereas females have a far more limited supply of large egg cells produced at a rate of about one per month during a shorter period of fertility. Added to this is the burden on the female of gestation and lactation, requiring years of investment. For males, whose contribution to offspring may be as little as sperm, the most effective strategy may be to take advantage of any and all sexual opportunities with fertile partners. The female, for whom every sexual encounter is potentially very costly, would be expected to be far more choosy in her copulations. Once her egg is fertilized, she will have to forgo other reproductive opportunities for a relatively long time. Her most effective strategy is to limit her sexual encounters to males who possess and appear willing to share valuable resources with her and the offspring she will have to nurture.

In a survey of 37 cultures, Buss (1989) found that sex differences in mate selection criteria are consistent with male–female differences in parental investment. For example, males assign greater importance than females to the physical appearance of potential mates, preferring females who look youthful and healthy—among the best indices of fertility in our species (Buss, 1989; Symons, 1979). In contrast, females care more than males about the social status and earning power of potential partners, a sensible mate selection strategy for ensuring that he will provide well for the offspring. As for attracting potential mates, males tend to display their resources, sometimes even exaggerating them, and females try to make themselves appear as young and healthy (i.e., fertile) as possible, using deception if necessary. Apparently, when it comes to pair-bond relationships, evolutionary theory predicts, and empirical evidence confirms, that men and women are different, at least in what they seek in a mate and how they go about enticing one.

If human reproductive fitness required nothing more than successful mating, reproductive partners could part ways as soon as a viable pregnancy had been ensured. But, in actuality, the vast majority of human males and females opt to remain with the same partner for a more extended period of time (Eibl-Eibesfeldt, 1989). This trend of forming lasting bonds is not new in human phylogenetic history (Mellen, 1981), nor is it likely to change anytime soon. The norm for our species is enduring pair-bonds and, from the perspective of evolution and reproductive fitness, they are highly adaptive.

The Evolution of Pair-Bonds

The apparent evolutionary trend among mammals, and primates in particular, was to bear increasingly immature offspring (Trevathan, 1987). In humans, this trend followed a birthing crisis in which the infant's large head (with its more fully developed brain) could not easily pass through the birth canal of our bipedal female primate ancestors. Infants born prematurely, with less developed brains (and smaller heads) were more likely to survive, as were their mothers. Immaturity at birth also afforded the advantage of a longer period of learning during a time of neural plasticity. This would have been a distinct advantage in a species with complex social organization, such as our own. However, along with the multiple advantages of a protracted period of immaturity came new risks and challenges. The effort associated with feeding, carrying, and protecting an infant, along with the tasks of socializing and training, necessitated paternal investment. Without an adequately strong force to keep fathers around and involved, exceptionally helpless and vulnerable offspring would have had very poor chances of surviving to reproductive age or developing the necessary skills for their own eventual mating and parenting roles.

As we see it, the central question of interest is what psychological mechanism(s) evolved to solve this adaptive problem of keeping parents together. Implicit in this question are several related issues. One concerns the nature of the mechanism itself—how it works, what activates it, whether it is specialized or serves multiple functions, and so on. Another concerns the length of a coparenting period needed to ensure that offspring would, themselves, be reproductively fit. Yet another related issue is whether the adaptive problem differed significantly for males and females, resulting in qualitatively different mechanisms. We address each of these issues in turn.

Serial Monogamy

It has been proposed that humans are, by nature, serial monogamists (Fisher, 1992). That is, we do tend to form pair-bonds but, rather than remaining with the same partner for life or even a significant proportion of life, we develop multiple sequential bonds with different partners, each of a relatively limited duration. According to this theory, the natural duration of pair-bonds corresponds roughly to the length of the human reproductive cycle. It includes not only conception and gestation, but also nursing—in all, approximately 4 years. The length of the cycle is hypothesized to be indirectly responsible for the fact that, worldwide, more marriages are dissolved in the fourth year than at any other time, and the pattern of marital breakups is consistent across cultures with hugely varying rates of divorce (Fisher, 1987). The theorized mechanism that evolved to keep parents together for the benefit of their offspring is romantic infatuation. It is effective in drawing and holding lovers together temporarily, but as a psychological state that fades as surely as the sun sets each day, it does not bind them to each other for longer than necessary.

The theory rests on the assumption that serial monogamy evolved because it enhanced the reproductive fitness of offspring beyond what could be achieved in the absence of paternal investment. Otherwise, a male strategy of impregnating as many females as possible would likely have won out in the competition. We question whether short-term involvement on the part of the father would have significantly improved the fitness of his offspring. Granted, a four-year-old could be more easily cared for by other group members than an infant not yet weaned and not yet mobile. But such a young child would still require a great deal of care in the form of monitoring, protection, assistance in obtaining food, some carrying, and so on, and would remain relatively vulnerable for several additional years. Beyond issues of survival, for humans to successfully mate, parent, and exist cooperatively in social groups, children must undergo extensive socialization, most of which they are not cognitively equipped to assimilate until they are older than 3 or 4 years of age (Piaget, 1952).

There is also the practical matter of a probable subsequent pregnancy. Nursing is reasonably effective in suppressing ovulation, but only so long as it occurs frequently and without long interruptions (Short, 1984). As infants mature and begin to consume solid foods, they naturally nurse less and less often. Thus, ovulation would typically resume before weaning. Assuming a couple continued to have sexual intercourse, a subsequent pregnancy would be likely, and, we might add, not evident for several months. If the infatuation by then had subsided, which it most surely would, the couple would separate shortly after having conceived. If the female then quickly became infatuated and began having sex with a new male, he would end up investing his resources, at least for a short while, in another man's progeny. Moreover, this type of recoupling increases the risk to offspring of abuse at the hands of a stepparent (Daly & Wilson, 1988). Finally, the degree of genetic relatedness among siblings would be expected to correspond to their motivation to promote one another's survival.

For a number of reasons, serial monogamy might not appear to provide sufficient improvement over serial copulation to have been chosen by natural selection to solve the problem posed by the extreme immaturity of human young. Whether something like infatuation has an important role to play in pair-bonding is an issue to which we return. We also return to the data on which the theory of serial monogamy is based—the universal phenomenon of marriages being most susceptible to breakup in their fourth year—and offer a slightly different explanation.

Sexual Jealousy

An alternative perspective on the adaptive problem of keeping reproductive partners together focuses on the various evolved strategies for retaining a mate. One mate-keeping tactic frequently used and judged to be effective is for each mate to continue supplying whatever relational provisions initially attracted the other (Buss, 1988). For males, this translates into continued provision of resources; for females, it means keeping themselves physically attractive. Other strategies include emotional manipulation of a mate (e.g., threatening to leave, inducing guilt), limiting a mate's access to rivals (e.g., not allowing the mate to socialize with attractive alternatives), and making it clear to potential rivals that a mate is already taken (e.g., publicly displaying affection).

We have evolved a special mechanism for alerting us that mate-keeping maneuvers are in order: sexual jealousy. Consistent with sex differences in mating preferences, men and women differ with respect to the situations that evoke a jealous reaction. Males are more bothered by a mate's real or imagined sexual infidelity, which calls paternity into question and raises the possibility of investing in another man's offspring. Females are more troubled by signs of a mate's emotional infidelity, which could be an indication that he plans to take his resources elsewhere (Buss, Larsen, Westen, & Semmelroth, 1992). Feelings of jealousy are emotional signals that corrective, mate-guarding action is needed.

The various evolved tactics and strategies for ensuring that a mate, once attracted, is retained are further indications that keeping a mate has important reproductive payoffs. If enduring bonds between reproductive partners did not enhance fitness, it is doubtful that humans would devote so much time and energy to maintaining them. Such mechanisms are clearly part of human nature, yet they tell us little about the nature of the bond whose continuance they serve. We propose that an understanding of the bond itself is not only an essential component of any model or theory of pair-bonds, but that it reveals a powerful additional mechanism for keeping mates together.

Pair-Bond Attachment

When the adaptive problem of exceptionally immature offspring arose in the course of human evolution, our species, by virtue of its altricial nature, already had available a well-designed, specialized, flexible, but reliable mechanism for ensuring that two individuals would be highly motivated to stay together and resist attempts to separate them; the mechanism was attachment. In light of the generally conservative tendencies of evolution and natural selection, it seems likely that this preexisting mechanism would have been exploited for the purpose of keeping reproductive partners together. Further evidence that this available mechanism was utilized can be found in the many similarities between infant-caregiver relationships and adult pair-bonds.

A Similar Dynamic.   The dynamics of the attachment mechanism were first evident to Bowlby in the reactions of infants and young children to being separated from their primary caregivers. The first extensions of attachment theory into the realm of adult relationships also began with separation and loss. A review of prior research (Gorer, 1973; Marris, 1958; Palgi, 1973) revealed that, despite tremendous cultural variation in associated customs, rituals, and attitudes, there is universal human response to the breaking of a pair-bond (Parkes & Weiss, 1983; Weiss, 1975).

The initial reaction, whether due to voluntary or involuntary separation, is anxiety, often to the point of panic. Individuals also report being preoccupied with thoughts of the lost partner and experiencing a strong compulsion to reestablish contact. The eventual realization that the loss cannot be recovered precipitates a period of deep sadness, during which intense activity and rumi-nation give way to lethargy and depression. Gradually, sadness subsides, and most people achieve a sufficient degree of emotional detachment to return to ordinary living. Thus, the sequence of reactions to the loss of a pair-bond relationship is essentially the same as the protest–despair–detachment se-quence observed in infants and children separated from their attachment figures (Bowlby, 1980; Hazan & Shaver, 1992; Parkes & Weiss, 1983). Even temporary separations from partners, such as those associated with military service and work-related travel, are at least mildly distressing for most adults (Vormbrock, 1993).

The similar dynamic suggests that the same mechanism is at work in both types of relationships. It also suggests that the bond, in both cases, serves a similar affect-regulation (i.e., security-promoting) function.

Similar Selection Criteria.   The attachment mechanism also seems to be implicated in mate preferences. Although sex differences in the relative impor-tance of such traits as physical beauty and social status are robust, neither trait is assigned highest priority by either sex. For both men and women, the most highly valued traits in a potential mate are "kind-understanding" and "intelli-gent" (Buss, 1989). Furthermore, being kind is judged, by men as well as women, to be the most effective strategy for retaining a mate (Buss, 1988). However, the most robust factor, by far, in mate selection is similarity. Although in evaluating the desirability of various traits in potential mates, we may express a desire for wealthy or attractive ones, we actually end up marrying individuals who are similar to us in numerous dimensions, including socioeconomic status and attractiveness (Berscheid, 1984; Hinsz, 1989; Rubin, 1973).

We see a connection between the attachment mechanism's sensitivity to familiarity and the overwhelming evidence for assortative mating on the basis of similarity. In the case of infants, attraction to a particular caregiver is enhanced through repeated exposure and interaction that serves to increase familiarity. In the case of adults, it is debatable whether preexisting similarities draw potential mates into the same activities and social circles, thereby increasing familiarity, or whether similarity creates a false sense of familiarity. (The word familiar comes from the Latin *familia*, which connotes family or household. Individuals who are similar may seem like family, and thus safe to approach.) At any rate, familiarity appears to be an important selection factor for adults as well as infants, and, desiring a partner who is kind, understanding, and intelligent is also reminiscent of the "preferences" evident in the "selection" of an attachment figure during infancy. Again, the similarities suggest to us that the same mechanism may be operating.

A Similar Process. When a potential mate is identified, most adults experience a complex of thoughts and feelings colloquially referred to as *infatuation* (Tennov, 1979). This state is characterized by heightened arousal, preoccupation with the target person, hypervigilance for signs of responsiveness, and a strong desire for close physical proximity. In addition, moods become somewhat dependent on the perceived reactions of the other: Responsiveness triggers elation, unresponsiveness evokes anxiety.

In adulthood, this combination of cognitions and emotions is directed exclusively toward individuals who are or have the potential to become sexual partners (Tennov, 1979), that is, one who is or could become the partner in a pair-bond relationship. In infancy, such emotions and behaviors are directed toward the primary caregiver. In both instances, such feelings are more characteristic of newly forming, or newly formed bonds than of well-established ones. As such, they characterize the initial phase in the process of bond formation. The cognitive and emotional features of this process in nascent pair-bonds are markedly similar to those known in infancy to result in an attachment (Hazan & Shaver, 1994; Shaver, Hazan, & Bradshaw, 1988).

Attachments can be distinguished from other types of social relationships in terms of four defining features: proximity seeking, safe haven, secure base, and separation distress. Recall that the accepted marker of attachment formation in infancy is a resistance to being separated from the primary caregiver and expressed distress when separations do occur. Relationships characterized by these four features, and especially separation distress, qualify as attachment bonds.

We developed measures of these components of attachment and administered them to a large group of individuals representing the age range from late adolescence to late adulthood (Hazan & Zeifman, 1994). We found that full-blown attachments (i.e., relationships in which all four components were present) were almost exclusively limited to bonds with parents or sexual partners. The results also revealed something about the time course of attachment formation. Pair-bond relationships tended not to contain the separation distress component unless the partners had been together for at least 2 years. Pair-bonds of shorter duration were characterized by the presence of only two components: proximity seeking and safe haven.

The results suggest that adult relationships that may eventually develop into attachment bonds begin with an extended period of interactions involving close physical proximity and mutual alleviation of distress. In infancy, the antecedents of attachment are the same. The fact that pair-bond relationships are typified by the defining features of attachment and seem to follow the same developmental course as attachment formation in infancy (in terms of the sequence in which various components are added), provides additional evidence that the attachment mechanism is at work in both types of relationships.

Physical Contact of a Similar Nature.   The proximity-seeking tendencies of the human infant appear to be largely motivated by fear and the desire for contact comfort. However, when two adults experience an intense urge to be in close physical proximity, sexual desire is likely what fuels it. Our hunch is that sexual attraction is primarily responsible for getting two adults into an attachment-promoting situation.

Earlier, we described the intimate nature of physical contact between infants and their primary caregivers. With the important exception of sex, the description applies equally well to lovers. Like caregivers and their infants, adult sexual partners (at least initially) spend a lot of time engaged in mutual gazing, cuddling, nuzzling, sucking, and kissing in the context of prolonged face-to-face, skin-to-skin, belly-to-belly contact, and the touching body parts otherwise considered private.

Although some forms of intimate contact may occur in isolation within other types of social relationships (e.g., kissing among friends), their joint occurrence in infancy is usually restricted to the infant–caregiver relationship. When this complex behavioral package later reemerges in adulthood, it typically is restricted to pair-bond relationships. The possibility that intimate physical contact plays a key role in attachment formation is bolstered by the finding that attachment bonds tend to develop only out of relationships in which such intimate contact is the norm (Hazan & Zeifman, 1994).

Prohibitions against sexual behavior in other social relationships are rampant in Western society and religion, and seem to be present in one form or another in all cultures (Alcock, 1989). The universal existence of prohibitions against sex outside of a recognized pair-bond (at least for females) has generally been attributed to the fact that extrapair copulations reduce confidence in paternity.

However, such restrictions may be owing, in part, to an implicit understanding that close physical contact with another could lead to an emotional bond that would jeopardize the primary one. Even in subcultures where extrarelationship sexual contact is permitted, efforts to avoid emotional involvement are common. For example, prostitutes commonly refuse to engage in kissing, nuzzling, and other forms of intimate face-to-face contact with their clients (Nass & Fisher, 1988). Gay male couples who consensually engage in extrapair sexual activities tend to reserve kissing and cuddling for the primary partner (Blumstein & Schwartz, 1983). Ground rules among so-called "swinging" heterosexual couples often forbid repeated sexual contact with the same person (O'Neill & O'Neill, 1972). A plausible interpretation of these findings is the implicit understanding that repeated and intimate physical contact naturally fosters attachment formation. If an emotional bond is not desired, special steps must be taken to protect against its formation.

The chemical basis for the effects of close physical contact may be the same for lovers and mother–infant pairs. Oxytocin, a substance released during suckling–nursing interactions, and thought to induce maternal caregiving, is also released at sexual climax and has been implicated in the cuddling that often follows sexual intercourse (i.e., "afterplay"; Carter, 1992). Cuddling, or contact comfort as was demonstrated by Harlow, is important in the establishment of emotional bonds.

Many unique features of human sexuality appear to have evolved to enhance the development and maintenance of emotional bonds between sexual partners (Morris, 1972; Symons, 1979). The most striking change in our reproductive physiology in comparison to other mammalian species is the loss of outward signs of estrus in the female. Most mammals mate only during the short estrus periods of the female, whereas human sexual desire is not limited to certain seasons or periods in the reproductive cycle. Women are sexually receptive during all phases of their cycle, despite the fact that conception is possible only during a small fraction of it (although desire does increase somewhat around the time of ovulation; Alcock, 1989). This physiological adaptation enables the couple to maintain a continuous tie on the basis of sexual reward (Eibl-Eibesfeldt, 1975).

The evolution of genital differences between us and our closest primate relatives also suggests the important role of sex in maintaining the integrity of the human pair-bond. It has been hypothesized that the large size of the human penis, in marked contrast to that of all other great apes, made possible a wide variety of copulatory positions (including more intimate face-to-face, mutually ventral positions) which may have served to enhance the sexual pleasure of both partners and increase the probability of female orgasm (Short, 1979). The ability of the human female to experience an orgasm comparable to that of the male may have served to increase the female's readiness to engage in sexual behavior and thereby strengthen emotional ties to her mate (Eibl-Eisbesfeldt, 1975).

Similar Effects on Overall Functioning. The notion that attachment is a very real biological need was established in studies of infants reared in orphanages and other institutional settings (Robertson, 1953; Spitz, 1946). Although adults are clearly less dependent on an attachment bond for basic survival, there is ample evidence that they incur health benefits from having one and suffer health decrements as a result of the loss or lack of an attachment relationship. Attachment disruption, especially through divorce, makes one more susceptible to everything from automobile accidents to alcohol abuse to admission into a psychiatric facility (Bloom, Asher, & White, 1978). In addition, the grieving and lonely are more vulnerable to cardiac disease (Lynch, 1977) and impaired immune functioning (Keicolt-Glaser et al., 1984), and at greater risk of death from cancer (Goodwin, Hurt, Key, & Sarret, 1987). Social losses jeopardize, not only health and happiness, but job performance and achievement as well (Baruch, Barnett, & Rivers, 1983; Lee & Kanungo, 1984; Vaillant, 1977).

The effects of attachment disruption on well-being and adjustment during all phases of development provide strong support for Bowlby's claim that the attachment system is functioning throughout life. Beyond the benefits accrued by offspring who have a stable home life and the care of two parents, a pair-bond relationship brings advantages to the adult partners as well. Through its stress and distress alleviation functions, the pair-bond relationship helps to keep adults healthy and actively engaged in the world during their reproductive, parenting, and grandparenting years.

The Biology of Adult Attachment. In the development of pair-bonds, two qualitatively different phases can be discerned: an attraction phase and an attachment phase (Mellen, 1981). The two phases are hypothesized to correspond to distinct neurotransmitter and limbic system activity (Liebowitz, 1983). Evidence suggests that the experience of infatuation is mediated by naturally occurring amphetaminelike substances, especially phenylethylamine (PEA). Like other stimulants, PEA produces heightened arousal, alertness, and activity. It may be implicated in the "high" often experienced in the early stages of pair-bond formation. It also has mild hallucinogenic effects, which may be partially responsible for the well-known tendency of new lovers to have unrealistically positive, almost idealized views of each other (Tennov, 1979).

Eventually, the brain habituates to the high PEA levels, at which time it is hypothesized that endogenous opiates—endorphins—take over and the so-called attachment phase begins. Endorphins have long been associated with the formation of affectional bonds between infant and caregiver (Panksepp, Siviy, & Normansell, 1985), and their release is accompanied by feelings of security and contentment, a sense that all is right with the world. In diverse species, opioid administration ameliorates the disorganizing effects of separation whereas opioid blockage exacerbates them. The effects of opioids on attachment-related emotions and behaviors suggest that they are implicated in the formation of such bonds.

In addition to alleviating anxiety, opioids are powerful conditioning agents. Through classical conditioning, stimuli paired with opioid drugs rapidly become associated with their calming effects, are strongly preferred, and such preferences are extremely difficult to extinguish (Panksepp et al., 1985). A typical intimate encounter between lovers involves an initial increase in stimulation and arousal, followed by relief, satisfaction, and calm. We argued previously that attachment, in adulthood as well as infancy, involves the conditioning of an individual's opioid system to the stimulus of a particular other as a result of repeated calm-inducing (i.e., opioid-releasing) interactions (Zeifman & Hazan, in press). The experience of anxiety and distress during separations from one's primary source of comfort is thus understandable.

As noted earlier, the theory of serial monogamy postulates infatuation as the mechanism that evolved to hold pair-bonds together for the approximate duration of the human reproductive cycle. We propose a slightly different model. Attachment was the selected mechanism because, to enhance reproductive fitness of offspring, parents needed to remain together for more than a few years. Infatuation fueled by sexual attraction is likely what initially draws them to each other. The point at which the attraction (infatuation) phase ends—approximately 2 to 3 years into the relationship, according to our own data (Hazan & Zeifman, 1994) and findings reported by Tennov (1979) and Money (1980)—is the time when a relationship either begins to deteriorate (possibly resulting in a fourth-year breakup) or is transformed into a more enduring attachment bond.

Benefits for Offspring and Parents

Enduring pair-bonds evolved because, at some point in human evolution, it became reproductively advantageous for parents to remain together. As we have already argued, it seems unlikely that short-term paternal investment of resources would have resulted in sufficient improvements in offspring fitness. Thus, it is doubtful that a short-term bonding mechanism, or serial monogamy, would have been selected for. The multitude of similarities between pair-bond and infant–caregiver relationships provide strong evidence that the already available attachment mechanism was exploited for the purpose of keeping mates together.

Numerous benefits accrue to the offspring of stable pair-bonds, including increased chances that they will survive to reproductive age and be successful in the competition for mates. But, the reproductive fitness of the parents is also enhanced. Benefits in the form of more robust physical and mental health and the stress-relieving effects of having a reliable and trusted companion have been noted. There is also evidence that a stable pair-bond enhances a female's reproductive value. For example, women ovulate more regularly if they are in a stable sexual relationship (Cutler, Preti, Huggins, Erickson, & Garcia, 1985; Veith, Buck, Getzlaf, Van Dalfsen, & Slade 1983). They are also more likely to continue ovulating into their middle years and reach menopause later if they continue to engage in regular sexual activity (Cutler et al., 1985), which, for middle-age women, probably happens most often within the context of a long-term relationship.

Changes, of course, occur over the life of a pair-bond. One prominent and universal change is in partners' interest in and frequency of sexual activity. But even in couples who have reached an age when the female's reproductive potential is near zero, sexual intercourse still occurs an average of once per week (Udry, 1979). It appears that sexual exchanges have benefits beyond their reproductive value. Although the frequency of sexual interactions declines predictably, and its role in overall relationship satisfaction becomes less central (Reedy, Birren, & Schaie, 1981), it continues to serve the function of maintaining the emotional connection between partners.

Gradually, the arousal and passion of the attraction phase is replaced by calm and compassion. The decline in sexual desire and activity could signal a loss of interest in the relationship. Alternatively, it could be a sign that an attachment bond, which will go a long way toward keeping the couple together, is firmly established (Berscheid, 1983). The disruptive effects of separations and the strong emotions they evoke are an indication that the attachment behavioral system, though usually quiescent, is in place and still functional. The strong desire for frequent and prolonged physical contact that typifies the preattachment phase of a developing relationship fosters the formation of an attachment bond and would be expected to diminish once the bond had formed. Both human and nonhuman primate young show decreasing proximity- and contact-seeking behavior once a clear-cut attachment bond has been established, but continue to exhibit attachment behavior when stressed or frightened, and in response to prolonged involuntary separations (Bowlby, 1969; Suomi, 1990).

Despite their power to bind individuals, attachment bonds can be and, especially in adulthood, are sometimes severed voluntarily, but rarely without emotional pain. A mechanism that evolved to ensure an enduring bond would not be very effective or reliable if individuals could break it easily and without personal consequence. The currently high rate of divorce in some societies may overestimate the frequency of voluntary attachment dissolution. The world census data, on which serial monogamy theory is founded (Fisher, 1992), indicate that the probability of marital breakup increases steadily throughout the early years of marriage, peaking in the fourth. In light of the data on the duration of the attraction phase and the time course of attachment formation, it seems fair to conclude that many of these broken marriages may not represent the severing of attachment bonds. Attachments take time to form, and the dissatisfaction that presumably precedes marital dissolution may preclude their development. Consistent with this interpretation, Weiss (1988) found that widows and widowers married for less than 2 years did not show the same sequence of reactions as those grieving the loss of longer-term bonds.

The various mate-keeping tactics that have been identified, along with the mechanism of sexual jealousy, have proven effective in helping to maintain pair-bonds. They constitute individual strategies for keeping a valued relationship going. Attachment, in contrast, represents an emergent structure: a relational bond. It is a structure that transcends the boundaries of individuals. People may select mates on the basis of conscious or even unconscious calculations about such qualities as physical attractiveness or earning potential, but once they begin to engage in frequent and repeated intimate interactions, they may get more than they initially bargained for.

It makes good evolutionary sense that men would be concerned about the reproductive value of a potential mate and that women would be very selective in the high-risk, high-cost arena of sexual relations. It makes good sense, not only because securing these types of relational provisions improves their chances of producing reproductively fit offspring, but also because, through no deliberate intention or fault of their own, they may end up spending a very long time with that particular mate. Although human males and females differ in what it takes to get them to engage in sexual intercourse, once they do, circumstances are ripe for another evolved mechanism, attachment, to take over (Hazan & Zeifman, 1994).

INDIVIDUAL DIFFERENCES IN ATTACHMENT AND REPRODUCTIVE STRATEGIES

Our emphasis so far has been on the normative aspects of attachment: the human tendency to form attachment bonds, their predictable dynamics and developmental course, their effect on overall functioning, and so on. We have not yet addressed the vast differences among individuals in the ability or desire to establish and sustain attachment relationships. Bowlby's ethological attachment theory provides a lexical tool for classifying related individual differences and an explanation of their ontogeny. To date, differences in individuals' styles of relating to attachment figures have been the focus of most research and theorizing in the areas of adult and infant attachment.

Central to Bowlby's theory were hypotheses about how the evolved attachment mechanism would respond and adapt to variations in environmental input. Caregiver responses to the activation of their infant's attachment system, and the associated bids for proximity and contact comfort, influence the set-goal of the system, as well as the emotional triggers that activate it. Through repeated social exchanges, infants learn whether caregivers can be counted on to respond kindly, promptly, and effectively to their signals of distress. These experiences eventually form the basis of a mental model that guides future attachment-related behavior. Mental models of attachment, like all schemata, can be modified but tend toward stability, particularly if subsequent experiences are model consistent (Bowlby, 1979).

Bowlby's predictions concerning individual differences have been empirically tested and confirmed by scores of researchers (for reviews see Belsky & Nezworski, 1988; Greenberg, Cicchetti, & Cummings, 1990; Parkes, Stevenson-Hinde, & Marris, 1991). Three major patterns of infant–caregiver attachment were originally identified by Ainsworth and her colleagues (Ainsworth et al., 1978). Cross-culturally, the majority of infants display a pattern of behavior in relation to their primary caregiver that corresponds to what Bowlby considered the ideal. The caregiver responds consistently, warmly, promptly, and appropriately to the infant's distress signals. The infant seeks contact when distressed, is readily soothed by it, and engages in active environmental exploration in the safety of the caregiver's presence. Ainsworth et al. labeled this pattern of attachment *secure*.

In contrast, inconsistent responsiveness on the part of the caregiver has the effect of intensifying the infant's attachment behavior, which interferes with exploration. This pattern was labeled *insecure–ambivalent* because of its associated intermingling of desperate contact-seeking and angry resistance to soothing. When a caregiver consistently rebuffs an infant's bids for comfort, the result is often *insecure–avoidant* attachment, which is characterized by active avoidance of the caregiver in situations in which proximity would normally be sought, and corresponding exploratory behavior that appears compulsive or defensive.

Each of these patterns of attachment represents a reasonable adaptation to a particular caregiving environment. It is adaptive to rely on a caregiver who has proven dependable (secure attachment). It is equally adaptive to steer clear of a rejecting caregiver (avoidance), or to turn up the volume of demands on, and also be reluctant to venture far from, a caregiver who responds unpredictably (ambivalence). Similar patterns of interpersonal relating have been identified in adulthood and have proven useful for organizing individual differences in the way adults think, feel, and behave in established and potential pair-bond relationships (e.g., Collins & Read, 1990; Feeney & Noller, 1990; Hazan & Shaver, 1987; Kirkpatrick & Davis, 1993; Kobak & Hazan, 1991; Main, Kaplan, & Cassidy, 1985; Simpson, 1990; Simpson, Rholes, & Nelligan, 1992).

The Effects of Attachment Experiences on Reproductive Behavior

It is well-documented that biological sex plays an important role in determining the reproductive strategies that individuals will employ. There is mounting evidence that attachment-related experiences, even those that predate sexual maturation, can also have a profound effect on mating behavior. The evidence comes from studies of both human and nonhuman primates.

For example, rhesus monkeys deprived of an opportunity to form attachments during infancy developed gross abnormalities in their sexual behavior as adults (Harlow & Harlow, 1965). Although they were reproductively normal in the physiological sense (e.g., sperm production, estrus cycles), they showed severe deficits in the display and sequencing of the motor movements involved in copulation. When females deprived of early attachments were inseminated by artificial means (because they were sexually incompetent and incapable of being impregnated by natural means), they demonstrated highly abnormal and even violent caregiving responses toward their newborn young (Harlow & Harlow, 1962).

For obvious ethical reasons, there is no experimental evidence for the effects of early attachment experiences on human sexual or caregiving behavior, but there are data that bear on the issue. For example, according to retrospective accounts, dysfunctional early attachment relationships are a common precursor of adulthood sexual deviance (Sroufe & Fleeson, 1986). A correlation has also been found between retrospective reports of maternal rejection during childhood and adult sexual promiscuity (Brennan, Shaver, & Tobey, 1991). Relative to their insecure counterparts, securely attached adults are more skilled at providing care to their offspring (Main et al., 1985), more responsive to their infants' distress (Main, 1990), and more caring and supportive toward their adult partners (Kunce & Shaver, 1994).

Draper and Belsky (1990; see also Belsky, Steinberg, & Draper, 1991) proposed a more specific model of the influence of early attachment experiences on subsequent reproductive strategies. They apply the evolutionary concept of r- and K-selection to human mating (Pianka, 1970). The terms refer to the partitioning of an organism's effort between mating and reproduction. R-selected organisms tend to evolve in unstable environments in which rapid change and catastrophic events that decimate populations can be overcome only through a high rate of fertilization. They are characterized by rapid maturation, reduced parental care, and shortened life span. K-selected organisms usually evolve in more stable, densely populated environments. In such stable but competitive ecologies, lower fertility, delayed maturation, and greater parental care result in more efficient exploitation of resources.

The concept of r- versus K-selection has been used most often in reference to variability between species, but it may also apply to variability within the same species in terms of rates of development and reproductive strategies. According to Draper and Belsky, insecure attachment in childhood might predispose an individual toward an r-selected rather than a K-selected strategy. Based on the concept of mental models of attachment, they reason that securely attached children will anticipate enduring, trusting, and mutually rewarding close relationships with romantic–sexual partners. Insecurely attached children, in contrast, will come to regard close relationships as transient, opportunistic, and untrustworthy. These perceptions could result in drastically different mating and childrearing practices. Specifically, insecure attachment should predispose individuals to adopt short-term, r-selected strategies as a result of their belief that partners cannot be counted on for the long haul.

The ways in which early rearing environments, and attachment relationships in particular, can influence reproductive strategies are manifold. There is compelling evidence, for example, that the presence or absence of an enduring bond between parents is related to the rate of sexual maturation and preferred reproductive strategies of adolescent offspring. Draper and Harpending (1982) documented a link between father-absence, antagonistic attitudes toward the opposite sex, and precocious sexuality. Adolescents reared under father-present conditions, in contrast, are more likely to delay the onset of sexual activity and have more positive attitudes toward the opposite sex. Draper and Harpending reasoned that the two behavior patterns reveal greater or lesser interest in developing a stable pair-bond. They propose a developmental model in which there is a sensitive period for acquiring reproductive strategies. The mother's pair-bond status acts as the crucial switch in determining which strategy an individual will ultimately choose.

The word *choose* is not meant to imply conscious decision, and it is unclear whether alternative strategies are the result of psychological mechanisms or physiological ones. Girls in father-absent homes reach menarche earlier than those in father-present homes (Jones, Leeton, McLead, & Wood, 1972; Moffitt, Caspi, & Belsky, 1990), and the younger the girl when her father leaves home, the earlier her menarche (Surbey, 1990). One possible explanation for this effect is that the presence of related or familiar adult males delays puberty through a chemical signal or pheromone. It is also the case that the presence of unfamiliar males accelerates puberty. Father-absent girls who had stepfathers before menarche tended to mature earlier than those who did not have stepfathers (Surbey, 1990).

Attachment theory provides yet another possible explanation for the phenomenon of insecure attachment leading to earlier and more promiscuous sexual behavior. If attachment constitutes a biologically based and irrepressible need, and the need is not being met by parents, individuals may attempt to satisfy it elsewhere. Where could one better obtain the longed-for contact comfort than in a romantic relationship? In a study of adolescents, we found that full-blown attachments to peers were rare in the group as a whole, but quite common among those who were insecurely attached to their parents (Hazan & Zeifman, 1994).

A separate study suggested an additional link between attachment and reproductive strategies. We found attachment-related differences in the types of physical contact that individuals seek and enjoy or try to avoid (Hazan, Zeifman, & Middleton, 1994). For example, avoidant adults reported enjoying purely sexual contact (e.g., oral and anal sex), but found more emotionally intimate contact (e.g., kissing, cuddling, nuzzling) to be aversive. Ambivalents reported the opposite pattern of preferences and viewed sexual activity primarily as a means for gratifying intimacy and comfort needs. The secure group found pleasure in both types of physical contact, especially within the context of an ongoing relationship (as opposed to one-night stands, for example).

The physical contact preferences associated with each attachment style may help to explain their relative rates of success when it comes to establishing an enduring pair-bond relationship. The secure group, who are relatively more successful and significantly less likely to experience relationship dissolution (Hazan & Shaver, 1987; Kirkpatrick & Hazan, 1994), engage in the kinds of physical contact necessary to satisfy their own and their partners' sexual and attachment needs. Avoidants, in contrast, have an expressed aversion to the type of contact thought to foster attachment. The preferences of ambivalents suggest that they have somewhat excessive attachment needs and a marked lack of interest in sex, both of which could jeopardize their relationships. Again, insecure attachment appears to intefere with the development of stable pairbonds.

SUMMARY AND CONCLUSIONS

There is abundant and compelling evidence that stable pair-bonds are regulated by the same psychological mechanism that originally evolved to ensure that human infants would develop a strong and enduring attachment to a caregiver. The primary sources of evidence are the many fundamental similarities between infant–caregiver attachments and stable pair-bond relationships. Both qualify as a unique type of bond, distinguishable from other forms of social connectedness by virtue of their common and distinctive components and security-promoting function. Both are typified by a similar and qualitatively distinct type of physical contact. The time course and processes by which each type of relationship develops are essentially the same. The effects of both on overall physical and psychological well-being are profound. In short, there seems to be little room for doubt that the same mechanism that evolved to tie infants to their caregivers was exploited by natural selection for keeping adult partners together.

Reproductive fitness is achieved through solutions to the three adaptive challenges: surviving to reproductive age, successfully mating, and providing adequate care to offspring to ensure that they, too, reach sexual maturation and succeed in mating. It seems clear that the attachment mechanism was the selected solution to the first problem of survival through childhood. But it also plays a major role in the problem of mate selection.

The goal of mating is to identify and attract someone who is fit to serve as a reproductive partner. For both men and women, it forces consideration of a potential mate's suitability as an attachment figure and, consequently, a valuing of such qualities as kindness and intelligence. The attachment mechanism also supplies a solution to the third adaptive problem of ensuring the reproductive fitness of offspring. By helping to maintain the pair-bond, it increases the probability of long-term paternal investment and the kind of family stability that gives progeny an edge in the competition for mates and the skills needed to retain one. In these ways, the attachment mechanism directly and indirectly enhances the reproductive fitness of offspring. If pair-bond members fail in their parenting roles by providing insufficient stability and security, they risk producing progeny who are ill-equipped to meet the challenges of mate competition and retention.

Attachment is not a tactic that individuals employ to keep their mates, but rather an emergent property of their interactions. It is the bond that holds them together, that resides not only in the partners, but in the relationship between partners. From birth, humans are innately predisposed to seek out, develop, and maintain such bonds. This inclination and the need it fulfills does not end in infancy or childhood, but is constant throughout life. Once such a bond is established, the partner assumes a role of unique importance. In our beginning example, and in the case of the infants and children whose separation distress provided the original inspiration for attachment theory, the power to relieve the longing rests with one individual: the attachment figure.

Two recent and otherwise comprehensive treatments of the emerging field of evolutionary psychology (Buss, 1994; Wright, 1994) neglect to include attachment among the basic features of human nature. In our view, no theory of human nature or pair-bonds will be complete without it.

ACKNOWLEDGMENTS

We are grateful to Richard Canfield, Doug Kenrick, and Jeff Simpson for their thoughtful comments on earlier drafts of this chapter.

REFERENCES

Ainsworth, M. D. S., Blehar, M. C., Waters, E., & Wall, S. (1978). *Patterns of attachment: Assessed in the strange situation and at home.* Hillsdale, NJ: Lawrence Erlbaum Associates.

Alcock, J. (1989). *Animal behavior: An evolutionary approach.* Boston: Sinauer.

Baruch, G., Barnett, R., & Rivers, C. (1983). *Lifeprints: New patterns of love and work for today's women.* New York: McGraw-Hill.

Belsky, J., & Nezworski, T. (Eds.). (1988). *Clinical implications of attachment theory.* Hillsdale, NJ: Lawrence Erlbaum Associates.

Belsky, J., Steinberg, L., & Draper, P. (1991). Childhood experience, interpersonal development, and reproductive straegy: An evolutionary theory of socialization. *Child Development, 62,* 647–670.

Berscheid, E. (1983). Emotion. In H. H. Kelley, E. Berscheid, A. Christinsen, J. H. Harvey, T. Huston, G. Levinger, E. Mc Clintock, L. A. Peplau, & D. R. Peterson (Eds.), *Close relationships* (pp. 110–168). New York: W. H. Freeman.

Berscheid, E. (1984). Interpersonal attraction. In G. Lindzey & E. Aronson (Eds.), *Handbook of social psychology* (3rd ed., pp. 413–484). Reading, MA: Addison-Wesley.

Bloom, B. L., Asher, S. J., & White, S. W. (1978). Marital disruption as a stressor: A review and analysis. *Psychological Bulletin, 85,* 867–894.

Blumstein, P., & Schwartz, P. (1983). *American couples: Money, work and sex.* New York: Morrow.

Bowlby, J. (1944). Forty-four juvenile thieves: Their characters and home life. *International Journal of Psychoanalysis, 25,* 19–52.

Bowlby, J. (1958). The nature of the child's tie to his mother. *International Journal of Psychoanalysis, 39,* 350–373.

Bowlby, J. (1969). *Attachment.* New York: Basic Books.

Bowlby, J. (1973). *Attachment and loss, Vol. II. Separation: Anxiety and anger.* New York: Basic Books.

Bowlby, J. (1979). *The making and breaking of affectional bonds.* London: Tavistock Publications.

Bowlby, J. (1980). *Attachment and loss, Vol. III. Loss: Sadness and depression.* New York: Basic Books.

Bowlby, J. (1988). *A secure base: Parent–child attachment and healthy human development.* New York: Basic Books.

Brennan, K. A., Shaver, P. R., & Tobey, A. E. (1991). Attachment styles, gender, and parental problem drinking. *Journal of Personal and Social Relationships, 8,* 451–466.

Buss, D. M. (1988). From vigilance to violence: Tactics of mate retention. *Ethology and Sociobiology, 9,* 291–317.

Buss, D. M. (1989). Sex differences in human mate preferences: Evolutionary hypotheses tested in 37 cultures. *Behavioral and Brain Sciences, 12,* 1–49.

Buss, D. M., (1994). *The evolution of desire: Strategies of human mating.* New York: Basic Books.

Buss, D. M., Larsen, R., Westen, D., & Semmelroth, J. (1992). Sex differences in jealousy: Evolution, physiology, and psychology. *Psychological Science, 3,* 251–255.

Buss, D. M. & Schmitt, D. P. (1993). Sexual strategies theory: A contexual evolutionary analysis of human mating. *Psychological Review, 100,* 204–232.

Carter, C. S. (1992). Oxytocin and sexual behavior. *Neuroscience and Biobehavioral Reviews, 16,* 131–144.

Cutler, W. B., Preti, G., Huggins, G. R., Erickson, B., & Garcia, C. R. (1985). Sexual behavior frequency and biphasic ovulatory type menstrual cycles. *Physiology and Behavior, 34,* 805–810.

Collins, N. L., & Read, S.J. (1990). Adult attachment, working models and relationship quality in dating couples. *Journal of Personality and Social Psychology, 58,* 644–663.

Daly, M., & Wilson, M. (1983). *Sex, evolution, and behavior* (2nd ed.). Boston: Willard Grant.

Daly, M., & Wilson, M. (1988). *Homicide.* New York: Aldine de Gruyter.

Draper, P., & Belsky, J. (1990). Personality development in evolutionary perspective. *Journal of Personality, 58,* 141–163.

Draper, P., & Harpending, H. (1982). Father absence and reproductive strategy: An evolutionary perspective. *Journal of Anthropological Research, 38,* 255–273.

Eibl-Eibesfeldt, I. (1975). *Ethology: The biology of behavior.* New York: Holt, Rinehart & Winston.

Eibl-Eibesfeldt, I. (1989). *Human ethology.* New York: Aldine de Gruyter.

Feeney, J. A., & Noller, P. (1990). Attachment style as a predictor of adult romantic relationships. *Journal of Personality and Social Psychology, 58,* 281–291.

Fisher, H. E. (1987). The four-year itch: Do divorce patterns reflect our evolutionary heritage? *Natural History, 96*(10), 22–33.

Fisher, H. E. (1992) *Anatomy of love: The natural history of monogamy, adultery, and divorce.* New York: W.W. Norton.

Goodwin, J. H., Hurt, W. C., Key, C. R., & Sarret, J. M. (1987). The effect of marital status on stage, treatment, and survival of cancer patients. *Journal of the American Medical Association, 258,* 3125–3130.

Gorer, G. (1973). Death, grief, and mourning in Britain. In E. J. Anthony & C. Koupernik (Eds.), *The child in his family: The impact of death and disease* Interscience.

Greenberg, M., Cicchetti, D., & Cummings, M., (Eds.). (1990). *Attachment in the preschool years: Theory, research, and intervention.* Chicago: University of Chicago Press.

Harlow, H. (1958). The nature of love. *American Psychologist, 13,* 673–685.

Harlow, H., & Harlow, M. K. (1962). The effects of rearing conditions on behavior. *Bulletin of the Menninger Clinic, 26,* 213–224.

Harlow, H., & Harlow, M. K. (1965). The affectional systems. In A. M. Schrier, H. R. Harlow, & F. Stollnitz (Eds.), *Behavior of nonhuman primates* (Vol. 2, pp. 287–334). New York: Academic Press.

Hazan, C., & Shaver, P. R. (1987). Romantic love conceptualized as an attachment process. *Journal of Personality and Social Psychology, 52,* 511–524.

Hazan, C., & Shaver, P. R. (1992). Broken attachments. In T. L. Orbuch (Ed.), *Close relationship loss: Theoretical approaches.*

Hazan, C., & Shaver, P. R. (1994). Attachment as an organizational framework for research on close relationships. *Psychological Inquiry, 5,* 1–22.

Hazan, C., & Zeifman, D. (1994). Sex and the psychological tether. *Advances in Personal Relationships, 5,* 151–177.

Hazan, C., Zeifman, D., & Middleton, K. (1994, July). *Attachment and sexuality.* Paper presented at the 7th International Conference on Personal Relationships, Groningen, The Netherlands.

Hinsz, V. B. (1989). Facial resemblance in engaged and married couples. *Journal of Social and Personal Relationships, 6,* 223–229.

Jones, B., Leeton, J., McLeod, I., & Wood, C. (1972). Factors influencing the age of menarche in a lower socioeconomic group in Melbourne. *Medical Journal of Australia, 2,* 533–535.

Kagan, J. (1971). *Change and continuity in infancy.* New York: Wiley.

Kiecolt-Glaser, J. K., Garner, W., Speicher, C., Penn, G. M., Holliday, J., & Glaser, R. (1984). Psychological modifiers of immunocompetence in medical students. *Psychosomatic Medicine, 46,* 7–14.

Kirkpatrick, L. A., & Davis, K. E. (1993). Attachment style, gender, and relationship stability: A longitudinal analysis. *Journal of Personality and Social Psychology, 66*, 502–512.

Kirkpatrick, L. A., & Hazan, C. (1994). Attachment styles and close relationships: A four year prospective study. *Personal Relationships, 1*, 123–142.

Koback, R., & Hazan, C. (1991). Attachment in marriage: The effects of security and accuracy of working models. *Journal of Personality and Social Psychology, 60*, 861–869.

Kunce, L. J., & Shaver, P. R. (1994). An attachment-theoretical approach to caregiving. *Advances in personal relationships, 5*, 205–237.

Lee, M. D., & Kanungo, R. N. (Eds.). (1984). *Management of work and personal life: Problems and opportunities*. New York: Praeger.

Liebowitz, M. (1983). *The chemistry of love*. New York: Berkeley Books.

Lorenz, K. (1970). *Studies in animal and human behavior*. Cambridge, MA: Harvard University Press.

Lynch, J. J. (1977). *The broken heart: The medical consequences of loneliness*. New York: Basic Books.

Main, M. (1990). Parental aversion to infant–initiated contact is correlated with the parent's own rejection during childhood: The effects of experience on signals of security with respect to attachment. In K. E. Barnard & T. B. Brazelton (Eds.), *Touch: The foundation of experience* (pp. 461–495). Madison, CT: International Universities Press.

Main, M., Kaplan, N., & Cassidy, J. (1985). Security in infancy, childhood, and adulthood: A move to the level of representation. *Monographs of the Society for Research in Child Development, 50* (1–2), 66–104.

Marris, P. (1958). *Widows and their families*. London: Routledge & Kegan Paul.

Mellen, S. L. W. (1981). *The evolution of love*. Oxford, England: Freeman.

Miller, G. A., Gallanter, E., & Pribram, K. H. (1960). *Plans and the structure of behavior*. New York: Holt, Rinehart, & Winston.

Moffitt, T., Caspi, A., & Belsky, J. (1990, March). *Family context, girls' behavior, and the onset of puberty: A test of a sociobiological model*. Paper presented at the biennial meetings of the Society for Research in Adolescence, Atlanta, GA.

Money, J. (1980). *Love and love sickness: The science of sex, gender differences, and pair-bonding*. Baltimore: Johns Hopkins University Press.

Morris, D. (1972). *Intimate behavior*. New York: Random House.

Nass, G. D., & Fisher, M. P. (1988). *Sexuality today*. Boston: Jones & Bartlett.

O'Neill, N., & O'Neill, G. (1972). *Open marriage: A new lifestyle for couples*. New York: Evans.

Palgi, P. (1973). The socio-cultural expressions and implications of death, mourning and bereavement arising out of the war situation in Israel. *Israel Annals of Psychiatry, 11*, 301–329.

Panksepp, J., Siviy, S. M., & Normansell, L. A. (1985). Brain opioids and social emotions. In M. Reite & T. Field (Eds.), *The psychobiology of attachment and separation* (pp. 3–50). London: Academic Press.

Parkes, C. M., Stevenson-Hinde, J., & Marris, P. (1991). *Attachment across the life cycle*. London: Routlege.

Parkes, C. M., & Weiss, R. S. (1983). *Recovery from bereavement*. New York: Basic Books.

Piaget, J. (1952). *The origins of intelligence in children*. (M. Cook Trans.). New York: International Universities Press.

Pianka, E.R. (1970). On r- and K-selection. *American Naturalist, 104*, 592–597.

Reedy, M. N., Birren, J. E. & Schaie, K. W. (1981). Age and sex differences in satisfying love relationships across the adult life span. *Human Development, 24*, 52–66.

Robertson, J. (1953). Some responses of young children to the loss of maternal care. *Nursing Times, 49*, 382–386.

Rubin, Z. (1973). *Liking and loving: An invitation to social psychology*. New York: Holt, Rinehart, & Wilson.

Schaffer, H. R. (1971). *The growth of sociability*. Baltimore: Penguin Books.

Shaver, P. R., Hazan, C., & Bradshaw, D. (1988). Love as attachment: The integration of three behavioral systems. In R. J. Sternberg & M. L. Barnes (Eds.), *The psychology of love* (pp. 68–99). New Haven, CT: Yale University Press.

Short, R. V. (1979). Sexual selection and its component parts: Somatic and genital selection as illustrated in man and the great apes. *Advances in the Study of Behavior, 9*, 131–155.

Short, R. V. (1984). Breast feeding. *Scientic American, 250*, 23–29.

Simpson, J. A. (1990). The influence of attachment styles on romantic relationships. *Journal of Personality and Social Psychology, 59*, 971–980.

Simpson, J. A., Rholes, W. S., & Nelligan, J. S. (1992). Support-seeking and support-giving within couple members in an anxiety-provoking situation: The role of attachment styles. *Journal of Personality and Social Psychology, 62*, 434–446.

Spitz, R. A. (1946). Anaclitic depression. *Psychoanalytic Study of the Child, 2*, 313–342.

Sroufe, L. A., & Fleeson, J. (1986). Attachment and the construction of relationships. In W. W. Hartup & Z. Rubin (Eds.) *Relationships and development*. Hillsdale, NJ: Lawrence Erlbaum Associates.

Sroufe, L. A., & Waters, E. (1977). Attachment as an organizational construct. *Child Development, 48*, 1184–1199.

Suomi, S. J.(1990). The role of tactile contact in rhesus monkey social development. In K. E. Barnard & T. B. Brazelton (Eds.), *Touch: The foundation of experience* (pp. 129–164). Madison, CT: International Universities Press.

Surbey, M. K. (1990). Family composition, stress, and the timing of human menarche. *Socioendocrinology of primate reproduction* (pp. 11–32). New York: Wiley-Liss, Inc.

Symons, D. (1979). *The evolution of human sexuality*. New york: Oxford University Press.

Tennov, D. (1979). *Love and limerence: The experience of being in love*. New York: Stein & Day.

Trevathan, W. (1987). *Human birth*. New York: Aldine De Gruyter.

Trivers, R. (1972). Parental investment and sexual selection. In B. Campbell (Ed.), *Sexual selection and the descent of man, 1871–1971* (pp. 136–179). Chicago: Aldine.

Udry, J. R. (1979). Age at menarche, at first intercourse and at first pregnancy. *Journal of Biosocial Science, 11*, 433–441.

Vaillant, G. E. (1977). *Adaptation to life: How the best and the brightest came of age.* Boston, MA: Little, Brown.

Veith, J. L., Buck, M., Getzlaf, S., Van Dalfsen, P., & Slade, S. (1983). Exposure to men influences the occurrence of ovulation in women. *Physiology and Behavior, 31,* 313–315.

Vormbrock, J. K. (1993). Attachment theory as applied to war-time and job-related marital separation. *Psychological Bulletin, 114,* 122–144.

Weiss, R. S., (1975). *Marital separation.* New York: Basic Books.

Weiss, R. S. (1988). Loss and recovery. *Journal of Social Issues, 44,* 37–52.

Wright, R. (1994). *The moral animal: The new science of evolutionary psychology.* New York: Pantheon Books.

Young, J. Z. (1964). *A model of the brain.* London: Oxford University Press.

Zeifman, D., & Hazan, C. (in press). A process model of adult attachment formation. In S. Duck & W. Ickes (Eds.), *Handbook of personal relationships* (2nd ed.).

Zimmerman, S. J., Maude, M. B., & Moldawar, M. (1965). Frequent ejaculation and total sperm count, motility and forms in humans. *Fertility and Sterility, 16,* 342–345.

V

Kinship and Social Relations

10

Kinship: The Conceptual Hole in Psychological Studies of Social Cognition and Close Relationships

Martin Daly
Catherine Salmon
Margo Wilson
McMaster University

Kinship has been the central construct in evolutionary biological analyses of social phenomena since Hamilton (1964) extended the concept of Darwinian fitness (personal reproductive success) to encompass the actor's effects on the expected reproduction of collateral as well as descendant kin ("inclusive fitness"). Hamilton's theory replaced the classical Darwinian conception of organisms as evolved "reproductive strategists" with the more subtle notion that they have evolved to be "nepotistic strategists," instead. If each of a behaving animal's genes is just as likely to be duplicated in a sister as in a daughter, for example, then the evolution of sororal beneficence is not, in principle, more paradoxical than the evolution of maternal beneficence. No development in this century has more pervasively and fundamentally affected biologists' understandings of social influence and interaction.

Kinship has attained a central position in anthropological analyses of social phenomena as well. Here, the centrality of kinship emerged much earlier, not for theoretical reasons but because of its inescapable centrality in the lives and minds of human subjects. In the words of Edmund Leach (1966), "Human beings, wherever we meet them, display an almost obsessional interest in matters of sex and kinship" (p. 41). Of course, we grant that one may reasonably question whether the "kinship" to which Leach refers is quite the same thing as evolutionary biology's central construct of genetic relatedness. But they are, at the very least, sibling constructs, grounded in sexual reproduction and genealogical descent. Just how anthropology's and biology's kinship concepts are related to one another is another topic that we consider further.

Because kinship is so important both theoretically and phenomenologically, one might suppose that it would have attained a central position in social psychology, too. Remarkably, it has been virtually ignored. Human social life is dominated by interactions with relatives and acquaintances, but textbooks of social psychology are almost exclusively concerned with studies of stranger interactions. This seems a scandalous situation, especially when we consider that neither the textbook writers nor the experimentalists have offered any particular justification for this narrow focus. (But see Moghaddam, Taylor & Wright, 1993, who have lodged complaints that partly parallel our own.) We suspect that this narrow focus has arisen primarily as a result of the convenience of a captive subject pool of freshman psychology students who are strangers to one another, and secondarily because real social relationships are messy sources of just the sorts of complex "noise" that good experimental psychologists strive to eliminate. In any event, whatever the reason, the social psychology of even North American kinship has received more attention from anthropologists (e.g. Schneider and Homans, 1955) and, especially, from sociologists (e.g. Adams, 1968; Harris, 1970; Parsons, 1943) than from social psychologists themselves.

Since about 1980, social psychology's restrictive focus on stranger interactions has been somewhat alleviated, as the study of close relationships has become a significant subfield. However, research under this rubric has hitherto been largely confined to dating and marital relationships, with a secondary emphasis on friendships. (See, for example, Berscheid's, 1994, review of almost 300 recent references on the psychology of interpersonal relationships, in which familial relationships are conspicuous by their absence.) It is only very recently that social psychology journals have begun to include research, explicitly inspired by Hamilton's inclusive fitness theory, demonstrating the relevance of genealogical relatedness to human cooperation and altruism (Burnstein, Crandall & Kitayama, 1994; Petrinovich, O'Neill, & Jorgensen, 1993).

In this chapter, we introduce Hamilton's theory in a little more detail and address some common misconceptions about it. We then argue for the existence of a relationship-specific kinship psychology, in which specialized motivational and information processing devices cope with the peculiar demands of being a mother, a father, an offspring, a sibling, a grandparent, or a mate. Although human kinship systems exhibit interesting cross-cultural diversity, they share many universal features, several of which we proceed to discuss in light of evolutionary models such as Hamilton's.

What has yet to be fully appreciated by most psychologists who lack an evolutionary perspective is that sexual partnership, friendship, parenthood, and so forth, are qualitatively distinct kinds of close relationships that differ in many specific ways other than just in their degrees of intimacy. The attributes of an ideal mate, for example, are quite different from those of an ideal sibling or friend, and it is clear that the human mind processes information about these different sorts of intimates in different, specialized ways (Kenrick, Sadalla, & Keefe, in press; Krebs & Denton, chapter 2, this volume; Symons, 1995).

Even the several fundamental sorts of close genetic relationships require distinct analyses. Before addressing their differences, however, we must first consider what these kin relationships share that distinguishes them from other close relationships. Why and how is it that "blood is thicker than water"?

EVOLVED NEPOTISTS

According to Hamilton's (1964) analysis of "the genetical evolution of social behaviour," the ultimate arbiter of the evolutionary fate of a potentially heritable novel trait is its impact on the inclusive fitness of individuals who possess the trait. This inclusive fitness effect is the sum of the trait's effects, by any and all causal chains, on the survival and reproduction of the focal individual (its "direct" fitness effects, in Brown's 1975 terminology) plus whatever effects it may have on the survival and reproduction of the focal individual's relatives, weighted by the closeness of relationship ("indirect" fitness effects). Personal reproduction is, in a sense, just one form of kin-directed altruism—one of the ways in which an individual can contribute to the relative proliferation of her relatives and her genes within the interbreeding population to which she belongs.

Nepotism originally referred to the bestowal of patronage on the bastard sons, euphemistically called nephews, of popes and other high Vatican officials. It has come to mean the (usually illicit) use of one's social position to bestow benefits on relatives, both genetic and marital. Our meaning here is a little different again: Evolutionary biologists now refer to any sort of social discrimination on behalf of genetic relatives as nepotism, with no implication that such discrimination is reprehensible. It is simply what the evolved attributes of living creatures have been "designed" by the natural selective process to achieve. In the words of Richard Alexander (1979) living creatures "should have evolved to be exceedingly effective nepotists, and we should have *evolved* to be nothing else at all" (p. 46).

According to Hamilton's (1964) original formulation, "the social behaviour of a species evolves in such a way that in each distinct behaviour-evoking situation the individual will seem to value his neighbours' fitness against his own according to the coefficients of relationship appropriate to that situation" (p. 23). What are these "coefficients of relationship"? The elementary Hamiltonian analysis that has proven adequate in accounting for much of social evolution ignores such complications as sex chromosomes and mitochondrial DNA, and relies on Sewall Wright's (1922) coefficient of relatedness r : the probability that a particular autosomal allele in one individual will be identical to that in another by virtue of their being direct descendants of the same allele in a recent common ancestor. In a sexually reproducing diploid species such as *Homo sapiens*, Wright's $r = .5$ for (outbred) parent and offspring, .5 for full siblings, .25 for half siblings, .25 for aunt/uncle and nephew/niece, .125 for first cousins, and so forth. According to Hamilton, "altruistic" behavior in which the altruist incurs an average cost c (in units of expected future reproduction) and a beneficiary acquires an average benefit b (in the same currency) can proliferate under natural selection as long as $r > c/b$.

THE "GENETIC SIMILARITY" FALLACY

Hamilton's proviso that the relevant index of relationship is the probability of allelic identity "by descent" has been the object of some considerable confusion. Shared alleles are shared alleles whatever their origin. So should we not expect the extent to which natural selection favors altruistic behavior to depend simply on the beneficiary's genetic commonality with the altruist, regardless of "descent"?

This line of reasoning has been called "genetic similarity theory," and has been touted as a more general theory that encompasses Hamilton's inclusive fitness theory as a special case (Rushton, Russell & Wells, 1984; Russell, 1987). But it is no such thing. It is an instance of what Richard Dawkins (1979) has called "Washburn's fallacy," in honor of an earlier version of this attractive but flawed argument.

"This whole calculus upon which sociobiology is based is grossly misleading," anthropologist Sherwood Washburn (1978, p. 415) maintained, because most genes are identical regardless of descent. Parent and offspring are identical at 50% of their genetic loci by virtue of copying of an allele in that very parent, but they are sure to be alike at many other loci, too. If the alleles that two parties have in common were really the basis for altruism, we would have evolved to cherish all other humans as our close kin, and perhaps the great apes as well. By the same reasoning, selection should have favored our being nicer to monkeys than to other mammals, or, for that matter, being nicer to mosquitos than to marigolds. *Reductio ad absurdum.*

Obviously, something is wrong here. But it is not Hamilton's reliance on Wright's coefficient of relatedness. The proportion of genes that two individuals have in common may sound as if it captures what we mean when we refer to relatedness, but it is a red herring. The real issue is the evolutionary stability of trait states in competition with one another. As Dawkins (1979) explained:

> Let there be two strategies, Universal Altruist U, and Kin Altruist K. U individuals care for any member of the species indiscriminately. K individuals care for close kin only. In both cases, the caring behaviour costs the altruist something in terms

of his personal survival chances. Suppose . . . that virtually the entire population are universal altruists and a tiny minority of mutants or immigrants are kin altruists. Superficially, the U gene appears to be caring for copies of itself, since the beneficiaries of the indiscriminate altruism are almost bound to contain the same gene. But is it evolutionarily stable against invasion by initially rare K genes?

No, it is not. Every time a rare K individual behaves altruistically, it is especially likely to benefit another K individual *rather than* a U individual. U individuals, on the other hand, give out altruism to K individuals and U individuals indiscriminately, since the defining characteristic of U behavior is that it is indiscriminate. Therefore K genes are bound to spread through the population at the expense of U genes. . . .

[Now] assume that kin altruism has become common and ask whether mutant universal altruist genes will invade. The answer is no, for the same reason as before. The rare universal altruists care for the rival K allele indiscriminately with copies of their own U allele. The K allele, on the contrary, is especially unlikely to care for copies of its rival.

We have shown, therefore, that kin altruism is stable against invasion by universal altruism, but universal altruism is not stable against invasion by kin altruism (pp. 191–192).

In other words, nepotism is an evolutionarily stable state in competition with less discriminative social behavior, and this does not cease to be true when a nepotistic gene has become so prevalent that unrelated individuals are virtually as likely to share it as related ones.

There is a riposte available to the defender of "genetic similarity theory," but it is a riposte that restricts the theory to a domain so narrow as to be possibly nonexistent. A game theory argument like the one Dawkins uses to debunk Washburn's fallacy requires a complete specification of the "strategy set," and Dawkins' argument allows for only two alternatives, U and K. What about others? Would a population of either U or K individuals be evolutionarily stable against a mutant that neither bestowed altruism indiscriminately nor relied on kinship as an imperfect indicator of the likelihood that another individual carried the same mutant, but somehow detected copies of itself more directly and bestowed altruism accordingly?

Such a hypothetical mutant has been called a "green beard" gene (Dawkins, 1976), and it is probably obvious that it could indeed spread rapidly in a population of universal altruists or even in a population of kin altruists. But do such things exist? The requirements are formidable: The phenotypic effects of the mutant would have to include the unlikely combination of both self-recognition and some kind of discriminative impact on social interactions. A few candidate cases have been proposed, of which perhaps the most interesting is that of certain genes expressed in maternal endometrial tissue that effectuate the selective implantation of blastocysts carrying copies of themselves in preference to those carrying the mother's other allele (Haig, in press). But there is no reason to suppose that "green beard" genes are numerous, and in any event, they could not even in theory produce what Rushton et al.'s (1984) genetic similarity theory proclaimed: an evolved psychology that adjusts altruism in relation to an inferred degree of across-the-genome similarity rather than in relation to an inferred degree of genealogical relatedness. This is because no matter how many loci one uses in assessing overall genetic similarity, one is none the wiser about the probability of similarity at some other locus (such as a mutant gene affecting the rules for allocating altruism) except by virtue of what the similarity assessment implies about genealogical relatedness. Thus, if psyches that regulate altruism in response to gross similarity assessments do indeed exist, they have evolved because genetic similarity can be used as a cue indicative of genealogical relatedness, not the reverse. Discriminative altruism that is contingent on such an assessment is favoured by selection only to the degree that it constitutes discriminative nepotism.

A variant of Washburn's fallacy (or, if you like, a close relative) is the suggestion that it only makes sense to discriminate in favor of kin if there is significant genetic variability. Why should a "selfish gene" bother to engender nepotistic discrimination if nonrelatives were as likely to carry it as relatives? But to even ask this question is to take the metaphor of genes as replicative "strategists" too literally. One and the same gene cannot "change its mind" about its phenotypic expression. Genes that engender discriminative nepotism go to fixation and are stable against invasion by alternatives, and that is all there is to it. It is true that selection often favors phenotypes and hence genotypes that respond facultatively to environmental variations, but the entities that engage in flexible "strategizing," contingent on social and contextual variations, are entities at the level of whole organisms or complex subsystems thereof, not entities at the level of the "selfish gene."

More generally, there is an oddly prevalent misconception, even among biologists (e.g., Hrdy & Hausfater, 1984; Lewontin, 1979; Plomin, DeFries & McClearn, 1980), to the effect that any hypothesis that some phenotypic attribute is an adaptation requires that there be demonstrable variability at relevant genetic loci. If it is an evolved adaptation, goes the argument, then there must be genes "for" it, so show me that the trait is "genetic" first, and only then will I entertain the hypothesis that it might have been selected for. The trouble with this argument is that selection tends to eliminate heritability, with the effect that the attributes with the most direct and important effects on reproductive success are precisely the ones with the lowest heritabilities, not the highest (Falconer, 1960). The most highly heritable of the variable aspects of our eyes, for example, is the color of the iris, and this is because iris color and the genes affecting it are neutral (inconsequential) with respect to visual function. It follows that demonstrable, appreciable heritability should be considered prima facie evidence (though by no means conclusive evidence) against the hypothesis that the trait under consideration represents an adaptation (Daly, 1995). Nepotistic adaptations usually take the form of species-typical contingent decision rules: Sound the alarm when your neighbors are kin, for example, and stay mum when they are not (Sherman, 1977). It is therefore environmental rather than genetic sources of behavioral variation that usually provide the crucial tests of adaptationist hypotheses (Crawford & Anderson, 1989).

Oddly, arguments such as Washburn's have been advanced, not only as supposed critiques of Hamilton's theory, but as supposed validations of it. Consider, for example, Segal's (1984) claim that Hamilton's theory entitles us to expect a greater solidarity in monozygotic than in dizygotic twins. As it happens, Segal has convincingly demonstrated just such a difference, but this was never a straightforward prediction from inclusive fitness theory. Nepotistic inclinations are psychological adaptations, "designed" by the process of natural selection to promote inclusive fitness in the particular social environments in which they evolved. People living in circumstances like those of our ancestors probably seldom bore twins and were unlikely to rear both when they did (see Granzberg, 1973), so there appears to be little chance that we possess complex psychological adaptations whose specific function is to discriminate between our identical and fraternal twins and adjust our cooperativeness/competitiveness accordingly. A much more plausible hypothesis is that children possess psychological adaptations whose function is to assess whether a junior sibling has the same paternity as oneself, and to adjust the intensity of sibling competition accordingly. It is also possible that similarity to oneself is used more generally as one cue of relatedness, as Segal herself recognized. The exceptional solidarity of identical twins might then reflect a "supernormal" activation of psychological mechanisms whose normal function is to discriminate between lower values of r. Such solidarity can only be "predicted," however, from some such specific hypothesis about the nature of an evolved Hamiltonian adaptation. Rather than directly assessing Hamilton's analysis of the selective process, evolutionary psychology can progress by testing alternative specifications of the predictable "design" features of the psychological mechanisms of nepotistic discrimination under one versus another specific hypothesis about their primary functional contexts.

RELATIONSHIP-SPECIFIC
PSYCHOLOGICAL ADAPTATIONS

According to Haslam (1994b), "The study of social relationships lies at the heart of the social sciences, but psychologists' understanding of the cognitive structures that support them remains in the hinterlands" (p. 575). It is becoming clear, however, that those cognitive structures do not merely locate relationships in a space defined by a couple of dimensions such as "intimacy." They distinguish relationship categories. Haslam's (1994b) study of mental representations of relationships left both kin relationships and distinctions by sex out of consideration, yet he was still led to conclude, following Fiske (1991) and Fiske, Haslam, & Fiske, (1991), that "implicit knowledge of social relationships is modeled better by a small number of local, discontinuous representations, or categories, than by global laws and dimensions" (Haslam, 1994b, p. 582; see also Haslam 1994a).

When we include kinship (and gender), the categorical distinctions among relationships become inescapable. Genealogical kinship is not just one discrete category of relationship but several. The challenges that have faced human mothers, for example, are different from those confronting fathers or offspring or siblings or more distant relatives. It is certain that we possess distinct sets of evolved psychological adaptations for dealing with the peculiar demands of motherhood and offspringhood, and virtually as certain that we possess psychological adaptations for dealing with the challenges of being fathers and siblings as well. Specific psychological adaptations for grandparenthood and perhaps for other distal relationships remain plausible, too.

Motherhood

Consider first the psychology of motherhood. The most intimate of mammalian social relationships is that between mother and young. It is also the one with the largest inventory of special-purpose anatomical, physiological, and psychological machinery. But the task demands of motherhood are a good deal more complex than even a consideration of the component demands of conceiving, gestating, and raising a baby would imply. Because offspring are not all equally capable of translating parental nurture into increments in the long-term survival of parental genetic materials, there has been intense selection for subtle discriminations in the allocation of maternal effort. The result is that the evolved motivational mechanisms regulating maternal investment decisions are complexly contingent on variable attributes of the young, of the material and social situation, and of the mother herself including lifespan developmental changes from youth to menopause in the possibly exceptional case of *Homo sapiens* (Daly & Wilson, 1995a).

The problem of adaptive allocation of maternal investments is an especially subtle one because of the active intervention of other parties with conflicting interests, namely the offspring themselves. Parent–offspring conflict (Trivers, 1974) is endemic to sexually reproducing species because of a certain asymmetry of relationship: A mother is equally related to any two of her offspring, but each offspring is more closely related to self than to sib. It follows that mother and offspring are selected to see the relative fitness values of broodmates, and hence the optimal allocation of maternal investments, somewhat differently. Each offspring is selected to covet a little more from the mother than would be optimal for her own fitness. This conflict accounts for the otherwise puzzling existence of seemingly maladaptive aspects of mother–young interaction, including weaning conflict (Trivers, 1974), tantrums (Trivers, 1985), and the dangerously high levels of allocrine substances of fetal origin in the blood of pregnant women, including human chorionic gonadotrophin, which interferes with the mother's capacity to terminate pregnancies that are suboptimal from her perspective, and human placental lactogen, which upregulates the fetus's access to maternal glucose stores (Haig, 1993).

Of course, long before Trivers laid bare the logic of parent–offspring conflict, it was apparent to all who looked that the maternal relationship was special. Throughout human history, most women have devoted the majority of their waking hours to foraging for, educating, guarding, and otherwise nurturing their children. But try to find even a paragraph concerning this central domain of human social behavior in a textbook of social psychology, and you are likely to be disappointed. The sources of variability in maternal feeling and action have also been virtually ignored in scientific psychology's consideration of "motivation," apparently because theorists lacking an evolutionary perspective have had no framework for making sense of the vicissitudes of maternal inclinations (Daly & Wilson, 1988a). It is only developmental psychologists who have paid some attention to maternal behavior, but the interest here has been primarily in alleged impacts on the developing child (e.g. Cranley, 1993; Howes, Matheson, & Hamilton, 1994) and secondarily in maternal style as a personality attribute (e.g. Belsky, Fish, & Isabella, 1991) rather than as an adaptively contingent response. The neglect of motherhood is perhaps the single most telling indictment of a social psychology devoid of the concept of qualitatively distinct fundamental relationships.

Fatherhood

Now consider fatherhood. There are some obvious parallels with the maternal case, but also some crucial differences. In both mothers and fathers, parental solicitude evolves to vary adaptively in relation to phenotypic and situational cues affording information about the expected impact of any parental investments on the offspring's reproductive value (expected future fitness), so mother and father alike are selected to assess offspring quality and offspring need. However, a father cannot necessarily exploit the same information as a mother in making his parallel assessment, mainly because of the special avenues of communication (and manipulation) between a mother and her fetus or nursling (Haig, 1993). It is also the case that both maternal and paternal solicitude evolve responsiveness to cues of the fitness value of the available alternatives to present parental investment, but here it is even clearer that the specifics are different. In particular, the chronic possibility that extrapair mating effort might enhance male fitness lures male effort away from parental investment. Finally, both mother and father are selected to discriminate with respect to available cues that the offspring is indeed the parent's own, but the relevant information sources are again distinct. Female mammals generally identify their own babies on the basis of initial circumstantial cues soon after birth and subsequent learning of their young's distinguishing features, and there is no reason to doubt that this applies to women (Daly & Wilson, 1995a). But the mere fact that a particular baby emerged from a particular woman's birth canal has never been conclusive evidence of paternity, so a putative father must rely on additional sources of information about the woman's probable fidelity, the baby's pheno-typic resemblance to his relatives or himself, or both.

An evolutionary psychological hypothesis derived from these considerations is that the affection felt by fathers is likely to be influenced by their children's resemblance to themselves, whereas little or no such effect would be expected in mothers. As far as we know, this hypothesis has not been tested directly. However, there is evidence that all interested parties pay a great deal more attention to a baby's phenotypic resemblances on the paternal than on the maternal side, and that mothers and their kin actively promote perceptions of paternal resemblance (Daly & Wilson, 1982, 1988a; Regalski & Gaulin, 1993).

The issue that we are discussing has been referred to as one of the "uncer-tainty" of paternity, or variable paternity "confidence," especially in studies of nonhuman animals in which males invest in their ostensible offspring. The same terms have also been used in anthropological discussions of the avunculate, a social practice whereby men transmit titles and resources primarily to their sisters' sons rather than their own. Following a suggestion by Alexander (1974), evolution-minded anthropologists have discussed the incidence of extrapair paternity that would be necessary for putative fathers to actually be more closely related, on average, to their sisters' sons than to their wives' sons (Kurland, 1979). Avuncular inheritance is indeed cross-culturally associated with condi-tions conducive to low levels of paternity confidence (Flinn, 1981; Gaulin & Schlegel, 1980), and if those levels are seldom or never quite low enough to make the sisters' sons closer kinsmen than wives' sons, we should remember that these inheritance decisions are not the man's alone and that his parents can always be surer of their relationship to his sister's children than to his own children (Flinn, 1981; Hartung, 1985).

The use of the terms uncertainty and confidence has been unfortunate, however, because of their misleading psychological implications. Both terms seem to refer to a subjective degree of confidence or doubt of which the putative father is aware. But that is not the way these concepts are actually used. In practice, they have been synonymous ways of referring to a population parame-ter: the proportion of offspring actually sired by their ostensible fathers (or, if you like, 1.0 minus the incidence of "cuckoldry"). One problem with this usage is that it can mislead readers (and occasionally even the writer!) into imagining that the writer is actually speaking about men's beliefs. Another is that present terminology leaves no room for consideration of variations in confidence among the fathers within a population. Some men have children who resemble them and wives in whose fidelity they have every reason to believe; others do not. Thus, *confidence of paternity* is surely a variable whose determinants and consequences can be studied. In certain songbirds, for example, males attain a significant, albeit imperfect, match between actual paternity and the effort expended in feeding young by making their paternal efforts proportionate to how thoroughly they were able to monitor their mates during the fertile egg-laying period (e.g., Davies, 1992). Human males may be capable of comparable cognitive feats.

This points up yet another problem with the prevalent terminology of *uncertainty* and *confidence*, however: the implication of awareness is superfluous. A male songbird need not doubt his paternity, nor indeed have a concept of paternity at all, in order to adjust paternal investment adaptively. And neither need a man. It is perfectly possible that at least some of the mental mechanisms that instantiate paternity confidence adaptations are activated automatically in isolation from articulatable beliefs. We hypothesize, for example, that the affection of adoptive fathers may very well be more strongly affected by the adoptee's resemblance to self than is the affection of adoptive mothers, simply because such resemblance is a cue that the paternal psyche has evolved to respond to. This need not imply that the emotional aspects of parental feeling and commitment are utterly isolated from that stream of cognition of which we are aware, nor even from the influence of rational deliberation; indeed, there is anecdotal evidence that paternal affection can be shattered by a verbal revelation of nonpaternity (Daly & Wilson, 1988a). Here, as in other domains of kinship psychology, there is much to learn about the ways in which emotional responses can and cannot make contact with articulatable beliefs and knowledge.

Sibship

Sibling relations also warrant scrutiny from a selectionist perspective (Mock & Parker, 1996). Sisterhood is, of course, at the heart of Hamilton's (1964) analysis of the evolution of exceptional sociality and altruism in haplodiploid bees, ants, and wasps, but sibling relations are prominent in the sociality of diploid creatures, too. Such phenomena as cooperative courtship displays by pairs of male turkeys (Watts & Stokes, 1971) or delaying personal reproduction to help at the parental nest (Moehlman, 1986; Stacey & Koenig, 1990) are testimony to the relevance of sib relations to social evolution. But if siblings are major social allies by virtue of relatedness, they are even more surely major competitors, especially for crucial maternal resources. It is little wonder, then, that sibling relationships are so often ambivalent.

The sibships into which we are born are crucial social environments, with associated opportunities, costs and "niches," and it would be remarkable if our evolved social psyches did not contain features adapted to the peculiarities of sibling relationship. Sulloway (1995, 1997) has developed the idea of niche differentiation in an evolutionary psychological perspective, with principal reference to the ways in which one deals with one's ordinal position in a sibship. Evolutionary considerations suggest that parents would favor their eldest offspring, and in tough choices there is evidence that they do just that (Daly & Wilson, 1995a), so it is not surprising that firstborn children tend to be conservative supporters of the status quo (Sulloway, 1997). There is some theoretical and empirical support for the notion of parental indulgence of lastborns, too, which suggests that it may be the middle birth positions that derive the least benefit from nepotistic solidarity. In support of this conjecture, Salmon (1997) has found that both first- and lastborn Canadians differ from middleborns in measures of familial solidarity and identity: Middleborns are substantially less likely than either first- or lastborns to name a close genetic relative as the person to whom they feel closest, and are also significantly less likely than either first- or lastborns to assume the role of family genealogist.

Sibling conflict is the other side of the coin of parent–offspring conflict. Children are selected to try to manipulate mother to extend the interval before the next birth (Blurton Jones & daCosta, 1987; Trivers, 1974), but mother has her counterploys, and the next sibling often arrives too soon from the perspective of the toddler (Dunn & Kendrick, 1982). The costs that the toddler is willing to impose on its infant sibling in competition for maternal investment may vary adaptively in relation to nutritional status, the birth interval, the social situation, and perhaps even the phenotypic quality of each youngster.

Uncertainty of paternity engenders a problem for siblings as well as for fathers: Given common maternity, do we share a father ($r = .5$) or have different ones ($r = .25$)? A particularly interesting example of such discrimination comes from the study of nepotistic discrimination in ground squirrels. Because females often mate with more than one male in a single estrous period, even littermates might be only half siblings. Males disperse at maturity, but females do not, with the result that if two female littermates both live to adulthood, they may occupy adjacent territories. Such littermate sisters vary in the extent to which they cooperate in mutual defense of young or are mutually hostile, and it turns out that a significant determinant of that variability is relatedness, with full sisters the more cooperative neighbors (Holmes & Sherman, 1982). This is a remarkable finding because there is apparently no circumstantial cue that would enable females to make this discrimination. However, if individual aspects of odor are genetically based (as they surely are) and if the squirrels habituate to their own odors, then there is a simple heuristic that they might use to make this and other discriminations: Be most hostile to those who smell strongest.

Could there be analogous discrimination in human beings? It is certainly plausible that the distinction between full and half sibs was selectively significant in our evolution. Studies of contemporary foraging peoples suggest that a succession of more or less monogamous mateships may have been common among our ancestors. Children fare better when they have the benevolent attention of both genetic parents (Daly & Wilson, 1988a; Hurtado & Hill, 1992; Voland, 1988), and this fact provides a strong disincentive against taking divorce lightly, but nevertheless both marital breakups and (perhaps more importantly) untimely deaths probably assured that half sibship was common. It could well be the case that in human prehistory it was virtually a toss-up whether successive children of the same woman were full or half-siblings, and the distinction between ($r = .5$) and ($r = .25$) is by no means trivial when the decision to cooperate or to compete is a close call. It therefore seems to us very plausible that the psychology of toddlers has evolved to adjust the intensity of competitive tactics toward newborn siblings in relation to either phenotypic cues or direct evidence of male turnover.

Grandparenthood

Is grandparenthood, too, a relationship status for which we possess specific adaptations? This question is harder to settle. The very fact that women experience menopause has seemed to many writers to suggest that our female ancestors may have attained with some regularity a life stage in they could serve their fitness interests better by prolonged investment in their extant young and perhaps in their grandchildren than by further reproduction. However, an evaluation of this hypothesis in light of the known demographic and life history characteristics of contemporary foragers led Hill and Hurtado (1991) to question whether women in such populations could really attain greater fitness from helping kin in old age than from continuing to reproduce (see also Rogers, 1991). But even if menopause is not itself a grandmaternal adaptation, it is a cross-culturally general fact that postmenopausal women contribute significantly to their grandchildren's welfare (Lancaster & King, 1985), and it is therefore at least plausible that mental processes specific to the task of adaptive allocation of grandparental investment have been targets of natural selection (see also Turke, in press; Smith, 1988).

Mateship

Although mates are not typically close genetic relatives, kinship is often conceived of as encompassing this relationship, too. There is a certain logic to this conflation of genealogical and mating relationships: In both cases, the two parties have a commonality of interest grounded in the fact that the fitness of both is promoted by the reproductive success of their common kin. A long-standing mateship becomes increasingly like a genetic relationship because as children are produced and mature, it is more and more the case that the exigencies and resource allocations that would be ideal for promoting the fitness of one party are optimal for the other too. Indeed, as Alexander (1987) noted, if mating partners are faithfully monogamous and their efforts are channeled predominantly into reproduction rather than collateral nepotism, their commonality of interest, and hence of perspective, may become nearly total. This would seem to explain why established couples may become more solidary in their approach to the world around them than even the closest genetic kin.

There is an important difference between mateship and genetic kinship, however: The former can be more readily and irredeemably betrayed (Wilson & Daly, 1992). Whatever failures of reciprocity and other provocations may strain blood-kin relations, shared interests in the welfare of common relatives have provided a countervailing force selecting for a kin-specific readiness to forgive and reconcile. The correlation between the fitness interests of husband and wife, by contrast, can be abolished if one or both parties engage in extrapair mating effort. Moreover, if a husband is cuckolded and unwittingly invests his parental efforts in a rival's young, then the very acts that promote the wife's fitness are positively damaging to the husband's. These considerations would seem to account for the fact that suspected or actual infidelity is a uniquely potent source of severe marital conflict and violence (Daly & Wilson, 1988b; Wilson & Daly, 1993).

Steprelationship resembles cuckoldry in that a child raised by a couple is a potential vehicle of fitness for one party but not the other. It is different, however, in that this asymmetry is out in the open and has ideally entered into the negotiation of entitlements and reciprocities in the remarriage. Nevertheless, the presence of stepchildren is an important risk factor for marital disruption and violence (Daly & Wilson, 1996), and the stepchildren themselves incur greatly elevated risk of severe assault (Daly & Wilson, 1988a, 1988b, 1995b). It is apparently stepparenthood itself that is the relevant risk factor and not some correlate or "confound" (Daly & Wilson, 1996), reinforcing the point that the motivational mechanisms of parental feeling are designed to channel affection and investment preferentially toward one's own offspring. Stepparenthood is one of several forms of "fictive" or nominal kinship in which people find themselves placed, not always altogether willingly, in interpersonal statuses that are artificial analogues of kinship statuses. It would be very surprising if the appropriate relationship-specific psychology were fully activated by such experiences, because the interests of our interactants are seldom identical with our own and selection has presumably acted to buffer us against being the manipulanda of others' social agendas (a subject to which we return). It is not at all surprising, on the other hand, that the genetic parent in a stepfamily should do what she or he can to induce stepparent and stepchildren to feel and act more like genetic relatives than their inclinations might dictate.

The solidarity even of faithful couples is apt to wane if children are not forthcoming (Rasmussen, 1981). A mated pair's separate interests in their separate kindreds is a potential source of conflict for any couple, but these "in-law problems" may be more acutely felt when there are no children to cement the marital relationship itself and the broader alliance between the two parties' families of origin that the marriage represents. For marriage is indeed an alliance between kin groups, as many anthropologists have stressed. Marriage is a cross-culturally ubiquitous feature of human societies, notwithstanding variations in social and cultural details of the marital relationship (Flinn & Low, 1986; Murdock, 1967; van den Berghe, 1979): Women and men everywhere enter into publicly acknowledged one-on-one reproductive alliances, with mutual obligation to invest biparentally in the union's joint progeny, and this reproductive purpose appears to be felt at least as acutely by interested kin as by the marriage partners themselves. Many writers have tried to argue that the economic and/or political aspect of marriage is fundamental, yet all interested parties, including the kin groups allied by the union, understand the first purpose of marital union to be reproductive.

SOME UNIVERSAL ASPECTS OF HUMAN KINSHIP

The human brain/mind contains an as yet unspecifiably large number of special-purpose modules. The notion that some general-purpose cognitive device does all our information processing has been shown to be unworkable (Miller & Todd, 1990; Pinker & Prince, 1988). Consider language: There is no longer a reasonable doubt that human language is a complex functional aspect of our evolved human nature (Pinker, 1994). The hallmarks of evolutionary adaptation are apparent in the large number of analytically separable but functionally integrated elements of neuroanatomy, of the peripheral speech apparatus, and of the elements of natural languages themselves, which perform astonishingly efficient encoding and decoding of linguistic materials and of nothing else. Some considerable fraction of our human nature, and especially of our mind/brain, is designed specifically to generate and comprehend speech.

Can anything comparable be claimed about human kinship cognition? We think it can, but study of the evolutionary psychology of the family is not nearly so advanced as psycholinguistics in identifying the constituent adaptations. Human kinship systems are dauntingly diverse, so much so that social anthropologists have devoted more attention to their variations than to any other aspect of human society (Fox, 1967). Clan (or lineage or moiety) memberships may be single or multiple, and they may have pervasive social consequences or

virtually none. Kin relationships may dictate scarcely any limitation on potential marriage partners, or they may proscribe thousands of complete strangers.
Descent reckoning may be strictly patrilineal (47% of 857 societies according
to Murdock, 1967), or matrilineal (14%), or "bilateral" (36%, for which a better
label would be "multilineal"), or even a "double descent" system (3%; these are
genuinely bilineal in that each person is considered to be descended from a
female line extending back from the mother and a male line from the father).
Most notably, the domains of kin terms are cross-culturally variable. The
Yanomamö studied by Chagnon (1974), for example, employ an Iroquois
terminology, such that a man would address not only his brother as *Abawä* but
also his patrilateral parallel cousin (his father's brother's son) and even, at least
in principle, more distant same-generation male kin related through strictly
patrilineal links. To some anthropologists, these facts have seemed to imply that
kinship is "cultural rather than biological."

To evolutionists, of course, there can be no such antithesis: "culture" is
produced by a living species. Moreover, cross-cultural diversity is not arbitrary,
and its orderliness has provided evolutionists with opportunities for statistical
tests of adaptationist hypotheses about what kinship means and how it is used
by self-interested actors. Matrilineal descent reckoning and inheritance, for
example, are systematically associated with residential and subsistence practices
that threaten paternity confidence (Flinn, 1981). Patriliny is associated with
concentrated bride price, blood feud, and severe penalties for adultery (Daly &
Wilson, 1988b). These and other correlations are readily interpreted as the
compromise outcomes of reproductive and nepotistic struggles under different
socio-ecological conditions (Alexander 1974, 1977, 1979; Flinn & Low, 1986;
Gaulin and Boster, 1990; Hartung 1985).

If we wish to identify the core adaptations underlying human kin classification, we will need to burrow under this diversity and identify that which is
cross-culturally universal (Brown, 1991). When the French say "chien" where
we say "dog," we can infer that it is not these lexical specifics that are the general
features of language; better candidates for linguistic universals are more abstract
attributes such as the existence of discrete words and their assemblage out of a
short list of language-specific phonemes, or the existence of nouns and the fact
that such entities as animal species are among their referents (Pinker, 1994).
Can we identify any comparably abstract but well-specified universal features
of human kinship?

We propose the following list, without imagining that it is anywhere near
complete. The first few principles are well established, but farther along the list
are attributes of kinship that have been demonstrated in only a few societies.
Their universality is an hypothesis. As far as we know, there are no known
counterexamples to any of them, and although one or more may yet be
overturned, we won't be holding our breath.

1. *Ego-centered kindred terminologies are universal.* In all societies, kin relationships are classified with reference to each focal individual: My mother is not the same person as your mother. One could imagine a society without this feature. A man's kinship status might be fully specified, say, by some combination of his stable attributes (e.g., he was born a member of the Raven clan) and his ephemeral attributes (e.g., he is a prepubertal, unmarried male), without any reference to his relationships to particular others. However, no such society exists. Such nonrelational attributes are important elements of social identity in some societies, but never so important as to obviate the ego-centered kindred relations.

2. *Parent-offspring relationships are the fundamental building blocks of this ego-centered structure, so that the terminology implies a genealogy.* Again, one could imagine a society with no such notion of kinship. Special relations functioning in a manner analogous to kin relations might, in principle, be erected on some other basis. One's "brothers" could, for example, be defined as all those men who were born in the same lunar month as oneself. But again, this is not the way *Homo sapiens* conducts its social life.

3. *All kinship systems include terminological (and practical) distinctions according to sex.* No system lacks words to distinguish daughters from sons or mothers from fathers, for example. Neither is there anywhere a society lacking the notion that there are qualitative differences between these sexually distinguished relationships and attendant differences in the social behavior characteristic of and appropriate to each.

4. *All kinship systems include terminological (and practical) distinctions according to generation.* No system lacks words to distinguish daughters from mothers or sons from fathers, for example, and parental roles and filial roles are not reversible.

5. *Kin relations are universally understood to be arrayed along a dimension of closeness.* Once again, one could imagine a society in which this was not so, a society whose participants perceived no such ordinality in the qualities that they considered essential to their kinship categories. We can imagine such a system, but it would not be human kinship.

6. *This dimension of the characteristic closeness of kinship categories is always negatively correlated with the characteristic number of genealogical links defining them, and hence positively correlated with genetic relatedness* (r). The first five universals are apparently conceded by even the most biophobic of commentators. Where cultural determinists have tried to draw the line is on this sixth point. In a famous attempt to refute the applicability of Hamiltonian theory to human affairs, Sahlins (1976) claimed to have demonstrated "that the categories of 'near' and 'distant' vary independently of consanguineal distance and that these categories organize actual social practice" (p.112). He had, of course, demonstrated nothing of the sort. His evidence consisted entirely of typological descriptions of alleged practices in certain societies which, if verified, would indicate only that the correlation between "closeness" and genetic relatedness is sometimes less than perfect. Nobody ever doubted that there are mismatches in detail between conceptions of the charac-

teristic closeness of particular kinship categories in particular societies and the number of genealogical links involved (see Farber, 1981). But the categories of "near" and "distant" do not "vary independently of consanguineal distance," not in any society on earth.

It probably is not even true that kinship systems treat certain close genealogical relationships as socially significant while obliterating other equally close relationships, or at least not to the degree that Sahlins and other cultural determinists imagine. It is true that the relatives of one's mother are nominally excluded from one's kindred in a system with strict patrilineal descent reckoning, whereas in a matrilineal system, the male role in reproduction may be "unknown" or denied. But we would be naive to suppose that this represents psychological reality. As Meyer Fortes argued in many writings (e.g., 1953, 1969) and with many ethnographic examples, actual sentiments of attachment to kin are always bilateral. Articulated kinship systems are to a considerable degree ideologies, and like all ideologies, they are the arenas of contest and social manipulation, as we discuss in additional detail with respect to proposed universals 10 and 12.

The universality of a correlation between closeness and r is simply inexplicable for those who would divorce "culture" from "biology" and then place kinship in the former domain.

7. *There is everywhere a strong positive correlation between the average or characteristic r of kinship categories and the levels of solidarity and cooperation among those so related.* This is not quite the same proposition as point 6, for although solidarity and subjective closeness are strongly associated with one another, they are not synonymous. It is not merely that one considers one's brother a closer relative than one's cousin; they really do cooperate more. In situations of cooperative labor exchange, for example, brothers tolerate imbalances of reciprocity that would be considered unacceptably exploitative in a friendship not based on kinship (Hames, 1988). Close kin do violence to one another less and collaborate in violence against third parties more than one would expect on the basis of the opportunities provided by their proximity and frequency of contact (Daly & Wilson, 1988b). Even in a relatively nonkin-based society such as ours, people turn to close relatives when in need, and are increasingly likely to do so the greater the imposition or demand (Essock-Vitale & McGuire, 1985; Hogan & Eggebeen, 1995).

Documenting and measuring the relevance of genealogical relatedness to cooperative action has been one of the principal achievements of evolution-minded anthropologists. An important limitation of the ethnographic record is its typological nature: In this society inheritance is matrilineal; in that society cross-cousins are preferred as marriage partners. What we cannot tell from such claims is how well they describe practice. Is such a characterization a valid generalization, an ideal, or what? Although ethnographers have often noted imperfect adherence to professed rules and even conflicts of interest, few have bothered to quantify their incidence or resolutions. Why worry about who really marries whom, how the resources actually flow, or who honors or reneges on what obligations, if your agenda is descriptive and "interpretative" rather than hypothesis testing? An experienced fieldworker such as Sahlins could persist in believing that merely nominal kinship categories "organize social practice" because he never felt the need to evaluate the fit between a prediction and data. But the anthropologists who began using Hamiltonian theory to generate hypotheses about commonalities and conflicts of interest within societies, and about social manipulation and ideology, needed behavioral data, and so they collected it. If evolutionists had contributed nothing else to ethnographic practice, anthropology would be in their debt for their insistence on behavioral observation and quantification. But of course, they have contributed more than methodological rigor: What these studies have consistently shown is that cooperative and conflictual interactions are best predicted from genealogical relatedness, even when ideology and lip service say otherwise (Betzig & Turke, 1986; Chagnon, 1981; Flinn, 1988; chapters in Chagnon & Irons, 1979).

8. *In all societies, persons related by marriage are deemed to be in a sort of quasi-kinship relationship.* The marital relationship brings one's spouse into one's kindred—after a sort. As we noted earlier, the marital relationship is analogous to genetic related-ness in the fact that people who reproduce together have genetic relatives in common (their descendants) with resultant shared interests. The same applies, albeit more weakly, to the two kindreds: As children of a marriage are produced, the relationship between the two marriage partners' separate kindreds becomes qualitatively more like a degree of kinship. One's spouse's brother becomes a sort of figurative or partial brother (as is hinted at by such qualifiers as "brother-in-law"), but he never becomes indistinguishable from a "real" brother.

9. *In all societies, people are motivated to inquire how strangers and new acquain-tances might be genealogically linked to people they already know, and feel that they have acquired useful social information when such links are uncovered.* Insofar as kinship links are predictive of social sentiments and action, as they certainly are, it makes good sense that we should wish to situate people who are unknown and therefore unpredictable in matrices of kinship. And we do.

10. *In all societies, certain people make it their business to know genealogies and to educate others, especially their relatives, about exactly how they are related to one another.* Knowing our genealogical links serves our own interests and those of our kin, by helping us to manage our social affairs as effectively nepotistic. But more than this, the fact that kinship imparts only a partial congruence of interests means that there is room for those engaged in familial socialization and education to exert self-inter-ested influence. It is a predictable corollary of Trivers's (1974) parent–offspring conflict theory, for example, that senior family members will pressure those in subsequent generations to take a stronger benevolent interest in collateral kin than would otherwise be their inclination. For a focal female, for example, a sister's

son ($r = .25$) represents a vehicle of potential fitness with half the value of an otherwise equivalent son of her own ($r = .5$), but from the son's perspective, his mother's sister's son is a mere cousin ($r = .125$) for whom any sacrifice would have to yield 8 times what it cost the son for it to be worthwhile. Thus the old have special reason to remind the young of their links of collateral kinship and to urge that they base their cooperative undertakings on those links rather than on reciprocal ties of friendship. This sort of pressure can be at least partly effective, both because it comes from a communicator with a genuine interest in the listener's welfare and because it contains an element of valid wisdom.

In our society and many others, the interests of firstborn young may be especially well served by familial solidarity, and it is of interest that firstborns are disproportionately inclined to assume the role of family genealogist (Salmon, 1997). Ours is certainly not what anthropologists would call a kin-based society, and kinship networks and knowledge are evidently truncated in comparison to those in many traditional societies. Alexander (1979) claimed, citing Schneider and Cottrell (1975), that "most people in a modern technological society may know of the existence, at least, of all of their first cousins but few could count, let alone name, all of their second or third cousins" (p. 148). However, interest in genealogical research is considerable. Approximately 3,000 people visit the Mormon Genealogical Library in Salt Lake City each day (Shoumatoff, 1985), and it is estimated that in 1980, there were half a million North Americans who were active genealogists (Taylor, 1986).

Both sexes appear to be interested in genealogy as a hobby, but several studies suggest that women are particularly inclined to maintain active kin networks (Hogan & Eggebeen, 1995; Schneider & Cottrell, 1975). Troll (1987) provides evidence that men's kinship bonds operate through the influence of their wives or parents, and that older women typically adopt the role of "kinkeeper," providing family news updates, organizing get-togethers, maintaining contacts among family members, and training daughters or granddaughters for the role. And North American women do indeed know their own genealogies better than men (Salmon & Daly, 1996; Schneider & Cottrell, 1975).

This sex difference is probably not a universal, however. Although women in our society continue to rely heavily on relatives (Hogan & Eggebeen, 1995; Komarovsky & Philips, 1962), men's need for kin support may be substantially reduced in a modern nation state in comparison with societies in which kinsmen are crucial allies in intergroup conflict. In other social ecologies, therefore, men may well be keener genealogists. Among the Yanomamö, for example, knowledge of genealogy is a valuable social tool for negotiating male–male alliances and marital entitlements, and men appear to know more (or at least to be quicker in accessing) genealogical information than do women (Chagnon, 1988). Whether the evolved basis of kinship cognition is in any way sexually dimorphic thus remains an open question.

11. *In all societies, one's beliefs about one's genealogical links are core components of the phenomenology of self.* In a society organized into patrilineal fraternal interest groups, people grow up to perceive clan identity as paramount. Born into a Montenegrin tradition of blood feuds, for example, Milovan Djilas (1958) recalled:

> My forebears were drummed into my head from earliest childhood, as was the case with all my countrymen. I can recite ten generations without knowing anything in particular about them. In that long line, I am but a link, inserted only that I might form another to preserve the continuity of the family (p. 6).

In the context of more individualistic ideologies, of which the contemporary North American version may represent an extreme, this degree of familial subordination is alien. When asked "who are you?" one's kinship status is high on the list of responses for some people, but not for others (Salmon & Daly, 1996). However, familial links remain profoundly important to identity even here. To appreciate this, one need only consider the psychological impact of the discovery that a trusted link is a social artifact rather than a genetic fact, as in revelations of adoptive status (Hoopes, 1990; Kirk, 1981) or of donor insemination (Scheib & Daly, 1997).

12. *In all societies, some kinship terms incorporate more than one genealogical relationship, but people are nowhere oblivious to the distinctions that terminology obscures or ignores.* Although some primary kinship terms, especially those of the parent–offspring relationship, apply to one and only one genealogical relationship, others are extended to encompass more than one. Terminological extension and ambiguity have repeatedly been invoked against the proposition that kinship has a "biological basis" or that the primary meanings of kinship terms are genealogical (Sahlins, 1976; Schneider, 1984; Hirschfeld, 1986). After all, if a Yanomamö man, for example, places his brother ($r = .5$), and his patrilateral cousin ($r = .125$), and perhaps some more distant relatives as well, all in the same kinship category, well then, the cultural thing that we call *kinship* must not be about genealogical relatedness at all.

Chagnon's Yanomamö research provides two rejoinders to this argument. The first is that the terminological conflation of distinct relationships does not bespeak a failure to discriminate between them. Both my brother and my cousin may address me as *Abawä*, but the former is more likely to come to my aid in a conflict than the latter (Chagnon & Bugos, 1979). Terminological kinship is a predictor of who allies himself with whom, but genealogical relatedness is a significantly better predictor (Chagnon, 1981).

The second point is more subtle. Cultural determinists maintain that even to translate the word *Abawä* as *brother* and then to speak of its "extension" to patrilateral parallel cousins is to commit the error of ethnocentrism. There is no English translation of *Abawä*, according to this line of reasoning, because the way the Yanomamö partition their social universe is incommensurate with the way we partition ours. Chagnon's riposte is simplicity itself: Ask a native speaker. Having shown an informant some photographs of his brother and his cousin and having determined that both are *Abawä* to the speaker, Chagnon then asks (in Yanomamö, of course) "which one is your real *Abawä?*" To a cultural determinist, this would appear to be a vacuous question, compounding the ethnocentrism. However, the Yanomamö informant understands it perfectly well, and points to the brother. In other words, it is not just the western ethnographer who considers common parenthood to define the primary referent of *Abawä* and the inclusion of certain kinds of cousins to be a figurative extension; this is the Yanomamö's own understanding as well.

At least one critic of the proposition that kinship is fundamentally a matter of genealogy has rejected even this sort of evidence, arguing as follows (Hirschfeld, 1986):

> My *younger brother* and *my older brother* have distinct referents and as phrases mean different things, but the component term *brother* in both cases has a unique meaning. Similarly, qualifiers and hedges like "real" or "true" are readily attached to kin terms and frequently accompanied by a term or gesture associated with the womb and notions of procreation. But from the fact that it represents a culturally salient (perhaps even universal) distinction it does not follow that it is a necessary part of the meaning of a term any more than the recognition of *younger* and *older* entails that these always are part of the meaning of a term. (pp. 220–221)

In other words, "real brothers" are just one of an infinite number of arbitrarily delimited subsets of real "brothers"! If this argument were taken seriously, one would have to conclude that it is impossible to identify the primary or literal meaning of any word, as distinct from its figurative or metaphorical meanings.

In claiming that people remain sensitive to the distinctions among terminologically conflated categories of kin, we do not mean to imply that terminology has no behavioral impact. The extension of brotherhood to a group of men with patrilineal links, for example, reinforces the salience of certain connections and diminishes others. People make terminological distinctions that meet their social purposes. Many would find it bizarre that English speakers should use the single term *cousin* to encompass matrilateral and patrilateral parallel and cross-cousins of both sexes, but usage here both reflects and reinforces the absence of significant differentiations of social roles among these types of cousins in our society. (Note that we also call more distant relatives cousins too, a situation more analogous to the *Abawä* case, and that no one is deceived into assuming that the relationships are equally close psychologically.)

Insofar as kinship terminology affects social perceptions and behavior, however, the implication is not the sort of simple one-way cultural determinism to which many anthropologists still subscribe. People pursuing self-interested agendas exert at least as much influence on the use of kin terms as vice versa. Kinship terminology reinforces and advertises social entitlements and obligations and can be a contested domain for that reason. An example occurs in our society when a remarried woman calls her new mate "your dad" in speaking to her child of a former union. In tribal societies, more than one genealogical connection to a potential interactant may be known, and modes of address are then chosen for social advantage, such as in order to transfer someone into a marriageable category, often to the chagrin of other interested parties who stress an alternative relationship (Chagnon, 1982; Fredlund, 1985). People are obviously not completely bamboozled by such terminological manipulations, as Fortes (1969) showed by documenting bilateral kin attachments in unilineal descent systems, but they apparently are not completely unaffected either.

The fact that kinship terminology is a manipulable social device engenders one more putative universal, the last one that we discuss.

13. *In all societies, kinship terms are extended further still, being deployed figuratively rather than literally, for evocative and propagandistic purposes.* "Brother, can you spare a dime?" Speakers who wish to emphasize or promote kinlike beneficence commonly address nonrelatives with kin terminology. Perhaps even more common (and more effective) is the metaphorical brothering of a potential ally in a joint venture, when the speaker wishes to focus attention on a genuine (or at least plausible) shared interest and promote kinlike solidarity (Johnson, 1986).

Metaphors have been called "those sometimes explicitly acknowledged but often unconsciously or tacitly employed conceptual systems of images through which social life is interpreted and around which social life is organized" (Turner, 1987, p. 56). If kinship is the primary organizing principle in human relations and the bedrock of most altruism, it is perhaps unsurprising that one should invoke kinship metaphorically in the negotiation of nonnepotistic cooperation. Indeed, Alexander (1974) and other writers have suggested that the cognitive abilities necessary for nonnepotistic reciprocal altruism must have evolved previously in the context of nepotistic sociality.

We are aware of only one experimental investigation suggesting the efficacy of such usage. Johnson, Ratwick, and Sawyer (1987) made students listen to political speeches in which the audience was addressed either with kin terminology or as "fellow citizens"; results were in the direction of the kin terms being both more physiologically arousing and more persuasive. This preliminary study was less than conclusive and needs replication, but it is hard to doubt that addressing someone as *brother* works at least to some degree, sometimes, or it would not be so prevalent. Yet it also seems likely that natural selection should have equipped us with psychological defenses against being manipulated by easily faked words from the mouths of persons whose self-interests are not necessarily compatible with our own. Perhaps saying *brother* achieves little more than to signal to the listener that a claim of common cause is about to follow, a claim that the listener may still reject, but that he has at least been prepared to consider.

Metaphors evoking solidarity are most often those of sibship. "The Brotherhood of Free Masonry" or the feminist slogan "Sisterhood is powerful" are declarations of common cause (and implied threat against common foes) by ostensible equals. But asymmetrical kin relationships are also invoked metaphorically, especially by those laying claim to authority, as when kings and priests style themselves "fathers." Any implied threats in this case are mainly against the "children." Aptly, we call this style of imposing authority paternalism, even if kinship terminology is not prominent. According to van den Berghe (1985),

> If power is to be justified (so as to be more readily exercised), the aim of power must be hidden or denied. The best denial of the effect of power is that oppression is in the interest of the oppressed. . . . Paternalism mimics the genuine *concern* of the parent for the child, which is founded on the real overlap of interest inherent in genetically based nepotism, and thus hides the overwhelmingly conflictual basis of the ruler–subject relationship. Paternalism models itself on a relationship of genuine *dependence* and incapacity, in which the helpless child's survival and well-being is contingent on adult care, and extends it to a situation in which the dependence is *reversed*. The ruler who parasitizes the subject disguises parasitism as altruism. (p. 262)

In developing his Oedipal theory of "primal parricide," Freud (1913/1950) turned this manipulative metaphor on its head, maintaining that it is the subjects who make the ruler into a symbolic father to appease their guilty psyches. But despite Oedipal theory's tenacity in literary criticism and pop psychology, there is not and never was any evidence for it (see review by Daly & Wilson, 1990).

The metaphor of a religious denomination as family has wide appeal, perhaps especially among beleaguered sects. Religious leaders quite rightly perceive genealogical loyalties as rivals and threats to their own dynastic ambitions (Betzig, 1986; Goody, 1976), and have commonly offered a purported substitute. One discussion of the pseudofamilial structure and function of a Christian church described this phenomenon as follows (Anderson & Guernsey, 1985):

> The church as the new family of God, however, is not formed by mere consensuality between its members. Through spiritual rebirth we each become a brother or sister of Jesus Christ through adoption into the family of God. Consequently, we are all brother and sister to each other. (p. 81)

As genealogical kin share a common identity by virtue of the circumstances of their birth, the religious "family" forges a common identity through "rebirth" with all its solidary and authoritative implications.

People use the relatively egalitarian fictive kinship of nominal brotherhood and sisterhood in individualized contexts as well as in the context of groups such as freemasons and feminists. Close friends may invent secret rituals that make them "blood brothers," for example, and African American women "make family" by establishing functionally sororal partnerships with "play sisters," perhaps especially when helpful blood kin are scarce (Johnson & Barer, 1990; Stack, 1974; Staples, 1985; Taylor, Chatters & Mays, 1988). Kinship is evidently the dominant mental model for helpful social interaction (Bailey, 1988), and hence provides an almost inescapable metaphor for establishing, describing, or explaining cooperative relationships with nonrelatives.

UNRESOLVED ISSUES IN THE SOCIAL COGNITION OF KINSHIP

Cognitive psychologists have concerned themselves with ostensible general processes of categorization/similarity (Brooks, 1990). But is kinship a special domain with its own rules? One reason it might be is that children acquire an understanding of kinship terminology in ways that cannot be accounted for by the hypothesis of domain-general inductive processes, but that seem to require an "innate theory" of the nature of human social relationships (Hirschfeld, 1989).

There is also an a priori reason to think that kin classification might require its own mental processes: Unlike most other categorizations, kinship cannot be represented as a nested hierarchy because of sexual reproduction. In a phylogenetic descent diagram, all branching ("radiation") is downward: Several species may belong to a genus and several genera to a family, but no entity at one level in the hierarchy can have multiple ancestors at a higher level. Our ancestors, by contrast, double at each generation, making "family" an altogether different computational problem. Yet people often seem determined to bend their conceptualization of family links to fit the inappropriate nested hierarchy model, as when they belong to one or another named family (usually a patrilineage).

Why do human beings so often resort to this inappropriate mental model? Or do they? Are clan memberships and shared patrilineal names really such powerful elicitors of familial feeling as they sometimes appear? Fortes's (1969) claim that even the most extreme systems of unilineal descent reckoning do not suppress bilateral affiliation and sympathy is relevant here. How do people retain their sensitivity to genealogical relatedness while operating within a clan structure?

Hughes's (1988) reanalysis of Bryant's (1981) genealogy of a Tennessee mountain community is intriguing in this context. All Bryant's subjects were nominal members of one of four named "families," and when she discovered that patrilineal descent from the putative male founders of the four families could not explain membership status, she concluded that genealogical relatedness was not the basis of kinship at all. Hughes (1988) showed, however, that "family" membership was perfectly accounted for by one's degrees of relatedness to all the as-yet-unmarried young people in the community:

> They appear to use a metaphor of descent to describe groups that are in actuality based on relatedness to focal offspring. ... As long as the four "founding ancestors" are remembered by name, groups that contain at least some of their descendants can use their names as convenient labels even if group membership is in fact based on a principle other than descent. If the actual focus of family groups is dependent offspring, the offspring themselves will grow up and may move out of the area or even join other families. (p. 83)

There would seem to be some logic to defining family with reference to those close relatives who have yet to marry and whose eventual reproductive careers are therefore most vulnerable to kin influence. But what might these results imply about how the human mind computes kin ties? Unfortunately, Hughes (1987, 1988) has obscured the psychological interest of this and his other analyses by declaring an extreme version of the view that inclusive fitness maximization is a proximate cause of behavior, a view that has been cogently criticized by Symons (1989) and Tooby and Cosmides (1990).

Kin recognition and assessment are germane both to effective nepotism, as we have discussed at length, and to mate choice, which we have ignored. Evolutionists have been greatly interested in inbreeding avoidance, and the possible role of cosocialization of children in inducing sexual indifference (the "Westermarck effect" ; see Wolf, 1993; Wolf & Huang, 1980). But are inbreeding avoidance and nepotism subserved by common evolved psychological mechanisms and processes by which we infer kinship, or are they achieved by independent means?

The study of social cognition is ripe for an infusion of evolutionary theory and a serious consideration of kinship.

REFERENCES

Adams, B. N. (1968). *Kinship in an urban setting*. Chicago: Markham.
Alexander, R.D . (1974). The evolution of social behavior. *Annual Review of Ecology and Systematics, 5*, 325–383.
Alexander, R. D. (1977). Natural selection and the analysis of human sociality. In C. E. Goulden (Ed.), *Changing scenes in the natural sciences 1776–1976* (pp. 283–337). Philadelphia: Philadelphia Academy of Sciences.
Alexander, R. D. (1979). *Darwinism and human affairs*. Seattle, WA: University of Washington Press.
Alexander, R. D. (1987). *The biology of moral systems*. Hawthorne NY: Aldine de Gruyter.
Anderson, R. S., & Guernsey, D. B. (1985). *On being family: A social theology of the family*. Grand Rapids MI: Eerdmans.

Bailey, K.G. (1988). Psychological kinship: implications for the helping professions. *Psychotherapy, 25*, 132–141.

Belsky, J., Fish, M., & Isabella, R. A. (1991). Continuity and discontinuity in infant negative and positive emotionality: Family antecedents and attachment consequences. *Developmental Psychology, 27*, 421–431.

Berscheid, E. (1994). Interpersonal relationships. *Annual Review of Psychology, 45*, 79–129.

Betzig, L. (1986). *Depotism and differential reproduction: A Darwinian view of history.* Hawthorne NY: Aldine de Gruyter.

Betzig, L., & Turke, P. (1986). Food sharing on Ifaluk. *Current Anthropology, 27*, 397–400.

Blurton Jones, N. G., & daCosta, E. (1987). A suggested adaptive value of toddler night waking: delaying the birth of the next sibling. *Ethology and Sociobiology, 8*, 135–142.

Brooks, L. R. (1990). Concept formation and particularizing learning. In P. Hanson (Ed.), *Information, language, and cognitive science* (Vol. 1, pp. 141–165). Vancouver BC: University of British Columbia Press.

Brown, D. E. (1991). *Human universals.* New York: McGraw-Hill.

Brown, J. L. (1975). *The evolution of behavior.* New York: W. W. Norton.

Bryant, F. C. (1981). *We're all kin: A cultural study of a mountain neighborhood.* Knoxville: University of Tennessee Press.

Burnstein, E., Crandall, C., & Kitayama, S. (1994). Some neo-Darwinian decision rules for altruism: Weighing cues for inclusive fitness as a function of the biological importance of the decision. *Journal of Personality and Social Psychology, 67*, 773–789.

Chagnon, N. A. (1974). *Studying the Yanomamö.* New York: Holt, Rinehart & Winston.

Chagnon, N. A. (1981). Terminological kinship, genealogical relatedness and village fissioning among the Yanomamö Indians. In R. D. Alexander & D. W. Tinkle (Eds.), *Natural selection and social behavior* (pp. 490–508). New York: Chiron Press.

Chagnon, N. A. (1982). Sociodemographic attributes of nepotism in tribal populations: Man the rule-breaker. In King's College Sociobiology Group (Eds.), *Current problems in sociobiology* (pp. 291–318). Cambridge: Cambridge University Press.

Chagnon, N. A. (1988). Male Yanomamö manipulations of kinship classifications of female kin for reproductive advantage. In L. Betzig, M. Borgerhoff Mulder & P. Turke (Eds.), *Human reproductive behavior: A Darwinian perspective* (pp. 23–48). New York: Cambridge University Press.

Chagnon, N. A., & Bugos, P. E. (1979). Kin selection and conflict: an analysis of a Yanomamö ax fight. In N. A. Chagnon & W. Irons (Eds.), *Evolutionary biology and human social behavior: An anthropological perspective* (pp. 213–249). North Scituate MA: Duxbury Press.

Chagnon, N. A., & Irons, W. (Eds.). (1979). *Evolutionary biology and human social behavior: An anthropological perspective.* North Scituate MA: Duxbury Press.

Cranley, M. S. (1993). The origins of the mother–child relationship: A review. *Physical and Occupational Therapy in Pediatrics, 12*, 39–51.

Crawford, C. B., & Anderson, J. L. (1989). Sociobiology: An environmentalist discipline? *American Psychologist, 44*, 1449–1459.

Daly, M. (1995). Evolutionary adaptationism: Another biological approach to criminal and antisocial behavior. In G. Bock & J. Goode, (Eds.), *Genetics of criminal and antisocial behavior* (pp. 183–195). CIBA Foundation Symposium #194. Chichester: Wiley.

Daly, M., & Wilson, M. I. (1982). Whom are newborn babies said to resemble? *Ethology and Sociobiology, 3,* 69–78.

Daly, M., & Wilson, M. I. (1988a). The Darwinian psychology of discriminative parental solicitude. *Nebraska Symposium on Motivation, 35,* 91–144.

Daly, M., & Wilson, M. I. (1988b). *Homicide.* Hawthorne NY: Aldine de Gruyter.

Daly, M., & Wilson, M. I. (1990). Is parent-offspring conflict sex-linked? Freudian and Darwinian models. *Journal of Personality, 58,* 163–189.

Daly, M., & Wilson, M. I. (1995a). Discriminative parental solicitude and the relevance of evolutionary models to the analysis of motivational systems. In M. Gazzaniga, (Ed.), *The cognitive neurosciences* (pp. 1269–1286). Cambridge MA: MIT Press.

Daly, M., & Wilson, M. I. (1995b). Some differential attributes of lethal assaults on small children by stepfathers versus genetic fathers. *Ethology and Sociobiology, 15,* 207–217.

Daly, M., & Wilson, M. I. (1996). Evolutionary psychology and marital conflict: the relevance of stepchildren. In D. M. Buss & N. Malamuth (Eds.), *Sex, power, conflict: Feminist and evolutionary perspectives* (pp. 9–28). New York: Oxford University Press.

Davies, N. B. (1992). *Dunnock behaviour and social evolution.* Oxford: Oxford University Press.

Dawkins, R. (1976). *The selfish gene.* New York: Oxford University Press.

Dawkins, R. (1979). Twelve misunderstandings of kin selection. *Zeitschrift für Tierpsychologie, 51,* 184–200.

Djilas, M. (1958). *Land without justice.* New York: Harcourt.

Dunn, J., & Kendrick, C. (1982). *Siblings.* Cambridge MA: Harvard University Press.

Essock-Vitale, S., & McGuire, M.T. (1985). Women's lives viewed from an evolutionary perspective: II. Patterns of helping. *Ethology and Sociobiology, 6,* 155–173.

Falconer, D. S. (1960). *Introduction to quantitative genetics.* New York: Ronald Press.

Farber, B. (1981). *Concepts of kinship.* New York: Elsevier.

Fiske, A. P. (1991). *Structures of social life: The four elementary forms of social relationships.* New York: Free Press.

Fiske, A. P., Haslam, N., & Fiske, S. T. (1991). Confusing one person with another: What errors reveal about the elementary forms of social relations. *Journal of Personality & Social Psychology, 60,* 656–674.

Flinn, M. V. (1981). Uterine vs. agnatic kinship variability and associated cousin marriage preferences: an evolutionary biological analysis. In R. D. Alexander & D. W. Tinkle (Eds.), *Natural selection and social behavior* (pp. 439–475). New York: Chiron Press.

Flinn, M. V. (1988). Step- and genetic parent/offspring relationships in a Caribbean village. *Ethology and Sociobiology, 9,* 335–369.

Flinn, M. V., & Low, B. S. (1986). Resource distribution, social competition, and mating patterns in human societies. In D. I. Rubenstein & R. W. Wrangham (Eds.), *Ecological aspects of social evolution* (pp. 217–243). Princeton, NJ: Princeton University Press.

Fortes, M. (1953). The structure of unilineal descent groups. *American Anthropologist, 55,* 17–41.

Fortes, M. (1969). *Kinship and the social order.* Chicago: Aldine.

Fox, R. (1967). *Kinship and marriage: An anthropological perspective.* New York: Penguin.

Fredlund, E. V. (1985). The use and abuse of kinship when classifying marriages: a Shitari Yanomamö case study. *Ethology and Sociobiology, 6,* 17–25.

Freud, S. (1950). *Totem and taboo*. (J. Strachey, Trans.). New York: Norton. (Original work published 1913)

Gaulin, S. J. C., & Boster, J. S. (1990). Dowry as female competition. *American Anthropologist, 92*, 994–1005.

Gaulin, S. J. C., & Schlegel, A. (1980). Paternal confidence and paternal investment: A cross-cultural test of a sociobiological hypothesis. *Ethology and Sociobiology, 1*, 301–309.

Goody, J. (1976). *Production and reproduction: A comparative study of the domestic domain*. New York: Cambridge University Press.

Granzberg, G. (1973). Twin infanticide—a cross-cultural test of a materialistic explanation. *Ethos, 1*, 405–412.

Haig, D. (1993). Genetic conflicts in human pregnancy. *Quarterly Review of Biology, 68*, 495–532.

Haig, D. (in press). Gestational drive and the green-bearded placenta. *Proceedings of the National Academy of Science*.

Hames, R. (1988). Relatedness and garden labor exchange among the Ye'kwana. *Ethology and Sociobiology, 8*, 354–392.

Hamilton, W. D. (1964). The genetical evolution of social behaviour. I and II. *Journal of Theoretical Biology, 7*, 1–52.

Harris, C. C., (Ed.). (1970). *Readings in kinship in urban society*. Oxford: Pergamon Press.

Hartung, J. (1985). Matrilineal inheritance: New theory and analysis. *Behavioral and Brain Sciences, 8*, 661–688.

Haslam, N. (1994a). Categories of social relationship. *Cognition, 53*, 59–90.

Haslam, N. (1994b). Mental representations of social relationships: dimensions, laws, or categories? *Journal of Personality and Social Psychology, 67*, 575–584.

Hill, K., & Hurtado, M. (1991). The evolution of premature reproductive senescence and menopause in human females: An evaluation of the "grandmother hypothesis." *Human Nature, 2*, 313–350.

Hirschfeld, L. A. (1986). Kinship and cognition: Genealogy and the meaning of kin terms. *Current Anthropology, 27*, 217–242.

Hirschfeld, L. A. (1989). Rethinking the acquisition of kinship terms. *International Journal of Behavioral Development, 12*, 541–568.

Hogan, D. P., & Eggebeen, D. J. (1995). Sources of emergency help and routine assistance in old age. *Social Forces, 73*, 917–936.

Holmes, W. G., & Sherman, P. W. (1982). The ontogeny of kin recognition in two species of ground squirrels. *American Zoologist, 22*, 491–517.

Hoopes, J. L. (1990). Adoption and identity formation. In D. M. Brodzinsky & M.D. Schecter, (Eds.), *The psychology of adoption* (pp. 144–146). New York: Oxford University Press.

Howes, C., Matheson, C. C., & Hamilton, C. E. (1994). Maternal, teacher, and child care history correlates of children's relationships with peers. *Child Development, 65*, 264–273.

Hrdy, S. B., & Hausfater, G. (1984). Comparative and evolutionary perspectives on infanticide: introduction and overview. In G. Hausfater & S. B. Hrdy (Eds.), *Infanticide: comparative and evolutionary perspectives* (pp. xiii–xxv). Hawthorne NY: Aldine de Gruyter.

Hughes, A. L. (1987). Social and antisocial behavior. *Quarterly Review of Biology, 62,* 415–421.

Hughes, A. L. (1988). *Evolution and human kinship.* New York: Oxford University Press.

Hurtado, M., & Hill, K. R. (1992). Paternal effect on offspring survivorship among Ache and Hiwi hunter-gatherers: implications for modeling pair-bond stability. In B. Hewlett, (Ed.), *Father-child relations* (pp. 31–35). Hawthorne, NY: Aldine de Gruyter.

Johnson, C. L., & Barer, B. M. (1990). Families and networks among older inner-city blacks. *Gerontologist, 30,* 726–733.

Johnson, G. R. (1986). Kin selection, socialization, and politics: an integrating theory. *Politics and the Life Sciences, 4,* 127–154.

Johnson, G. R., Ratwik, S. H. & Sawyer, T. J. (1987). The evocative significance of kin terms in patriotic speech. In V. Reynolds, V. Falger, & I. Vine (Eds.), *The sociobiology of ethnocentrism* (pp. 157–174). London: Croom Helm.

Kenrick, D. T., Sadalla, E. K., & Keefe, R. C. (in press). Evolutionary cognitive psychology: The missing heart of modern cognitive science. In C. Crawford & D. Krebs (Eds.), *Evolution and Human Behavior: Ideas, issues and applications.* Mahwah, NJ: Lawrence Erlbaum Associates.

Kirk, H. D. (1981). *Adoptive kinship.* Toronto: Butterworth & Co.

Komarovsky, M., & Philips, J. H. (1962). *Blue-collar marriage.* New Haven, CT: Yale University Press.

Kurland, J. A. (1979). Paternity, mother's brother, and human sociality (pp. 145–180). In N. A. Chagnon & W. Irons (Eds.), *Evolutionary biology and human social behavior: an anthropological perspective.* North Scituate, MA: Duxbury Press.

Lancaster, J. B., & King, B. J. (1985). An evolutionary perspective on menopause. In J. K. Brown & V. Kern (Eds.), *In her prime: A new view of middle aged women* (pp. 13–20). Boston, MA: Bergin & Garvey.

Leach, E. (1966). Virgin birth. *Proceedings of the Royal Anthropological Institute of Great Britain and Ireland, 1966,* 39–49.

Lewontin, R. C. (1979). Sociobiology as an adaptationist program. *Behavioral Science, 24,* 5–14.

Miller, G. F., & Todd, P. M. (1990). Exploring adaptive agency. I: Theory and methods for simulating the evolution of learning. In D. S. Touretskz, J. L. Elman, T. J. Sejnowski, & G. E. Hinton (Eds.), *Proceedings of the 1990 Connectionist Models Summer School* (pp. 65–80). San Mateo CA: Morgan Kaufmann.

Mock, D. W., & Parker, G.A . (1996). *The evolution of sibling rivalry.* New York: Oxford University Press.

Moehlman, P. D. (1986). Ecology of cooperation in canids. In D. I. Rubenstein & R. W. Wrangham (Eds.), *Ecological aspects of social evolution* (pp. 64–86). Princeton, NJ: Princeton University Press.

Moghaddam, F. M., Taylor, D. M., & Wright, S. C. (1993). *Social psychology in cross cultural perspective.* New York: W.H. Freeman.

Murdock, G. P. (1967). *Ethnographic atlas.* Pittsburgh: University of Pittsburgh Press.

Parsons, T. (1943). The kinship system of the contemporary United States. *American Anthropologist, 45,* 22–38.

Petrinovich, L., O'Neill, P., & Jorgensen, M. (1993). An empirical study of moral intuitions: Toward an evolutionary ethics. *Journal of Social and Personality Psychology, 64,* 467–478.

Pinker, S. (1994). *The language instinct: How the mind creates language*. New York: Morrow.

Pinker, S., & Prince, A. (1988). On language and connectionism: Analysis of a parallel distributed processing model of language acquisition. *Cognition, 28*, 73–193.

Plomin, R., DeFries J. C., & McClearn, G. E. (1980). *Behavioral genetics: A primer*. San Francisco: Freeman.

Rasmussen, D. R. (1981). Pair-bond strength and stability and reproductive success. *Psychological Review, 88*, 274–290.

Regalski, J. M., & Gaulin, S. J. C. (1993). Whom are Mexican infants said to resemble? Monitoring and fostering paternal confidence in the Yucatan. *Ethology and Sociobiology, 14*, 97–113.

Rogers, A. R. (1991). Conserving resources for children. *Human Nature, 2*, 73–82.

Rushton, J. P., Russell, R. J. H., & Wells, P. A. (1984). Genetic similarity theory: beyond kin selection. *Behavior Genetics, 14*, 179–193.

Russell, R. J. H. (1987). Genetic similarity as a mediator of interpersonal relationships. In V. Reynolds, V. Falger, & I. Vine (Eds.), *The sociobiology of ethnocentrism* (pp. 118–130). London: Croom Helm.

Sahlins, M. (1976). *The use and abuse of biology*. Ann Arbor: University of Michigan Press.

Salmon, C. (1997). Birth order and the salience of family. Manuscript submitted for review.

Salmon, C., & Daly, M. (1996). On the importance of kin relations to Canadian women and men. *Ethology and Sociobiology, 17*.

Scheib, J., & Daly, M. (1997). Donor insemination and the evolutionary psychology of parenthood. Manuscript in preparation.

Schneider, D. M. (1984). *A critique of the study of kinship*. Ann Arbor, MI: University of Michigan Press.

Schneider, D. M., & Cottrell, C. B. (1975). *The American kin universe: A genealogical study*. Chicago: University of Chicago Press.

Schneider, D. M., & Homans, G. C. (1955). Kinship terminology and the American kinship system. *American Anthropologist, 57*, 1194–1208.

Segal, N. L. (1984). Cooperation, competition, and altruism within twin sets: A reappraisal. *Ethology and Sociobiology, 5*, 163–177.

Sherman, P. W. (1977). Nepotism and the evolution of alarm calls. *Science, 197*, 1246–1253.

Shoumatoff, A. (1985). *The mountain of names: A history of the human family*. New York: Simon & Schuster.

Smith, M. S. (1988). Research in developmental sociobiology: parenting and family behavior. In K. MacDonald (Ed.), *Sociobiological perspectives on human development* (pp. 271–292). New York: Springer.

Stacey, P. B., & Koenig, W. D. (1990). *Cooperative breeding in birds: Long-term studies of ecology and behavior*. New York: Cambridge University Press.

Stack, C. (1974). *All our kin*. New York: Harper & Row.

Staples, R. (1985). Changes in black family structure: The conflict between family ideology and structural conditions. *Journal of Marriage and the Family, 47*, 1005–1015.

Sulloway, F. J. (1995). Birth order and evolutionary psychology: A meta-analytic overview. *Psychological Inquiry, 6*, 75–80.

Sulloway, F. J. (1997). *Born to rebel: Radical thinking in science and social thought*. New York: Pantheon.

Symons, D. (1989). A critique of Darwinian anthropology. *Ethology and Sociobiology, 10,* 131–144.

Symons, D. (1995). Beauty is in the adaptations of the beholder: The evolutionary psychology of female sexual attractiveness. In P. R. Abramson & S. D. Pinkerton (Eds.), *Sexual nature sexual culture* (pp. 80–118). Chicago: University of Chicago Press.

Taylor, R. J., Chatters, L. M., & Mays, V. M. (1988). Parents, children, siblings, in-laws, and non-kin as sources of emergency assistance to black Americans. *Family Relations, 37,* 298–304.

Taylor, R. M. (1986). *Generations and change: Genealogical perspectives in social history*. Macon, GA: Mercer Press.

Tooby, J., & Cosmides, L. (1990). The past explains the present: Emotional adaptations and the structure of ancestral environments. *Ethology and Sociobiology, 11,* 375–424.

Trivers, R.L. (1974). Parent–offspring conflict. *American Zoologist, 14,* 249–264.

Trivers, R.L. (1985). *Social evolution*. Menlo Park, CA: Benjamin/Cummings.

Troll, L. E. (1987). Gender differences in cross-generation networks. *Sex Roles 17,* 751–763.

Turke, P. L. (1997). Did menopause evolve to discourage infanticide and encourage continued investment by agnates? *Evolution and Human Behavior.*

Turner, M. (1987). *Death is the mother of beauty: Mind, metaphor, criticism*. Chicago: University of Chicago Press.

van den Berghe, P. (1979). *Human family systems*. New York: Elsevier.

van den Berghe, P. (1985). Comment on G. L. Goodell's "Paternalism, patronage and potlatch". *Current Anthropology, 26,* 262–263.

Voland, E. (1988). Differential infant and child mortality in evolutionary perspective: data from late 17th to 19th century Ostfriesland. In L. Betzig, M. Borgerhoff Mulder & P. Turke (Eds.), *Human reproductive behavior* (pp. 253–261). Cambridge: Cambridge University Press.

Washburn, S. L. (1978). Human behavior and the behavior of other animals. *American Psychologist, 33,* 405–418.

Watts, C. R., & Stokes, A. W. (1971). The social order of turkeys. *Scientific American, 224*(6), 112–118.

Wilson, M. I., & Daly, M. (1992). The man who mistook his wife for a chattel. In J. Barkow, L. Cosmides & J. Tooby (Eds.), *The adapted mind* (pp. 289–322). New York: Oxford University Press.

Wilson, M. I., & Daly, M. (1993). An evolutionary psychological perspective on male sexual proprietariness and violence against wives. *Violence and Victims, 8,* 271–294.

Wolf, A. P. (1993). Westermarck redivivus. *Annual Review of Anthropology 22,* 157–175.

Wolf, A. P., & Huang, C. S. (1980). *Marriage and adoption in China, 1845–1945*. Stanford, CA: Stanford University Press.

Wright, S. (1922). Coefficients of inbreeding and relationship. *American Naturalist, 56,* 330–338.

11

Four Grammars for Primate Social Relations

Nick Haslam
New School for Social Research

The recent rise in the fortunes of evolutionary approaches to psychology has been especially kind to social psychologists, who find themselves in a vanguard position. Many of the animating concerns of evolutionary psychology overlap the traditional content domains of social psychology, and both disciplines share a methodological commitment to understanding the individual in context. Besides having this basic affinity, the evolutionary perspective offers social psychologists a unifying explanatory framework that often departs intriguingly from folk intuitions, an unaccustomed pleasure. As a result, social psychological questions are now central to the mission of evolutionary psychology.

It is now generally accepted, for instance, that many of the most crucial selective pressures operating over the course of primate and hominid evolution arose from the complexities of group living. The complexity of social organization appears to have increased in tandem with increases in cognitive capacity, and we are more and more willing to grant that the former may have driven the latter (Humphrey, 1976; Byrne & Whiten, 1988). We have come to see primate and hominid social life as a more demanding arena for problem solving than the technical challenges of subsistence (Quiatt & Kelso, 1985) and have deduced that social competence must call upon an impressive array of adapted skills and propensities (Cosmides & Tooby, 1995). Understood in this way, social intelligence fragments into an assemblage of mental modules dedicated to the tasks of face recognition, cooperation, mental state attribution, affect perception, reciprocity, kin recognition, mate choice, deception, cheater detection, and so forth.

Two aspects of this emerging program are particularly important. First, evolutionary social psychology explicitly concerns aspects of human sociality that are

297

generally understood to be cultural, or at least culturally elaborated. Although some elements of social competence, such as the detection of facial affect, are concrete, perceptual, and in the service of moment-to-moment attunement in social interaction, other elements are of a more abstract and conceptual kind. The complicated social figuring involved in keeping track of favors and obligations, for instance, must rely on a tacit understanding of the abstract patterning of social relations within a group, a patterning that constitutes a central aspect of social organization and culture (Jackendoff, 1992; Tooby & Cosmides, 1992). Evolutionary social psychology therefore aims to map those aspects of the capacity for culture that might be in some sense universal, and recognizes that this aim requires it to deal seriously with alternative traditions within anthropology and sociology.

A second noteworthy aspect of evolutionary social psychology's program is the degree to which it is deeply cognitive. The adaptations that it invokes are information processing capacities specialized for restricted domains of social problem solving (Hirschfeld & Gelman, 1994). By specifying social competence in computational terms, evolutionary psychology aims to understand the proximate mechanisms underlying social behavior, rather than functionally accounting for behavior in terms of its presumed adaptiveness. Although they may have adaptive consequences, proximate cognitive mechanisms clearly cannot operate directly on the currency of inclusive fitness. Although we would not expect most social behavior to deviate very systematically from expected fitness maximization, we can be confident that the computations involved do not always closely resemble economic analyses of costs and benefits.

Specifying what forms proximate cognitive mechanisms might take is another matter. At this point, it seems sensible to make a conceptual distinction between cognitive processes and the representations on which they operate. A cognitive mechanism is a certain kind of operation on a certain kind of representation (Chomsky, 1980). Cognitive mechanisms are generally referred to in terms of rules, grammars, algorithms, heuristics, or variations on these (e.g., condition–action rules, Quiatt & Reynolds, 1993; decision rules, Burnstein, Crandall, & Kitayama, 1994; action grammars, Greenfield, 1991; cognitive–affective heuristics, Kenrick, Groth, Trost, & Sadalla, 1993; social decision heuristics, Allison & Messick, 1990). Representations have been dubbed *representational frames, formats,* or *currencies* (Cosmides & Tooby, 1995). Economic cost–benefit analysis, in these terms, would refer to addition and multiplication rules operating on quantitative representations of monetary values, probabilities, and discount rates. One task of evolutionary social psychologists, then, is to describe the rules and representational frames employed in human social behavior.

Pulling together the two elements of evolutionary social psychology introduced earlier, we can see that a central task of the enterprise is to characterize the representational frames and rules that underlie social cognition at the abstract level of social relations. This chapter presents one such characterization, which was developed by Alan Fiske (1991, 1992) as an account of the cognitive underpinnings of social organization. Although the relational models theory was developed as a synthesis of anthropological and sociological thinking on elementary forms of human sociality, and was initially put to service in an ethnographic study of the Moose people of Burkina Faso, I hope to show its evolutionary relevance through an examination of continuities with nonhuman primates. By exploring regularities and patterns across species, it becomes possible to make evolutionary claims about the possible origins of some elements of human social cognition, and to put its distinctiveness into clearer relief (Cartmill, 1990).

In addition to offering insights into human distinctiveness, however, an approach based on parallels and continuities can challenge sharply discontinuous views of human social cognition. According to such views (Tomasello, 1994), distinctively human social–cognitive capabilities such as language and the attribution of mental states make human sociality and culture completely *sui generis*. These capabilities are undeniably fundamental for human social life, but the discontinuity position has dubious implications. By claiming that the truly important adaptations on which human sociality is based are mostly without precedent in the primate world, it suggests that human culture and social life should be understood exclusively in terms of symbolic language and intersubjectivity. This chapter argues that human social cognition has important dimensions besides these crowning abilities, dimensions in terms of which human culture can be apprehended. Relational models theory describes a part of human social cognition and culture that reveals interesting evolutionary continuities with our primate ancestors, and sheds light on human distinctiveness. My approach is to explore primatological evidence for parallels to the relational models. I then make some tentative claims about how the human social–cognitive apparatus differs from that of our closest evolutionary relatives, and about what factors in hominid evolution might have brought these differences into being. Prior to doing so, it is necessary to describe the relational models and their empirical support.

THE RELATIONAL MODELS

In bald summary, Fiske's fundamental claim is that everyday social living is undergirded by four elementary cognitive models. The relational models organize a broad range of social activities: the distribution and contribution of resources, work, social identity, interpretations of misfortune, political ideologies, moral judgment, social motivation, and much more besides. The models are in some sense innate and universal and emerge in an orderly sequence over the course of development. This sequence corresponds to the increasing order of formal complexity of the models, whose operations can be given abstract mathematical expression. The models do not emerge simply by induction, but provide schematic frames without which observational learning would be chaotic. Consequently, cultural learning corresponds more to parameter setting in Chomskyan linguistics; each culture sets which social activities will be organized by which components of a "universal grammar" of models, and socialization consists of learning the culture's distinctive rules of implementation of the species' shared social–cognitive inheritance. Socialization is therefore as much a matter of the externalization of highly constraining inner patterns as of the internalization of social organization.

Fiske's four relational models are *Communal Sharing, Authority Ranking, Equality Matching,* and *Market Pricing.* These models govern social relationships singly or in combination, so that although actually existing relationships may prototypically resemble one model, the models are not simply types of relationships. The *Communal Sharing* model organizes relationships in terms of collective belonging or solidarity. Members of an in-group are treated as equivalent and undifferentiated elements of a bounded, categorical set, and consequently individual distinctiveness is ignored. Grounding principles are unity, identity, intimacy, conformity, interdependence, and ideas of shared substance. By contrast, the *Authority Ranking* model organizes relationships in terms of asymmetrical difference. Parties to relationships governed by this model are hierarchically ordered, ranked transitively according to status markers such as age, skill, knowledge, class, or social position. Higher-ranked individuals are authorized to command, protect, dominate, bestow, precede, and show largesse; lower-ranked individuals defer, obey, show loyalty and respect, and yield precedence. Low-ranking individuals are not simply dominated or exploited; they typically benefit from protection, advice, leadership, and intervention in disputes.

The *Equality Matching* model organizes the construction and interpretation of relationships into an egalitarian framework. This model is manifested most distinctly in turn-taking, distributions of equal shares, democratic voting, and tit-for-tat retaliation. *Equality Matching* focuses close and unforgiving attention on reciprocity and upholds even balance in social exchange as the primary value. Departures from balance create degrees of obligation that can be added or subtracted.

The *Market Pricing* model, organizes relationships with reference to a common scale of ratio values such as money. Emphasis is on proportions; earning a wage based on hours worked, getting a good return on an investment of effort, or making efficient use of time. Social transactions are reckoned as rational calculations of cost and benefit. Relationships are implicitly or explicitly understood in terms of commodities, with autonomous participants seeking benefit through joint trading.

By Fiske's account, the four relational models are not commensurable or reducible to one another. Although the kinds of social reckoning implied by them can be arranged, as shown, in an order of increasing formal or mathematical complexity—nominal, ordinal, interval, and ratio—no model supercedes another. Rather, each model is discrete and autonomous, especially fitted, perhaps, for its own field of social life. The models represent four distinct ways to organize social activities and relationships, and those that emerge later in development and are more computationally complex are not necessarily or generally superior. *Market Pricing*, for example, is unlikely to be a satisfactory way to conduct a romance. By ordering the models "horizontally," in this way, Fiske makes the important point that to privilege one above others is itself a culturally saturated value judgment. Fiske's theory gives no comfort to those who would argue for any sovereign principle of social organization or cognition; cultures forge successful social organization from quite different admixtures of the models.

How the models are concretely realized and combined in social arrangements is governed by a culture's implementation rules. For instance, the harvesting of crops might be realized by a distinctive implementation of *Equality Matching*, in which it must be decided who will be required to perform the equal shares (perhaps low-status postpubertal females only), and whether these equal shares will be organized by turn-taking or synchronous activity. Because the relational models are abstract frames that are free of content, they can be used to compare cultures without the proposal of substantive universals (Brown, 1991). This is not to say, of course, that there are no substantive universals or social–cognitive adaptations of a less abstract sort than the models. Indeed, much of the best work in evolutionary psychology has been devoted to discovering relatively concrete and domain-specific adaptations that are complementary to the relational models. In theory, the models organize shared understandings of social relations, their abstractness providing flexibility but little guidance in the implementation of actual relationships, whereas concrete adaptations such as facial affect perception assist in the implementation of face-to-face interaction. Because the models can be implemented in widely varying ways, cultures can be respectfully understood in terms of elaborations on a limited vocabulary of elementary structures. This distinction between invariant, universal structures and relativizing implementation rules is a well-precedented theoretical move that shows the Chomskyan matrix from which Fiske's theory has emerged (Ekman, 1989; Jackendoff, 1992, 1994). The relational models theory, in other words, sketches a universal grammar of social relations, or perhaps four distinct grammars, out of whose rules and representations the myriad local forms of social life can be generated.

Empirical studies of the relational models theory have demonstrated the empirical coherence, social–cognitive centrality, cross-cultural generality, cognitive accessibility, and discreteness of the models. Exploratory and confirmatory factor analytic investigations (Haslam, 1995b; Haslam & Fiske, 1996b) have shown that features of the relational models germane to a wide variety of social domains systematically co-occur as predicted in social relationships. Fiske, Haslam, and Fiske (1991) demonstrated that the relational models govern naturally occurring slips of naming, person memory, and social action: People inadvertently substitute others with whom they relate in terms of the same model, confusing spouses with lovers, teachers with bosses, and so on. Fiske (1993) later replicated this finding on social errors in four disparate non-European cultures. In addition, Fiske (1995) showed that the models organize the free recall of acquaintances, such that acquaintances with whom the person has a certain kind of relationship tend to be recalled in runs.

Although the aforementioned studies indicate that the relational models give pattern to social cognition when they are not explicitly primed and in the absence of conscious decision, the models also appear to be somewhat cognitively accessible (Rozin, 1976; Rozin & Schull, 1988). Fiske and Haslam (1995) replicated the social error findings for intentional social substitutions: When an originally intended interaction partner is unavailable, people choose replacements with whom they relate in terms of the same model. Haslam and Fiske (1992) demonstrated that people's free sortings of their social relationships map well onto their classification of these relationships according to written descriptions of the models. Finally, a series of studies has offered strong support for Fiske's claims that the relational models are discrete cognitive structures, and do not simply represent poles on continuous relational dimensions (Wish, Deutsch & Kaplan, 1976). Haslam (1994a) supported the discrete boundedness of all four models using taxometric procedures, a set of results subsequently replicated with a novel feature set (Haslam & Fiske, 1996a). These studies indicate that individual relationships categorically are or are not governed by each relational model, and that many relationships are governed by some combination of two or more. Haslam (1994b) yielded further support for the categorical representation of social relationships, finding that prototypicality judgments for abstract relationship descriptions varying in authority and communality were better modeled by discrete Fiskean categories than by continuous dimensions or complementarity/symmetry laws.

All in all, the relational models theory has garnered considerable empirical support as an account of human social relations and their mental representation. Not only has it been supported in a wide variety of tasks and social–cognitive domains, but it has consistently predicted responses at least as well as alternative relational taxonomies, and usually with effect sizes that exceed those owing to the individual-level characteristics of persons—sex, age, race, and personality—that have been the focus of most social–cognitive research (Fiske, 1995; Fiske & Haslam, 1995; Fiske, Haslam, & Fiske, 1991; Haslam & Fiske, 1992). Nevertheless, despite the relational models theory's empirical support, its links to forms of primate social behavior have yet to be drawn, and it is to that task that we now turn.

Communal Sharing and Kinship

As the least formally complex of the relational models and by Fiske's account the first to appear developmentally, we should perhaps be especially confident about finding precedents for *Communal Sharing* in the primate world. Indeed, we only have to look as far as kinship relations. The theory of inclusive fitness (Hamilton, 1964) argues the importance of distinguishing between kin and nonkin, and directing help preferentially to the former, for all social creatures. As Cheney and Seyfarth (1990) noted, kin recognition has been much studied and established even in social insects. Whereas most species appear to employ heuristics based on familiarity or phenotypic similarity to establish kinship, some nonhuman primates appear to make more categorical distinctions (Dasser, 1988), to recognize kinship relations among other conspecifics (Cheney & Seyfarth, 1990), and to use distinct calls to communicate referentially about kin and nonkin (Gouzoules, Gouzoules & Marler, 1984). Unconditional grooming between kin is commonly observed (Cheney & Seyfarth, 1990), as is tolerance of payoff asymmetries.

All of these phenomena closely parallel the organization of *Communal Sharing relations*, such as the relative undifferentiation of kin group members, the concern with and categorical marking of in-group/out-group boundaries, and the lack of enforcement of symmetry or reciprocal obligation. At the same time, it is important to recognize that some aspects of primate kinship do not fully parallel *Communal Sharing*: Often clear differentiations are made among kin, and sometimes these can be modeled in terms of degrees of relatedness. Nevertheless, it can be argued that these differentiations are best understood in the context of a more fundamental kin/nonkin distinction, and that they reflect an unformalized intensity dimension that is quite compatible with the cognitive discreteness of the kin group.

Human *Communal Sharing*, of course, is not restricted to kin; it is one way to organize intimate relationships of many kinds, and characterizes some aspects of many close group identifications, from gangs to nations. Whether nonkin groupings in some higher primate species also follow a *Communal Sharing* logic is not clear (Harcourt, 1992). Nevertheless, in the human case it is reasonable to see this logic of equivalence at work, decoupled from kinship in some instances (in a way that will be discussed later in this chapter), but directly traceable to it in ways of thinking, acting, and speaking (e.g., the ideal of fraternity, the "Motherland," the "human family," and so on).

Authority Ranking and Dominance Hierarchy

Just as the equation of *Communal Sharing* and kinship relations in primates is fairly straightforward, so is the link between *Authority Ranking* and dominance. Dominance hierarchies are ubiquitous in primates, varying across species in their rigidity, pervasiveness, and spheres of operation (Hand, 1986; Vehrencamp, 1983). The hierarchies of our nearest relatives, the chimpanzees, appear to be relatively flexible and less based on individual competitive ability than those of other species, and their agonistic exchanges are notorious for their political complexity (de Waal, 1982, 1989). The possibility that dominance is represented in a categorically distinct way from other relational elements is supported by the finding that distinct referential calls in monkeys discriminate dominance and submission (Cheney & Seyfarth, 1990). Although hierarchies are evident in other organisms, primates appear to be unique in their capacity to make the transitive inferences by virtue of which they can infer the relative positions of their peers (McGonigle & Chalmers, 1977; Premack, 1983). Moreover, this capacity appears to depend on normal socialization, with socially deprived rhesus monkeys being unable to assess status differences (Anderson & Mason, 1974). In primates, then, the recognition and enaction of dominance relations may depend on a distinct, emerging mode of social cognition.

Although the marking of status in humans is highly complex, and plainly transitive hierarchies are most evident in traditional societies and formal organizations, authority relationships represent a distinct and underrecognized class of social relations (Haslam, 1994a). Just as kinship and status form the preeminent dimensions of primate social organization, so are related dimensions preeminent in a variety of domains of human sociality. Communality and authority are the principal dimensions of interpersonal behavior (Wiggins, 1979), interpersonal problems (Horowitz, Rosenberg, Baer, Ureno, & Villasenor, 1988), and interpersonal language (White, 1980), and similar dimensions are prominent in the domains of personality (agreeableness and extraversion; McCrae & Costa, 1989) and emotion (Kemper, 1978). People appear to be sensitive to physiognomic cues to these dimensions (Keating, 1985), variations on which also figure in many accounts of gender differences (Sidanius, Cling & Pratto, 1991; de Waal, 1993). Once again, there appear to be clear continuities between humans and nonhuman primates regarding the realization and representation of dominance. Although in humans the evidence for a distinct form of mental representation is stronger, and the ways in which authority is realized may be more varied, flexible, and sometimes unique (e.g., pedagogy, Premack, 1984; Tomasello, Kruger & Ratner, 1993), the case for a parallel social–cognitive apparatus is suggestive.

Equality Matching, Alliances and Coalitions

The reciprocity central to the *Equality Matching* model is at the very core of the theory of reciprocal altruism (Trivers, 1971), which was developed to account for cooperation between nonkin. Significantly, the *Equality Matching* principle of balanced egalitarian exchange, realized in equal outcomes and turn-taking, corresponds very closely to the simple tit-for-tat mechanism shown by Axelrod (1984) to be a superior solution to the repeated prisoner's dilemma game on which the problem of cooperation is commonly modeled. Also significantly, reciprocal altruism appears to be especially suited to social organizations composed of long-lived, frequently interacting organisms living in large groups (Hemelrijk & Ek, 1990; Messick & Liebrand, 1995), conditions met in many primate species and among hominids. Although reciprocity has been demonstrated in a variety of species, it seems to be manifest most fully and certainly in its most calculated forms among the primates (Harcourt, 1992), with important differences between species.

Among primates, reciprocity that is not simply based on symmetrical association among kin is chiefly observed in the formation of short-term coalitions and longer-term alliances, commonly in the context of striving for "leverage" (Hand, 1986) in struggles for social dominance. By this means, de Waal and Harcourt (1992) noted, "alliance formation links the vertical and horizontal components of social organization" (p. 5) in a way that greatly complicates social problem solving. Indeed, one theory of the evolutionary rise in primate intelligence, the Machiavellian intelligence hypothesis (Byrne & Whiten, 1988), argued that it was driven by the tactical maneuvering called into play by the cultivation of allies. True reciprocity calls upon advanced cognitive abilities, including keeping track of obligations sometimes over substantial time delays, the calculation of approximate balance, and the detection of others' failure to reciprocate equally (i.e., "cheating;" Cosmides, 1989). Cognitive limitations may explain why reciprocation appears to reach its fullest expression in chimpanzees. Only they have demonstrated retaliation against those who form alliances against them (de Waal & Luttrell, 1988), indicating a revenge or "moralistic aggression" system. De Waal (1991) went so far as to ascribe prescriptive rules to chimpanzees, corresponding to rudimentary justice or fairness norms, thereby significantly narrowing the gap between human and primate implementations of reciprocity.

Market Pricing and Human Distinctiveness

The forms of reciprocity displayed by at least some ape species appear to map rather cleanly onto the operation of the *Equality Matching* model in humans in such phenomena as turn-taking, friendship, punishment of cheating, heuristic attention to approximate balance, and prescriptive fairness rules. Their cognitive complexity is attested by the lack of their development in young children (Grammer, 1992), at a time when kinship and dominance relations appear to be well understood. However, an ambiguity in the concept of reciprocity stands in the way of establishing deeper correspondences between its human and nonhuman primate forms. As Cheney and Seyfarth (1990) noted, this ambiguity concerns the sense of equality, which must be understood from the animal's inscrutable point of view. If we are to invoke certain forms of cost/benefit computation underlying the phenomena of primate reciprocity, we must know how these costs and benefits are calculated.

To the extent that the *Equality Matching* model or a close analogue underlies the phenomena of primate reciprocity, calculation should be somewhat approximate and based on conventionalized equivalence, rather than on "rational" calculations of proportions and an implicit market mechanism. Whereas the former represents a simpler kind of computation, the latter is economic in a stronger, more sophisticated, sense. In *Equality Matching* relationships there is likely to be some latitude for small departures from equivalence, which Trivers (1971) allowed for and Boyd (1989) showed not to constitute serious threats to the evolutionary stability of reciprocal altruism. At any point in time, such a relationship either is or is not in balance, and imbalances are reckoned in relatively primitive additive interval form. Costs and benefits are only represented, that is to say, in terms of net advantage, disadvantage, or equality, normally within the time window of the most recent exchange of turns, rather than in terms of rates of return on investment of time, effort or goods, or considerations of proportional equity. What is at stake, in brief, is equality of amount versus equality of proportion. Without knowing how primates figure equality, this comparison is difficult to draw.

This issue is made all the more vexing by the fact that much primate reciprocity is not in kind. For example, grooming is commonly exchanged for support within an alliance (Cheney & Seyfarth, 1990), raising the possibility that a shared cognitive currency exists into which favors and obligations can be translated and traded. Hemelrijk and Ek (1990) distinguished this form of exchange from in-kind reciprocity and labeled it "interchange," inferring a degree of bargaining. Noë and his colleagues (Noë, 1992; Noë, van Schaik, & van Hooff, 1991) went even further in invoking explicit bargaining over profitable payoff divisions and complicated comparisons of the relative rates of return to be gained from alternative partners within baboon alliances. Within such alliances, Noë documented consistent payoff asymmetries based on the different options possible to players differing in power. Can we therefore assert an evolutionary continuity in the fourth and most complicated relational model?

There are reasons why we should be circumspect about ascribing to some primate species the capacity to enact marketlike exchanges using computations on proportions. Whether the outcomes of primate exchanges can be usefully analyzed with the tools of bargaining theory and the assumptions of economic analysis is another matter, of course, as we are interested here in proximate mechanisms rather than in outcomes. This distinction is vital here because, as Quiatt and Reynolds (1993) argued, references to cost–benefit accounting are often no more than a "layered metaphor for all that remains to be discovered about the relation between blind strategies and tactical decisions" (p. 74). I argue that there are four reasons against extending human *Market Pricing*-like social cognition into the nonhuman primate domain, and that the forms of cost–benefit thinking revealed in primate reciprocity are in essence primitive and not economic in the standard sense. First, there is no evidence to suggest that any primates lacking the benefit of language training are capable of solving problems requiring an understanding of proportionality (Woodruff & Premack, 1981) or of relations among relations (Premack, 1983). These capacities underlie market-style bargaining, which is carried out in the currency of proportions (i.e., prices and rates of return) and involves comparisons of rates in search of the most favorable. Second, it is difficult to disentangle phenomena such as those documented by Noë from the ubiquitous effects of dominance. Asymmetrical outcomes within alliances in the context of social hierarchy are probably better ascribed directly to the effects of precedence and power rather than indirectly to a complicated exercise of market reckoning. Demonstrations of market-style bargaining would be more compelling if they involved partners of equal rank or societies of a more egalitarian cast than those of baboons. Third, reciprocity that is not in kind does not require calculation of proportional value, but only a more rudimentary assessment of comparability, which may be embodied in ritualized or conventionalized exchanges rather than calculated afresh on each occasion. Certainly much human reciprocity is of this sort; the ledger of obligations is balanced by alternating exchanges of discrete acts that are both assimilated into an encompassing category, such as "small favor."

A fourth and crucial difference between human *Market Pricing* and primate interchange is related to this rough, categorical equalization of favors. As Cheney and Seyfath (1990) noted, primate reciprocity almost always involves exchange of acts rather than objects, whereas objects are commonly exchanged by humans. Although by Fiske's (1991) account acts might equally well be exchanged according to the *Market Pricing* model, it is important to recognize that objects, as tradeable goods, represent its prototypical field of operation. This may be no accident; external, substantial, and divisible things may be more readily represented in terms of quantity, number, and proportion than acts, which are probably construed more as matters of kind and quality. By this account, which needless to say, is speculative, the emergence of *Market Pricing* may be tied in hominid evolution to the exchange of goods. This emergence might be located within the egalitarian structure of the hunter–gatherer society (Cashdan, 1980), in which advantage might flow to those who could realize the profits of calculated, tactical departure from equality of exchange (e.g., basing distribution of a scarce good on assessments of what others can give in return and their state of need), or alternatively within the context of trade between social groups. In addition to yielding a fitness benefit, it need hardly be said that market calculation would benefit greatly from the capacity to take another's perspective, to attribute mental states such as desires and intentions, and to deceive on the strength of such attributions. Trade and bargaining may therefore have been one important domain in which selective pressure operated on the complex of abilities variously dubbed *the intentional stance* (Dennett, 1987), *mindreading* (Whiten, 1990), and *theory of mind* (Premack & Woodruff, 1978), abilities in which humans excel relative to apes. Once developed, whether or not in the context of market reckoning, these abilities would surely feed back into a great increase in the complexity of social activity governed by all of the relational models.

One problem with the scenario of *Market Pricing* arising endogenously within egalitarian hominid groups is that it would run into rigid fairness norms, possibly built into hominid psychology as cheater-detection rules (Cosmides, 1989), but certainly collectively enforced and punished. The second scenario based on trade between groups may avoid this difficulty, to the extent that fairness norms only operate within groups. Organized intergroup contact appears to have been a prominent and distinctive feature of early human societies (King, 1994; Rodseth, Wrangham, Harrington, & Smuts, 1991), and Stringer and Gamble (1993) argued that the existence of extensive trade networks among distant social groups of modern humans was one factor that played a role in their supplanting of the Neanderthals. Although the connection between organized intergroup contact and the emergence of *Market Pricing* is again highly speculative, it does seem plausible to argue that *Market Pricing*, as a calculative skill and as a representational frame, might constitute a powerful adaptation most likely to emerge under conditions of relaxed egalitarian norms and in concert with an array of "mindreading" skills.

THE RELATIONAL MODELS
AND HUMAN SOCIALITY

Review of the primate literature demonstrates a number of continuities with the relational models theory. Close correspondences can be drawn for three of the models, precedents for which can be observed in a progressively narrowing range of animal species. Kin-based discriminations, which are widespread in invertebrates, are more widespread than recognition of dominance relations, which are most developed in birds and mammals. Egalitarian alliance behavior and calculated reciprocity appear exclusively, perhaps apart from some marine mammals, among the primates, reaching full complexity among the apes. I have argued that *Market Pricing* phenomena, involving cost–benefit computations on rates and proportions, probably occur only in humans. Significantly, this ordering of cognitive exclusiveness corresponds to Fiske's (1991) description of the increasing formal complexity of the four models, and of their probable order of appearance in social–cognitive development. Whatever the explanation for this correspondence, two general points must be stressed.

First, social exchange takes several abstract forms with distinct cognitive underpinnings, and variations of some of these forms can be identified in humans and nonhumans. Second, these forms do not all involve explicit calculations of costs and benefits, and of the two that plainly do—*Equality Matching* and *Market Pricing*—only the latter approximates cost–benefit accounting in its commonly understood economic sense. Some primate reciprocity involves a great deal of calculation, to be sure, much of it over who is worth cultivating and who is owed, but there may be no shared cognitive currency for reciprocation, and no finely detailed reckoning of extent of obligation. Instead, there is the relatively simple recognition that imbalance and obligation exists, and no figuring of return on investment. In brief, exchange is not a singular phenomenon, cost–benefit calculation is not common to all exchange, and marketlike cognition is only one variant of cost–benefit calculation broadly conceived.

The presence of *Market Pricing* phenomena does not exhaust the distinctiveness of human social life, of course, and it is interesting to speculate on other differences in terms of the relational models theory. One uncontroversial way to characterize these differences is in terms of complexity. Although the "fission–fusion" social organization of chimpanzees—its frequent formation of new associations through division and combination—is very fluid and entangled, human social organization and culture are much more complicated still. This complexity might be divided, somewhat artificially, into two parts: the ramification of conventionalized social roles, and the ramification of social institutions. In regard to the first, human social groups show a proliferation of specialized social positions, aided, no doubt, by the advent of language and the allowances it makes for naming and the cultural transmission of norms and practices. As Quiatt and Reynolds (1993) commented, "In order to build a human social system, our ancestors needed an intelligence capable of inventing new categories, and then of labeling and reifying them" (p. 255). Regarding the second aspect of social complexity, human societies have a variety of organized forms of belief and behavior that arguably have no straightforward parallels in nonhuman primates, such as religion, war, cosmologies, codified moral systems, feuding, rituals, and so on. As with the multiplication of roles, these social institutions rely greatly on language and cultural transmission. Over and above such factors, what sense might the relational models theory be able to make of these two forms of social complexity?

One answer to this question regarding role complexity is that the relational models can be considered as discrete building blocks out of which conventional role categories are constructed (Fiske, 1991; Haslam, 1994a; Jackendoff, 1992), in combination with implementation rules set by the local culture. By such an account, social roles might be thought of as distinctively implemented admixtures of the models, such that some aspects of role performance in certain contexts are governed by one model and other aspects in other domains by other models. The role requirements of being a teacher, for instance, might be constructed in *Authority Ranking* terms in relation to students, *Equality Matching* terms in relation to fellow teachers, and *Market Pricing* terms in relation to school administrators. Relational models research documents a very high degree of model combination in social relationships (Haslam, 1994a), and to the extent that this finding reflects conventional role requirements as distinct from the flexibility of individual social behavior, it supports the claim that roles are derived from relational models. More compelling support for the derivativeness of conventional social roles comes from studies demonstrating that the relational models mediate effects of roles on social cognition and not the reverse (Fiske & Haslam, 1995; Fiske, Haslam & Fiske, 1991). For example, people tend to choose social replacements with whom they interact according to the same model that governed their interactions with their originally intended partner, even when the two substitutes enact different roles toward them, but people do not preserve the substitutes' role when the relational model is different. Moreover, conventional role terms are much less informative about relationship features than are the relational models (Haslam, 1994a): Knowing the relational model that governs an interactant's relationship enables more inferences about the person's probable social behavior than knowing the colloquial role term that describes their position in the relationship. Although this argument can only be sketched, its claim is that the diversification of human social roles is due to the generativeness of the relational models. The models enable this diversification by offering a collection of discrete relational elements out of which relationships can be built. Social complexity increases exponentially. In a hierarchical monkey society each monkey might occupy only one or a few single-model roles in relation to another—subordinate, superordinate, kingroup member, alliance partner and so on—whereas in human societies, not only are there more roles for acquaintances to inhabit, but each acquaintance may hold several in relation to another person, and each role may combine distinct relational elements.

Regarding the complexity added to human sociality by novel social institutions, another partial answer based on the relational models theory presents itself. Such institutions do not arise without precedent, and in many cases can be shown to bear very clear structural similarities to the relational models. Although Fiske (1991) offered a much more extensive catalogue of such connections, several might be mentioned. Relations with deities are often organized in *Authority Ranking* terms; ethnocentrism and xenophobia appear to reflect a sense of collective identity or metaphorically extended kinship governed by *Communal Sharing*; feuding between groups, a phenomenon apparently absent in our closest primate relatives (Boehm, 1992), reflects the tit-for-tat retaliation and preoccupation with balance characteristic of *Equality Matching*; and moral, legal, and ideological systems commonly bear the imprint of any one of the models. All of these phenomena appear to receive their form, at least in part, through the extension of the relational models into novel domains. The models, that is to say, although in some sense designed to organize social transactions among individuals, are capable in human social institutions of governing relationships between groups, with pets or divinities, with abstract principles of nature, and so on. In short, among humans the relational models appear to be readily decoupled from their original social functions.

I have argued that two aspects of the distinctive complexity of human social organization are role proliferation and the development of social institutions, and have claimed that the former may use the relational models as building blocks and the latter apply them to new cultural developments, such as religious beliefs and ideologies. It is interesting to ask how this might be cognitively possible. What is it about these elementary relational forms that enables them to serve novel developments in this fashion? Two tentative but linked specula-tions may be germane to this question: that the models represent discrete, abstract categories, and that they have become increasingly cognitively acces-sible in humans (Rozin, 1976; Rozin & Schull, 1988). Regarding categorization, Cheney and Seyfarth (1990) acknowledged the great benefits that relational categories offer for social perception and parsimonious representation of social information in memory and reviewed evidence for their existence in nonhuman primates (e.g., Dasser, 1988). Knowing the kind of relationship that one individual has with another supports a wide variety of social inferences and is much less demanding on memory than keeping track of the details of the pair's interactions (Haslam, 1994a). The cognitive flexibility and economy afforded by abstract relational categories would greatly facilitate the combination of relational elements in social roles in which obligations, rights, and responsibili-ties must be coordinated between individuals. In this connection it is worth noting the existence of adaptive specializations for categorization in two other domains crucial to social interaction, namely speech and facial affect perception (Etcoff & Magee, 1992; Liberman & Mattingley, 1989; cf. Haslam, 1995a). The findings of categorical representation of social relationships in humans (Haslam, 1994a, 1994b; Haslam & Fiske, 1996), of the frequency of social relationships being governed by more than one category (Haslam, 1994a), and of the apparent derivativeness of social roles from the relational models (Fiske, Haslam, & Fiske, 1991), make it tempting to attribute human social complexity to the discreteness of relational categories.

Categorization may also be linked to the second claim, that the relational models are more cognitively accessible in humans than in nonhumans. By Rozin and Schull's (1988) account, an important component of its evolution of human intelligence has been its increased ability to generalize initially domain-specific adaptations. The relational models represent just such adaptive specializations, and to the degree that they give shape to a variety of social institutions, decoupled from and generalized beyond the coordination of basic social ex-change, cognitive accessibility is a plausible explanation. Evidence for the accessibility of the models can be found in the strong associations between people's naive sortings of their relationships and their categorization in terms of the models (Haslam & Fiske, 1992) and in the tendency for people to make planned substitutions of acquaintances in social activities in ways predictable from the models (Fiske & Haslam, 1996). Although the concept of accessibility is somewhat vague, it offers one partial explanation for the degree to which the relational models permeate the more distinctive forms of human culture.

CONCLUSION

An analysis of the relational models theory from an evolutionarily informed standpoint yields several claims. First, there is considerable evidence for continuities between primate social organization and cognition and the elementary forms of human social relations posited by the theory. Second, marketlike cost–benefit accounting may, however, be a distinctively human means of social reckoning, and may have arisen in the context of trade between social groups and as a privileged domain for the selection of mental state attribution. Third, two further aspects of human distinctiveness—role proliferation and social institutions—may be understood in part as outgrowths of the capacity of the relational models to be combined and generalized, with these capacities depending, in turn, on the models' abstractness, discreteness, and increased cognitive accessibility.

All told, scrutiny of the relational models theory in the light of evolutionary psychology challenges the strongly discontinuous view of human social cognition, according to which "the social–cognitive adaptations on which human culture and cultural learning depend came only after the differentiation [of humans and chimpanzees]" (Tomasello, 1994, p. 315). Rather, the relational models may be understood as adaptively specialized grammars, more or less rudimentary forms of which can be inferred in a variety of primate species. With the flexibility afforded by their discreteness, abstractness, and accessibility, these grammars undergird many components of human culture (Fiske, 1991; Jackendoff, 1992). Apparently discontinuous capabilities such as language and "mindreading" do not exhaust the human social–cognitive apparatus.

Despite the sketchy and speculative character of some arguments that support these conclusions, it should be clear that the relational models theory offers a rich source of theoretical and empirical claims. Crucially, it reaches out in two directions: toward those cognitive scientists who seek a vocabulary for thinking abstractly and computationally about social relations and equally toward those social scientists who see the merit of understanding culture and social organization in terms of rules and representations (Fortes, 1983; Sperber, 1994). Evolutionary social psychology would do well to look both ways.

ACKNOWLEDGMENTS

This chapter benefitted greatly from the comments of Alan Fiske, Nick Humphrey and the editors and was supported by NIMH grant 2R01MH43857 to Alan Fiske.

REFERENCES

Allison, S. T., & Messick, D. M. (1990). Social decision heuristics in the use of shared resources. *Journal of Behavioral Decision Making, 3,* 195–204.

Anderson, C. O., & Mason, W. A. (1974). Early experience and complexity of social organization in groups of young rhesus monkeys (*Macaca mulatta*). *Journal of Comparative and Physiological Psychology, 87,* 681–690.

Axelrod, R. (1984). *The evolution of cooperation.* New York: Basic.

Boehm, C. (1992). Segmentary "warfare" and the management of conflict: Comparison of East African chimpanzees and patrilineal-patrilocal humans. In A. H. Harcourt & F. B. M. de Waal (Eds.), *Coalitions and alliances in humans and other animals* (pp. 137–173). Oxford: Oxford University Press.

Boyd, R. (1989). Mistakes allow evolutionary stability in the repeated prisoner's dilemma game. *Journal of Theoretical Biology, 136,* 47–56.

Brown, D. E. (1991). *Human universals.* Philadelphia: Temple University Press.

Burnstein, E., Crandall, C., & Kitayama, S. (1994). Some neo-Darwinian decision rules for altruism: Weighing cues for inclusiveness as a function of the biological importance of the decision. *Journal of Personality and Social Psychology, 67,* 773–789.

Byrne, R., & Whiten, A. (Eds.). (1988). *Machiavellian intelligence: Social expertise and the evolution of intellect in monkeys, apes, and humans.* Oxford: Clarendon Press.

Cartmill, M. (1990). Human uniqueness and theoretical content in paleoanthropology. *International Journal of Primatology, 11,* 173–192.

Cashdan, E. (1980). Egalitarianism among hunters and gatherers. *American Anthropologist, 82,* 116–120.

Cheney, D. L., & Seyfarth, R. M. (1990). *How monkeys see the world: Inside the mind of another species.* Chicago: University of Chicago Press.

Chomsky, N. (1980). Rules and representations. *The Behavioral and Brain Sciences, 3,* 1–15.

Cosmides, L. (1989). The logic of social exchange: Has natural selection shaped how humans reason? *Cognition, 31,* 187–276.

Cosmides, L., & Tooby, J. (1995). From function to structure: The role of evolutionary biology and computational theories in cognitive neuroscience. In M. Gazzaniga (Ed.), *The cognitive neurosciences* (pp. 1199–1210). Cambridge, MA: MIT Press.

Dasser, V. (1988). Mapping social concepts in monkeys. In R. Byrne & A. Whiten (Eds.), *Machiavellian intelligence: Social expertise and the evolution of intellect in monkeys, apes and humans* (pp. 85–93). Oxford: Clarendon.

Dennett, D. (1987). *The intentional stance.* Cambridge, MA: MIT Press.

de Waal, F. B. M. (1982). *Chimpanzee politics.* London: Jonathan Cape.

de Waal, F. B. M. (1989). *Peace-making among primates.* Cambridge, MA: Harvard University Press.

de Waal, F. B. M. (1991). The chimpanzee's sense of social regularity and its relation to the human sense of justice. *American Behavioral Scientist, 34,* 335–349.

de Waal, F. B. M. (1993). Sex differences in chimpanzee (and human) behavior: A matter of social values? In M. Hechter, L. Cooper, & L. Nadel (Eds.), *Towards a scientific understanding of values.* Stanford, CA: Stanford University Press.

de Waal, F. B. M., & Harcourt, A. H. (1992) Coalitions and alliances: A history of ethological research. In A. H. Harcourt & F. B. M. de Waal (Eds.), *Coalitions and alliances in humans and other animals* (pp. 1–19). Oxford: Oxford University Press.

de Waal, F. B. M., & Luttrell, L. M. (1988). Mechanisms of social reciprocity in three primate species: Symmetrical relationship characteristics or cognition? *Ethology and Sociobiology, 9,* 101–118.

Ekman, P. (1989). The argument and evidence about universals in facial expressions of emotion. In H. Wagner & A. Manstead (Eds.), *Handbook of social psychophysiology* (pp. 143–164). Chichester: Wiley.

Etcoff, N. L., & Magee, J. J. (1992). Categorical perception of facial expressions. *Cognition, 44,* 227–240.

Fiske, A. P. (1991). *Structures of social life: The four elementary forms of social relationship.* New York: Free Press.

Fiske, A. P. (1992). The four elementary forms of sociality: Framework for a unified theory of social relations. *Psychological Review, 99,* 689–723.

Fiske, A. P. (1993). Social errors in four cultures: Evidence about the elementary forms of social relations. *Journal of Cross-Cultural Psychology, 24,* 67–94.

Fiske, A. P. (1995). Social schemata for remembering people: Relationship and person attributes in free recall of acquaintances. *Journal of Quantitative Anthropology, 5,* 305–234..

Fiske, A. P., & Haslam, N. (1996). *Who will replace them? The structure of social substitutions.* Manuscript submitted for publication.

Fiske, A. P., Haslam, N., & Fiske, S. T. (1991). Confusing one person with another: What errors reveal about the elementary forms of social relations. *Journal of Personality and Social Psychology, 60,* 656–674.

Fortes, M. (1983). *Rules and the emergence of society* (Royal Anthropological Institute Occasional Paper #39). London: Royal Anthropological Institute.

Gouzoules, S., Gouzoules, H., & Marler, P. (1984). Rhesus monkey (*Macaca mulatta*) screams: Representational signalling in the recruitment of agonistic aid. *Animal Behaviour, 32,* 182–193.

Grammer, K. (1992). Intervention in conflicts among children: Contexts and consequences. In A. H. Harcourt & F. B. M. de Waal (Eds.), *Coalitions and alliances in humans and other animals* (pp. 259–283). Oxford: Oxford University Press.

Greenfield, P. M. (1991). Language, tools and brain: The ontogeny and phylogeny of hierarchically organized sequential behavior. *Behavioral and Brain Sciences, 14,* 531–551.

Hamilton, W. D. (1964). The genetical theory of social behaviour. *Journal of Theoretical Biology, 7,* 1–52.

Hand, J. L. (1986). Resolution of social conflicts: Dominance, egalitarianism, spheres of dominance and game theory. *Quarterly Review of Biology, 61,* 201–220.

Harcourt, A. H. (1992). Coalitions and alliances: Are primates more complex than non-primates? In A. H. Harcourt & F. B. M. de Waal (Eds.), *Coalitions and alliances in humans and other animals* (pp. 445–471). Oxford: Oxford University Press.

Haslam, N. (1994a). Categories of social relationship. *Cognition, 53,* 59–90.

Haslam, N. (1994b). Mental representation of social relationships: Dimensions, laws or categories? *Journal of Personality and Social Psychology, 67,* 575–584.

Haslam, N. (1995a). The discreteness of emotion concepts: Categorical structure in the affective circumplex. *Personality and Social Psychology Bulletin, 21*, 1012–1019.

Haslam, N. (1995b). Factor structure of social relationships. *Journal of Social and Personal Relationships, 12*, 217–227.

Haslam, N., & Fiske, A. P. (1992). Implicit relationship prototypes: Investigating five theories of the elementary cognitive forms of social relationship. *Journal of Experimental Social Psychology, 28*, 441–474.

Haslam, N. & Fiske, A. P. (1996a). *Categorical structure and distinctive content of the relational models.* Manuscript in preperation.

Haslam, N., & Fiske, A. P. (1996b). *Relational models theory: A confirmatory factor analysis.* Manuscript submitted for publication.

Hemelrijk, C. K., & Ek, A. (1990). Reciprocity and interchange of grooming and "support" in captive chimpanzees. *Animal Behaviour, 41*, 923–935.

Hirschfeld, L. A., & Gelman, S. A. (Eds.). (1994). *Mapping the mind: Domain specificity in cognition and culture.* New York: Cambridge University Press.

Horowitz, L. M., Rosenberg, S. E., Baer, B. A., Ureno, G., & Villasenor, V. S. (1988). Inventory of interpersonal problems: Psychometric properties and clinical applications. *Journal of Consulting and Clinical Psychology, 56*, 885–892.

Humphrey, N. K. (1976). The social function of intellect. In P. P. G. Bateson & R. A. Hinde (Eds.), *Growing points in ethology* (pp. 303–317). Cambridge: Cambridge University Press.

Jackendoff, R. (1992). Is there a faculty of social cognition? In *Languages of the mind: Essays on mental representation* (pp. 69–81). Cambridge, MA: MIT Press.

Jackendoff, R. (1994). *Patterns in the mind: Language and human nature.* New York: Basic.

Keating, C. F. (1985). Gender and the physiognomy of dominance and attractiveness. *Social Psychology Quarterly, 48*, 61–70.

Kemper, T. D. (1978). *A social interactional theory of emotions.* New York: Wiley.

Kenrick, D. T., Groth, G. E., Trost, M. R., & Sadalla, E. K. (1993). Integrating evolutionary and social exchange perspectives on relationships: Effects of gender, self-appraisal, and involvement level on mate selection criteria. *Journal of Personality and Social Psychology, 64*, 951–969.

King, B. J. (1994). *The information continuum: Evolution of social information transfer in monkeys, apes, and hominids.* Santa Fe, NM: School of American Research Press.

Liberman, A., & Mattingley, I. (1989). A specialization for speech perception. *Science, 243*, 489–494.

McCrae, R. R., & Costa, P. T. (1989). The structure of interpersonal traits: Wiggins's circumplex and the five-factor model. *Journal of Personality and Social Psychology, 56*, 586–595.

McGonigle, B. O., & Chalmers, M. (1977). Are monkeys logical? *Nature, 267*, 694–696.

Messick, D. M., & Liebrand, W. B. G. (1995). Individual heuristics and the dynamics of cooperation in large groups. *Psychological Review, 102*, 131–145.

Noë, R. (1992). Alliance formation among male baboons: Shopping for profitable partners. In A. H. Harcourt & F. B. M. de Waal (Eds.), *Coalitions and alliances in humans and other animals* (pp. 285–321). Oxford: Oxford University Press.

Noë, R., van Schaik, C. P., & van Hooff, J. A. R. A. M. (1991). The market effect: An explanation for pay-off asymmetries among collaborating animals. *Ethology, 87*, 97–118.

Premack, D. (1983). Animal cognition. *Annual Review of Psychology, 34,* 351–362.

Premack, D. (1984). Pedagogy and aesthetics as sources of culture. In M. S. Gazzaniga (Ed.), *Handbook of cognitive neuroscience* (pp. 15–35). New York: Plenum.

Premack, D., & Woodruff, G. (1978). Does the chimpanzee have a theory of mind? *Behavioral and Brain Sciences, 1,* 515–526.

Quiatt, D., & Kelso, J. (1985). Household economics and hominid origins. *Current Anthropology, 26,* 207–222.

Quiatt, D., & Reynolds, V. (1993). *Primate behavior: Information, social knowledge, and the evolution of culture.* Cambridge: Cambridge University Press.

Rodseth, L., Wrangham, R. W., Harrigan, A. M., & Smuts, B. B. (1991). The human community as a primate society. *Current Anthropology, 32,* 221–254.

Rozin, P. (1976). The evolution of intelligence and access to the cognitive unconscious. In J. N. Sprague & A. N. Epstein (Eds.), *Progress in psychology* (Vol. 6). New York: Academic Press.

Rozin, P., & Schull, J. (1988). The adaptive-evolutionary point of view in experimental psychology. In R. Atkinson, R. J. Herrnstein, G. Lindsey, & R. D. Luce (Eds.), *Handbook of experimental psychology: Motivation* (pp. 503–546). New York: Wiley.

Sidanius, J., Cling, B. J., & Pratto, F. (1991). Ranking and linking as a function of sex and gender role attitudes. *Journal of Social Issues, 47,* 131–149.

Sperber, D. (1994). The modularity of thought and the epidemiology of representations. In L. A. Hirschfeld & S. A. Gelman (Eds.), *Mapping the mind: Domain specificity in cognition and culture* (pp. 39–67). New York: Cambridge University Press.

Stringer, C. & Gamble, C. (1993). *In search of the neanderthals: Solving the puzzle of human origins.* London: Thames and Hudson.

Tomasello, M. (1994). The question of chimpanzee culture. In R. W. Wrangham, W. C. McGrew, F. B. M. de Waal, & P. G. Heltne (Eds.), *Chimpanzee cultures* (pp. 301–317). Cambridge, MA: Harvard University Press.

Tomasello, M., Kruger, A., & Ratner, H. (1993). Cultural learning. *Behavioral and Brain Sciences, 16,* 495–552.

Tooby, J., & Cosmides, L. (1992). The psychological foundations of culture. In J. H. Barkow, L. Cosmides, & J. Tooby (Eds.), *The adapted mind: Evolutionary psychology and the generation of culture* (pp. 19–136). New York: Oxford University Press.

Trivers, R. L. (1971). The evolution of reciprocal altruism. *Quarterly Review of Biology, 46,* 35–57.

Vehrencamp, S. (1983). A model for the evolution of despotic versus egalitarian societies. *Animal Behaviour, 31,* 667–682.

White, G. M. (1980). Conceptual universals in interpersonal language. *American Anthropologist, 82,* 759–781.

Whiten, A. (Ed.). (1990). *Natural theories of mind: Evolution, development and simulation of everyday mindreading.* Oxford: Blackwell.

Wiggins, J. S. (1979). A psychological taxonomy of trait-descriptive terms: The interpersonal domain. *Journal of Personality and Social Psychology, 33,* 409–420.

Wish, M., Deutsch, M., & Kaplan, S. B. (1976). Perceived dimensions of interpersonal relations. *Journal of Personality and Social Psychology, 33,* 409–420.

Woodruff, G., & Premack, D. (1981). Primitive mathematical concepts in the chimpanzee: Proportionality and numerosity. *Nature, 293,* 568–570.

VI

Groups and Group Selection

12

Groups as the Mind's Natural Environment

Linnda R. Caporael
Rensselaer Polytechnic Institute

Reuben M. Baron
University of Connecticut

Our evolutionary perspective represents a departure from the neo-Darwinist approach, which emphasizes inclusive fitness theory and reproductive-mindedness (Buss, 1995). During the last 15 years, an alternative to neo-Darwinism has been emerging under various names: expanded evolutionary theory (Gould, 1980), constructionist evolution (Gray, 1992), developmental systems theory (Griffiths & Gray, 1994; Oyama, 1991), process evolution (Ho, 1991), or group selection (Wilson & Sober, 1994). Proponents of the latter views argue that significant evolutionary units of analysis cannot be confined merely to the gene or the individual, as is the case for neo-Darwinism. Instead, there are multiple levels of organization upon which selection may occur, resulting in evolutionary dynamics that cannot be reduced to a summary of selection on alternative alleles. Leo W. Buss (1987), a biologist at Yale University, summarized the hierarchical perspective:

> The history of life is a history of transitions between different units of selection. . . . When a transition occurs in the units of selection, synergisms between the higher and lower unit act to create new organizations which may allow the higher unit to interact effectively in the external environment. However, the organization of the higher unit does not simply interact with the external environment; it is also the agent of selection on the lower unit. To the extent that control over replication of the lower unit is required for effective interactions with the external environment, organizations must appear in the higher unit to limit the origin or expression of variation at the lower unit. Any such organization will act to stabilize the higher unit, as it limits the capacity for variants to arise or be expressed. (p. 171)

Thus, genes must fit the environment of the cellular machinery; cells must fit the environment of their individual organism; and organisms must fit the environment of the next higher level of organization. There are, however, conflicts as well as synergisms between successive levels of organization. For example, there must be synergism among the cells of the body for an individual to function, but cell lineages can compete, as in the case of cancer. At the intraorganismic level, cancer cells demonstrate differential reproductive success. However, at the next higher level of organization, the success of the cancer cell lineage is the destruction of the organism—and with it, all of the competing cell lineages.

The difference between the neo-Darwinian and evolutionary systems approaches is particularly apparent in the treatment of groups. The sociobiological model predicts grouping on the basis of kinship (Hamilton, 1964), maintaining a balance of power (Alexander, 1979), self-interested interpersonal manipulation (Byrne & Whiten, 1988), or reproductive advantage and protective aggregation (Burgess, 1989). Psychologically, a group is an aggregate of individuals, each assessing (not necessarily consciously) the costs and benefits of social exchange. Evolution records the results of such individual choices as changes of gene frequencies within a population. In the hierarchical perspective, grouping is a feature of functional interdependence, which can result in the formation of a new entity or "individual." Self-replicating molecules can become incorporated in self-replicating gene complexes, which in turn can become incorporated in cell lineages, and finally in multicellular organisms, then in groups, and even ecosystems. Evolution is a historical trend in the expansion of interdependency, through conflicts and synergisms, across different scales of organization in both phylogeny and ontogeny (L. Buss, 1987). Where grouping, as opposed to mere aggregation occurs, there should be evidence of interdependence of function.

This chapter is an initial effort to apply the hierarchical perspective to the evolution of human social coordination. We begin with a discussion of evolution as "repeated assembly," as an alternative to nature–nurture dualism and dualistic gene–environment interactionist projects. This view of evolutionary and developmental processes provides the beginning of a vocabulary that more closely captures the claims and qualifications of evolutionary psychologists (Buss, 1995; Tooby & Cosmides, 1992; cf. Oyama 1985, Smith, 1992). It is also compatible with the reentry of group selection into the human sciences (Wilson, chapter 13, this volume). We then briefly introduce some concepts from dynamic systems theory, focusing on self-organizing systems to capture the contingent causal relationships characterizing ontogeny. Following these considerations of language and conceptual framework, we introduce four core configurations or specialized social-organizational domains, which we propose constitute the mind's natural environment. We then suggest how our framework lends itself to reinterpreting familiar research in personality and social psychology, as well as opening new domains of research.

EVOLUTIONARY DYNAMICS

Matters of Metaphor

Scientists rarely reflect on their metaphors, but Darwin (1896/1972) was a notable exception. His metaphor for natural selection was an intelligent agent that selects or favors. He deplored the mischief that this anthropomorphism caused and regretted not having a better alternative. We believe a digression here about our metaphors can forestall confusion later. In our opinion, the appeal to intentional processes, no matter how discreet, has become an obstacle in evolutionary thinking, and we strive to avoid them. Our preferred metaphors come from physical rather than intentional systems.

Endler (1986) emphasized that natural selection is not a cause of species-typical traits, but the statistical outcome of three conditions: phenotypic variation, correlated reproductive variation, and inheritance. He compared natural selection to a chemical reaction in which the frequency of reactants changes as a result of the properties of the molecules, not because the reactants have been pushed around by a mysterious force. Similarly we find the chemical metaphor more useful than a view of inclusive fitness and reproductive success as motivating forces of natural selection or as motors driving selective processes or putting them in motion. These metaphors introduce an extraneous and unnecessary causal process. Similarly, natural selection cannot be viewed as an agent that endows organisms with goals and purposes. (For plants and bacteria, goals and purposes are more blatantly anthropomorphic than they are for easily visible, animate creatures.) The chemical metaphor captures the view of natural selection as a contingent effect of extant biological and physical conditions and turns attention to those conditions.

The chemical metaphor also serves to negotiate the ubiquitous pitfalls of nature–nurture dualism. Despite wide agreement that genes and environment interact in the production of phenotypic characteristics, or even that the phenotype cannot be analyzed into separate genetically determined and environmentally determined components (Buss, 1995; Tooby & Cosmides, 1992), there are habits of mind that betray the grip of dualistic thinking and privileging of the gene. Too often, allegiance to interactionism merely shifts the question of nature versus nurture to one of relative influence: How much nature? How much nurture? In other instances, researchers may speak of genes "shunting" behavior in one direction or another as if responses to all possible environmental contingencies were preformed in the genes and could be merely elicited by environmental cues.

In contrast, people rarely think of chemical reactions as having blueprints or programs. There are no agents, motivating principles, nor goals in reactions (and the earliest stages of development from a fertilized cell are little more than biochemical reactions). Reactions automatically result from the co-occurrence of constituents, as, for example, combustion results from the co-occurrence of various materials against the enabling background of oxygen. There is little temptation to privilege one constituent over another. Without a combustible, such as paper, there is no fire; without the enabling condition of oxygen, there is no fire. Similarly, neither genes nor environment can result in an organism without the joint co-occurrence of both in relationships that repeatedly occur ontogenetically and generation-to-generation. In complex reactions, there are sequences of events, in which events at time$_1$ create conditions (i.e., "information") for the occurrence of events at time$_2$. Similarly, the co-occurrence of various resources—chemical, genetic, cellular, ecological, social, symbolic, and technological—are among the causally contingent, yet absolutely essential, resources entering into human development.

The chemical metaphor breaks down on the issue of heritable variation, but not without an important reminder. All biological systems are also physical systems. The events that recur in biological systems may be the results of natural selection, but they may also be a result of dynamic properties of physical systems and thus require no recourse to natural selection to explain their operation. Moreover, the dynamic properties of physical systems (e.g., temporality, size, and space) are significant contexts for evolutionary and developmental processes.

Repeated Assembly

Charles Darwin's (1859/1963) original notion of descent with modification was based on the idea of a fit or match between an organism and its environment. In the struggle for existence, heritable variations, however slight and no matter what their causes, would contribute to survival and reproduction if the variations were "in any degree profitable to an individual of any species, in its infinitely complex relations to other organic beings and to external nature . . ." (p. 61). Darwin was equivocal about the causes of heritable variation. Partly, he lacked an account of the mechanisms of inheritance; partly, he vacillated on the role of acquired habits in selective processes (Richards, 1987).

The neo-Darwinian synthesis between Mendelian genetics and Darwinism altered the notion of "fit" from the sense of match between organism and environment to "fitness" as the reproductive success of individuals. Theoretical population geneticists linked fitness to the relative representation of alleles in a population, a view that was more or less directly imported into the human sciences. Yet, as psychologists, we have little interest in gene frequency changes in populations, per se. In our view, genes are at a level of analysis different from, and inappropriate for, evolutionary psychology. Our interest and, we suspect, the interest of most evolutionarily inclined psychologists, is the relational properties between the organism and environment, an interest much closer to Darwin's original idea of descent with modification, regardless of the source of heritability and of fit.

We treat these two concepts in this and the next section. Descent with modification is elaborated as repeated assembly, and the fit or match between organisms, (i.e., their relational properties) is reinterpreted in terms of a familiar topic in psychology, J. J. Gibson's (1979) affordances.

The term *repeated assembly* (Caporael, 1995) refers to recurrent relationships between an entity (e.g., genes or organism) and its context (e.g., cellular machinery or habitat). For example, a human zygote is the repeated assembly of centrosomes from the sperm, two sources of DNA, maternal constituents, and various other cellular components. The zygote is not assembled by the genes. It is rather the automatic result of a set of genetic and epigenetic resources, all of which, if in the right place at the right time, reliably result in the zygote. Similarly, the continued development of the organism results from the reliable recurrence of appropriate contexts, from cellular machinery to social events to constancies of atmosphere (Griffiths & Gray, 1994). Neither genes nor environment are privileged in this view (although the microbiologist would still focus on genes, and the educational psychologist might focus on classrooms).

Repeated assemblies can be inorganic as well as organic; the substance of components in repeated assemblies may be irrelevant or critical. For example, humans have distinctive patterns of tool-use and manufacture that have changed markedly throughout their evolutionary history, but some sort of tool-use is repeatedly assembled by members of the species. At the other extreme, the absence or substitution of a single gene can have effects ranging from death to lifestyle change (e.g., the dietary requirements of individuals with pheynlketonuria). Among the many ways that recurrent resources differ is in temporality. The constituent components of repeated assemblies are heterogeneous partly because they recur over different time scales. DNA recurs over macro-evolutionary time; at the other extreme, social rituals can rise and fall within a lifetime (affecting phenotype, but having no selective effects). The conditions resulting in evolutionary processes are concretely situated in the set of relations between an entity and its environment. Hence, evolutionary repeated assemblies would recur within lifetimes (at least once, as in zygote formation) and across generations on a geologic timescale. That is, the conditions resulting in evolutionary change are the aggregation within lifetimes and across generations of life history events.

The difference between a dualist view and the one we advocate here can be illustrated with language. In the dualist view, the ability to learn a language is innate and the particular language one speaks is learned. Tooby and Cosmides (1992) suggested this view in their account of a child who is socially isolated and deprived of a language learning environment. According to them, such a child would have a species-typical language acquisition device, but would lack the variable environmental inputs for the device to work. If we view language as a repeated assembly, however, we might allow that such a child might have genetic resources for such a device, but lacking another critical resource—speakers of a language—that device never assembled. In the hierarchical view, both the genes and the language environment are inherited. Speaking English and speaking Kikuyu differ because the components involved in the repeated assembly of language have different cycles of repetition. One set, which includes genes, has a longer cycle of repetition relative to another set, which includes the language environment. The relevant resources recur, but on different scales of time: The cycle of language environment (English or Kikuyu) in cultural–historical time is nested within the cycle of other components, including genes, in evolutionary time. Nested within these two cycles of evolutionary time and historical–cultural time, is a third cycle, the sine qua non of variation and evolutionary process: the situated relation of "organism-in-setting," (e.g., a child raised in social isolation) in which the conditions for an assembly to be repeated, be it later in the day or in succeeding generations, take place or fail to take place.

Some repeated assemblies, and language is one of them, have "proper functions" (Millikan, 1984). A relationship repeatedly assembles ("is selected") because, in the past, the co-occurrence of a subset of organismic and environmental components in some particular relation (e.g., timing, frequency, contingency) contributed to the persistence of the assembly. If certain other relations had correlated better with these functions, then those relations would have proliferated instead. In other words, some feature of the repeated assembly contributed to the replication of new versions of the assembly. Of course, some repeated assemblies have no function; they result from illusory contingencies, such as the superstitious circling of pigeons in a Skinner box. Other repeated assemblies may have had a function at one time, but became dissociated from it.

The difference between proper functions and other categories of function can be illustrated with common tools (Millikan, 1984). A screwdriver's proper function is joining and unjoining objects held together by screws. However, the screwdriver can also be used to pry lids off paint cans or poke holes in tough materials. Screwdrivers are not repeatedly assembled because they are used to pry off paint can lids or poke holes. They are repeatedly assembled because they turn screws. The proper function of hearts is to pump blood (which contributes to the repeated assembly of hearts), not to make ticking noises (which make no contribution to new versions of hearts). It is important to mention that the proper function of an item is not necessarily the normative or most common outcome. The proper function of sperm is to fertilize, but the vast majority of sperm fail to fulfill this function. Millikan (1984) presented a theory and taxonomy of function that should prove useful to evolutionary psychologists.

Affordances

Repeated assembly gives us a way of capturing the complexity of descent with modification. Affordances (Gibson, 1979; McArthur & Baron, 1983; Springer

& Berry, chapter 3, this volume) give us a way of capturing the match between entity and environment, including the relational properties of the organism-environment system. *Affordances* are special cases of organism–environment fit, and they are referenced with respect the organism or agent: Only organisms with the appropriate species-specific sensorimotor sensitivities and motor competencies (in the case of motile organisms) can detect or use particular affordances. The environment is inherently complex, a plethora of nested affordances, only some of which are appropriate to any given organism (including humans) from its perspective. For example, a house may afford shelter for humans, and a hole in the wall may afford shelter for a mouse. Similarly, stairs may afford locomotion for humans, whereas a flat wall may do just as well for a spider or a fly.

McArthur and Baron (1983) extended the concept of affordances to include cultural and personal histories and motivational states as relevant attunements for the actor's discovery of affordances, and Baron and Misovich (1993a) allowed individual competencies to be augmented by group membership. For example, a boulder that does not afford moving by an individual may afford moving by a group of people acting in concert. Because affordances are available does not mean they will be attended to or utilized. Needs, goals, and other intentional states can affect utilization (Baron & Misovich, 1993a). Groups can affect both capabilities (as in the boulder example) and goals relevant to utilization (e.g., group-level goals). Motivational issues and groups issues combine: Just because a rock affords moving by a group does not mean that the group will actually move it. There could be a group goal to lunch at noon, leading the group to focus at noontime on the affordance of this flat rock for sitting as opposed to its social-level movability. Thus, any complex entity is a constellation of affordances that vary in availability depending on a wide range of user properties. Needs or goals can provide selection criteria in the sense that a mouse affords play or food to a cat, depending on how hungry the cat is. The mouse objectively affords both eatability and playability for this class of predators.

Lest readers have difficulty appreciating how user properties impact on the existential status of phenomena, we refer them to the affordances of electrons to function as either waves or particles. That is, electrons are in a state of flux involving "a superposition of states—hungry and full at the same time, as it were" (Horgan, 1992, p. 101). In effect, "Photons, neutrons, and even whole atoms act sometimes like waves, sometimes like particles, but actually have no definite form until measured" (Horgan, 1992, p. 96). The "measurement" of the environment is always part of the properties and situated activity of the user.

Complex Dynamic Systems

For our purposes, the basic theme of complex dynamic systems (Baron, Amazeen, & Beek, 1994) is that they are self-organizing. That is, interactions among discrete entities are capable of producing novel or emergent properties.

Emergent properties could not have been predicted from the individual properties of the components; and there is no resemblance between the individual components and their aggregation at the level of emergence. Concrete and glass are two simple inorganic examples. What particular state emerges depends on conditions intrinsic to the system (e.g., the bonding properties of water, sand, and cement; the melting point of glass). The nature of the emergent property (e.g., tensile strength) is not specified by external variables or internal scripts, blueprints, or executive programs. Rather, the nature of the property is the product of a contingent sequence of interactions.

Emergence (or when we speak of organic systems, development) is a phenomena of dynamic, self-organizing systems. Such systems produce collective–emergent properties through the internal circular causality of the components of the system. Dynamically, the actions of the system are influenced by feedback from previous states of the system. That is, outputs from one level of organization become inputs to further system functioning. For example, aggregates form groups that then influence subsequent member behavior.

More technically, the coordination of the microcomponents involves interactions of individual components (local dynamics), which give rise to the macrodynamics of downward causation (Campbell, 1990a), whereby the collective emergent property modifies the behavior of the individual components that gave rise to the property (see Fig. 12.1). That is, whether a system is chemical, physical, or human, there can be transitions from homogeneous to complex states of organization involving increasingly structured patterns of behavior among the components. For example, as the number of people participating in even an informal club increases beyond a certain size, the nature of interaction changes from individuals "doing their own thing" to individuals carrying out rule-regulated roles as when the club is coerced to enact bylaws and elect officers. At a global level, a field is created that allows interaction at a distance. For example, when an aggregate of workers forms a union, interaction is no longer restricted to people in the same assembly line, division, or even plant, because interaction is at the collective level of union members.

At a process level, as depicted in Fig. 12.1, we move from sidewise interactions to upward self-organization resulting in the formation of a global level of structuring that, in turn, allows for downward causation in which the output of the local interaction is transformed into input for the next level of organization. More broadly, the individual components have generated a group or collective level environment for subsequent functioning. A good example is Newcomb's (1943) Bennington study and its subsequent follow-up (Newcomb, Koenig, Flacks, & Warwick, 1967). As women became integrated into the college environment, they helped to create a liberal ethos that 25 years later was reflected in their forming liberal social networks and marrying liberal spouses.

Order-producing constraints on interaction are a general property of self-organizing social systems. For example, when too many people want to talk,

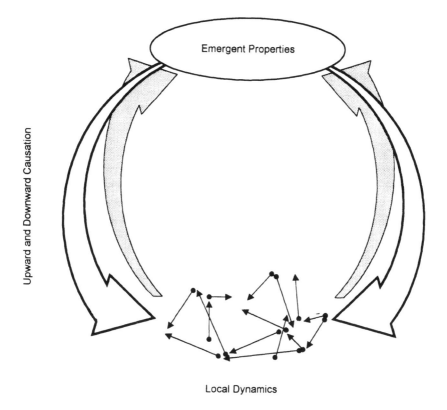

FIG. 12.1. Complex system. The small arrows are local dynamic systems. They are influenced by global dynamics (downward causation) and, in turn, influence global level systems (gray arrows).

unregulated shouting out does not work; we require some type of turn-taking rule to order the flow of communication. Similarly, when too many people started owning automobiles, devices such as stop signs and lights normatively order individual driving interactions. That is, individual attitudes, goals, and preferences become coordinated in ways that allow for new emergent properties at the group level, such as turn-taking norms. Those norms, in turn, allow an organization to emerge that has certain functional capabilities to exploit environmental affordances that the people as individuals do not have. For example, people working as a team, such as those navigating a ship, have a better chance to solve a problem of unknown dimensions than would individuals acting alone (Hutchins, 1990). Subsequent greater individual effectiveness, as a result of the downward causation effect of shared fate, is prototypic of a whole class of effects that reflect circular causality and emergence among system components.

Control and Order Parameters

The initial step in the production of a group norm involves the occurrence of events that shift people from individual modes of functioning to coordinated modes of social interaction. These triggering events are called *control parameters*. A simple social example is the proposed relationship between rising temperature and the occurrence of urban riots (see Forsyth, 1995, p. 561). Specifically, temperature may be assumed to increase rates of interaction among people in the course of generating deviant norms as the emergent states. In self-organization, however, the control parameter does not direct or determine the emergent organization. Thus, temperature does not shape the social organization. It merely increases rates of interactions among components. Such events are capable of generating nonlinear effects as the control parameter (in this case, temperature) passes through certain critical values. A small input change can produce a large change in mode of organization. For example, a small change in a stressor can change an aggregate into a group as when a study group coalesces into a "real" group days before an exam. (Here proximity to the exam is the control parameter.)

Familiar social psychological systems can be described in dynamic, complex systems terms. Consider Sherif's (1935) classic autokinetic paradigm of norm formation components of a circular causal system of self-organization. In his study an ambiguous stimulus, a pinpoint of light, appeared to move, but estimates of movement varied greatly. When a control parameter such as rates of interaction among subjects reached a critical threshold, the individual group fit was strained. New organizations occurred among components, and new affordances became available. By increasing sensitivity to discrepant judgments, rate of interactions spontaneously produced structures that ordered the interactions. The autokinetic problem instigated the emergence of the group vis-à-vis self-organization, one consequence of which was to modify the availability of affordances for dealing with an objective, intersubjectively available but unstable stimulus. The variance of estimated movement of the stationary light became considerably smaller, generating a norm that through downward causation resulted in individuals who functioned now as representatives of a group in the sense that they shared a common frame of reference. Furthermore, although the Sherif example involved a relatively short cycle, Newcomb et al.'s (1967) study of liberal Bennington students 25 years later, indicated both the persistence of a downward causation path and a further reorganization of the modified individual components as they created new social networks (friends or spouse selection) to perpetuate their individually changed judgments. In both

Sherif and Newcomb we can observe a self-organization mechanism showing how the whole can be in the parts and the parts in the whole (see discussion of self-identity, later).

Given dynamic relationships, efficient and stable phase relationships are expected to occur at a system level within a certain range of control parameters. These phase relationships are the *order parameters*, measures of "coordination, coherence, or cooperativity among the interacting components" (Baron et al., 1994, p. 119). In terms of a design metaphor, the order parameter is a measure of the extent to which a relationship exists that optimizes the fit between form and function. For example, in walking, a 180° discrepancy between the leading and the lagging leg is stable. Perturbed walkers resist disruption and/or rapidly attempt to repair the disruption, contingent on the speed of movement as a control parameter. In clumsy or awkward walking, the fit between form and function is visibly nonoptimal. At the group level a classic example of an order parameter at work is coordinating the time spent on task with time spent on socioemotional activities. Tuckman's (1965) proposed group development progression through "forming, storming, norming, performing, adjourning," can be viewed as a working out of this relation as groups oscillate back and forth between working hard and strengthening social bonds (Forsyth, 1995).

Control and order parameters also exist for establishing a turn-taking cycle for stable conversation. The cross-coupling of vocal amplitudes and nonverbal gestures (Newtson, 1994) coordinates the flow of information through the preferred frequency and amplitude of speaking. Such coordination builds order in the sense that initially independent actions can now be described by collective—a "couple" has emerged. Similarly, groups emerge from aggregates. We might picture a collection of oscillators resonating together much as a "wave" of spectators cheering occurs at a sporting event. With groups, such coordination persists because it is repeatedly assembled from the individual to the group at the level of roles. That is, in Fig. 12.1, the components at the level of local interaction can be the interacting of individuals, role occupants, groups, and so fourth. How these different units are related to each other is the next problem to which we turn.

THE MIND'S NATURAL ENVIRONMENT

Core Configurations

Like many other evolutionists, we believe that the face-to-face group is important to understanding human mental processes. However, we do not view all face-to-face groups as undifferentiated aggregates. Traditionally, a group is represented as an aggregate of individuals striving to optimize their reproductive fitness through pairwise social exchange relationships (Byrne & Whiten, 1988; Tooby & Cosmides, 1992; Winterhalder & Smith, 1992). Implicit in such individual selectionist views is an insensitivity to group size and structure. In contrast, we argue that throughout human evolutionary history, face-to-face groups have been structured as self-organizing systems by constraints imposed by both morphology and ecology.

Humans are an obligately interdependent species, unable to survive and reproduce outside a group context. The fit, or entity–environment relationships, that distinctively human mental systems achieve would be to group structure, and we should expect that there would be psychological correlates of that structure. Moreover, humans have been able to effect, and have been affected by, enormous changes in both their social organization and their habitats. Consequently, the psychological processes that we hypothesize must be stable through evolutionary time yet be flexible enough to engage in novel activity. In our view, the face-to-face group is the mind's natural environment.

The anthropological literature on nomadic hunter–gatherers identifies three levels of group structure: workgroups, bands and macrobands (Jarvenpa, 1993; Jarvenpa & Brumbach, 1988). Hunting and gathering is typically done by *workgroups*, subdivisions of a band. A band is composed of about 30 people. It serves domestic functions: butchering, preparing food for storage, childrearing. A *macroband* is a seasonal gathering of bands. Bands and macrobands are sources of group identity and function in the exchange of information, particularly about distant places, conditions, and resources.

To this structure, we add the dyad and hypothesize four evolutionary *core configurations* (Caporael, 1995). These are interdependent, face-to-face groups that afford different kinds of activity. Table 12.1 sketches the structural model. The labels given to the configurations are not intended to represent social roles; they represent kinds of interaction. A dyad is an interaction between two entities, one of which can be nonhuman. *Team* refers to "small-group, common task orientation" interactions, such as a family or workgroup. We have used the terms *deme* and *macrodeme* to indicate greater generality than that implied by anthropologists' use of *band* and *macroband*, and also to avoid inappropriate connotations of "primitive" social organization. (In biology, the term *deme* refers to a breeding population, but the word's origins are in ancient Greece. In Athens, demes designated geographic and political association and were part of an explicit policy to foster civic ties over kin ties.)

Each configuration is associated with a characteristic group size and task. Except for dyads, the numbers in Table 12.1 should be considered more as centers of gravity than as fixed sizes. Moreover, the combination of size and modal task, not just group size alone, constitute a configuration. Core configurations, in effect, are specialized domains for the evolution and development of psychologic i!l mechanisms. In the language we have developed so far, core configurations *repeatedly assemble*; they are developmental and social–cognitive

TABLE 12.1

Core Configurations

Core Configurations[a]	Group Size	Modal Tasks	Proper Funciton
Dyad	2	Sex, infant interaction with older children and adults	Microcoordination
Team (work/family group)	5	Foraging, hunting, gathering, direct interface with habitat	Distributed cognition
Deme (band)	30	Movement from place-to-place, general processing and maintenance, work group coordination	Shared construction of reality (includes indigenous psychologies), social identity
Macrodeme (macroband)	300	Seasonal gathering, exchange of individuals, resources, information	Stabilizing and standardizing language

[a]Core configurations are a function of both size and task. Except for dyads, these numbers should be considered as modal estimates.

Note. From "Sociality: Coordinating Bodies, Minds and Groups" by L. R. Caporael, 1995, *Psycoloquy [on-line serial]*, 6 (1). Available FTP: Hostname: Princeton.edu Directory: pub/harnad/psycoloquy/1995.volume 6 File: psycoloquy.95.6.01 griyo-selection.1.caporael. Copyright 1995 by L. R. Caporael. Adapted with permission.

affordances, and they have *proper functions*. There are, of course, many kinds of social configurations (e.g., crowds or parades), but they are not core. Just as a screwdriver can be used to pry the lids off paint cans, so the psychological correlates of core configurations can be used in the assembly of other kinds of configurations.

Obviously, more research is needed to establish core configurations empirically and to identify their affordances and functions. However, working under completely different assumptions, Hull (1988) observed similar configurations in the organization of scientific practice. He identified a "demic structure of science" consisting of small research groups, "conceptual demes," and seasonal society meetings, respectively. Group size at these three levels was fairly constant—about 3 to 5 individuals, 30 to 50 individuals, and 100 to 500 individuals. Dunbar (1993) also identified similar configurations, although he labeled them differently.

Table 12.1 represents a basic proposition: *Different size/task configurations afford different environments for the evolution and development of associated proper functions.* We use *proper functions* to refer to evolved psychological mechanisms and their affordances as a single relational unit. Each configuration is likely to have multiple affordances and proper functions, but we have limited the discussion to one per configuration. The dyad affords possibilities for microcoordination (e.g., facial imitation in a mother–infant dyad, the automatic adjustment of gait that occurs when two people walk together, the motoric skills of complex, hands-on laboratory tasks); the workgroup affords possibilities for distributed cognition (the sharing of memory, perception, and preferenda across a group; Hutchins, 1990); the face-to-face deme affords a shared, construction of reality (common knowledge), which also mediates interaction in macrodemes and other superordinate group configurations. Macrodemes afford the stabilization and standardization of language, signs, and symbols (including the technical languages of scientific specialties).

What Makes a Configuration "Core"?

In addition to anthropological research, there are other reasons for judging these four configurations to be touchstones for human evolutionary theory. Human bodies and habitats constrain, entrain, and enable typical boundary conditions for groups. A domestic band moves through a given terrain at a certain rate and requires certain resources. The available physics of body form, energy, time and space roughly bind configurations (Turvey, 1990). The mechanics of configurations are realized in activity or tasks. For example, given human morphology, only one person at a time can hold an infant. Moreover, infant-holding tends to place the infant and the holder in a face-to-face position. Such constraints are, on the one hand, simply mechanical; on the other hand, they constitute opportunities for the evolution of psychological processes.

Another reason these configurations are core is that they appear to be phylogenetically continuous and elaborated. Dyadic coordination among mammals is a prerequisite for sexual reproduction, which involves coordination between two bodies to achieve internal fertilization. As reproduction is extended to include parental care and other dyadic interactions (e.g., mutual grooming), novel evolutionary affordances can occur. Dyads are common among primates, and on rare occasions, even coordinated triadic interactions occur (Dunbar, 1993). Larger interacting coordinated units seem to be distinctively human.

Finally, because of the preceeding considerations, we believe that core configurations have been repeatedly assembled, generation to generation, over evolutionary time. Consider, for example, the ecology of a field that begins with grass, which is followed by shrubs, which are followed by oak trees. Each increasingly complex order of plant requires the existence of the prior simpler level. Moreover, once a stand of oaks is in place, its properties vis-à-vis downward causation affects the surrounding plants. Just as ecologies are repeatedly assembled, always variably, yet always recognizable, so human individuals and groups are repeatedly assembled, and in both instances, higher-level structure modifies lower-level outcomes. For example, the size of a band can influence infant birth weight and care-taking practices (Tronick, Winn, & Morelli, 1985).

Development: The Repeated Assembly
of the Mind's Natural Environment

We think of infants as becoming increasingly independent as they grow older. What happens in practice, however, is that they become increasingly interdependent. Very young infants interact with a single caregiver at a time; as they get older and achieve better control over their bodies, they interact with families and small playgroups. The classroom is typically an even larger group, about the size of a band, and by high school and the beginning of work life, students easily recognize themselves as part of larger group contexts, say, of the Washington High School student body macroband, which itself is part of a school district in a town in a county in a region in a country.

The repetition of core configurations in ontogeny is not a consequence of developmental programs or genetic maturation schedules. Rather, core configurations and their psychological correlates are contingent on the reliable co-occurrence of genetic and epigenetic resources. Development is the product of prior conditions of a self-organizing developmental system composed of genes and expanding circles of interactants (Oyama, 1985).

Infants develop microcoordination in dyads. As their coordination increases, they participate in workgroups (families), and through them, face-to-face groups (extended networks of kin, family friends, etc.). The basics of dyadic level microcoordination is the scaffolding for developing skills for small face-to-face group interaction. Similarly, at the small group level, one masters forming and internalizing norms, values, general knowledge, and specialized knowledge (including knowledge about gender roles, e.g., the distinction between a lipstick tube and a cylinder). This, in turn, readies people for larger groups that require role specialization and higher-level coordination.

These four configurations are the contextual domains for the evolution of all the essential social–cognition relationships and functions. Functions that evolved for one domain can be extended to and combined with functions in other domains. For example, some of the same proper functions (or cognitive modules) can be called into play while engaging in sex, interacting with an infant, chatting about food with a friend, or arguing with a partner about infidelity. The domains are dyadic, but the relevant functional relationships might include facial monitoring, movement synchrony, turn-taking and other nonverbal properties repeatedly assembled for coordinated behavior in a dyad. To illustrate, Perper (1985) in a study of the initiation of courtship sequences, identified movement synchrony as a critical and invariable stage of attraction to the opposite sex.

CORE CONFIGURATIONS AS SYSTEMS

The critical point for an evolutionary psychology is this: If core configurations, from the dyad to the macrodeme, are the structural dynamics to which human cognition was adapted in the past, we should expect human cognition to maintain the same structural dynamics in the present. That is, despite the complexity of modern life, humans will parse large-scale social organization into a small-group grammar when circumstances allow. People may be members of multiple independent demes (church choir, professional association, ski club), a historically recent reorganization of human possibilities, but their interactions in subgroups relative to superordinate groups will continue to engage and reflect basic configurations. For example, minority–majority relationships within a group may prefigure intergroup conflict, yet in each case a superordinate goal could promote conflict reduction (compare Aronson, Stephen, Sikes, Blaney, & Shapp's, [1978] jigsaw paradigm with Sherif's [1966] intergroup threat–conflict resolution strategy).

Social affordances are more reciprocal and dynamic than the affordances of the physical environment because they often unfold in mutual ways. For example, between a mother and child there are reciprocal opportunities for interaction and action. Generically, there are body positions that afford closeness (cuddling, etc.) that can be described as sociopetal or affording nurturance and others that discourage approach (sociofugal), such as being stiff or leaning away. More generally, social affordances are dynamically relational: They exist in the interphase or coordination of the actors. Their emergence can be indexed by imitative facial gestures, coordinated body leans, and eye contact.

There is no central control in this view. Higher-level configurations cannot be reduced to lower-level configurations, but once a proper function is available, it may be used in other higher- or lower-level domains. For example, there is no reason to propose that language evolved for dyadic interaction. However, once language evolved, it could be used in dyadic interactions. Similarly, dyadic microcoordination (involving perception of movement dynamics, emotional expressiveness, etc., whose proper function probably includes sex and infant–elder interactions) can be extended to interaction between animal trainer and animal, and human and artifact (where it becomes the basis for embodied knowledge, tinkering, or highly skilled performance; (cf. Caporael & Heyes, 1996; Caporael, Panichkul, & Harris, 1993).

Coordination at higher organizational levels is achieved by the extension of functions from lower levels. Group identity is illustrative. Although its proper function is coordination in face-to-face groups, it can create a shared identity for large groups. For example, if ethnicity, nationality, or scholarly discipline is used to create group boundaries, in-group–out-group differences are exaggerated in superordinate groups just as they are for face-to-face groups.

A dynamics perspective on core configurations opens the door for new types of research questions. We would first consider how each level of organization is stretched in terms of how many adjustments it can accommodate at a particular level before the stable type of person–environment relationship is destabilized. Specifically, a *phase transition* occurs when the entity–environment relationship breaks down at a particular level. Warren's (1984) research on stair climbers' accommodations to stair riser height changes illustrates such phase transitions. When the vertical height between steps becomes too steep relative to leg length, the climber is forced to shift from bipedal to quadrupedal movement. A new organization of limbs occurs to accommodate the shift in affordances occasioned by the lack of fit between the old bipedal organization and the new constraints imposed by increased riser height.

There are analogous examples in the social psychological literature that can be reframed in the preceding terms. For example, in the Asch's (1952) conformity paradigm we can ask how many others have to disagree before subjects switch from judgments based on stimulus constraints to judgments based on social pressure. Similarly, using Barker's (1968) theory of crowding as reformulated by Wicker (1979) to staffing theory, one can ask when behavior setting norms switch from correcting deviants to expelling them from the activity setting. At issue here is detecting when a behavior setting shifts from being understaffed to being overstaffed. Thus, the same principle holds whether we focus at an individual or group level relationship. In each case one can treat the source of the phase transition as having the properties of a control parameter in that a small change in input can produce a major qualitative shift in response mode. Furthermore, what changes in each case is a repeated assembly of person–environment relationships, be it experimenter–subject or deviant–group or the individual's limb adjustments to step height.

The Relational Status of Dispositions

In a relational view of social dispositional concepts, manifestations of mind are as much in the world as in the head, or in Shotter's (1983) apt terms, dispositions are in the dance as opposed to the dancers. In effect, dispositions are organized in time and space. For example, in Vygotsky (1962), mind is manifested in activity; for Gibson (1979), perception is for doing. From this perspective the structure of mind in the long run has to be isomorphic to the structure of the world including the social environment of groups. Thus, social dispositions become real when manifested in social relations. For example, Heider (1991) proposed that the cognitive units of his balance triad derive their compellingness from Newcomb's (1953) A–B–X model of what makes for stability in social communication. "For Heider, it is social interaction and interdependence in groups that gets attitudes out of head and into the world" (Baron, 1991, p. 567). We use dominance and social identity to illustrate the relational view.

Relationships can be assembled and selected because it is in joint activity that dispositions become actualized and perceivable, whether the focus is on personality traits, attitudes, or motives. For an analogy we used light, which also has a perspectival ontological status. By selecting suitable circumstances for testing, one can demonstrate that light has either wave or particle properties

(Horgan, 1992). It is in this relational sense that we can think of proper functions and affordances as relationships being repeatedly assembled. Concretely, we can expect a fit or match between people and settings (Kenrick, McCreath, Govern, Kinj, & Bordin, 1990). For example, athletic settings afford dominance, and dominant people select settings in which they can dominate (Baron & Misovich, 1993a). Such dispositions are organized and manifested as affordances and are realized in situated activity. In effect, attitudes, traits, and motives are not fixed individual properties any more than electrons are things as opposed to relationships.

Dominance.   One can specify certain invariant circumstances for the emergence of dominance as a form of social action rather than as a persistent trait internal to a person. For example, dominance is relationally instantiated when a context affords people taking charge or influencing or controlling the actions of other people.

The proposition that dominance is relationally specified in the world is consistent with the finding that children are able to accurately place themselves within a dominance hierarchy. That is, they can see how to place themselves in regard to affordances of bossing versus being bossed (Baron & Misovich, 1993b). At the small group level, dominance becomes embedded in taking a central position in a communication network, thereby constraining asymmetries in the flow of information. As group size grows, such role differentiation becomes formalized into leadership roles and castes. Finally, to propose that social dispositions have relational reality is necessary if we are to assume, for example, that females seek males who are high in dominance by observing their actions in suitable circumstances such as how they react to status threat (Ellyson & Dovidio, 1985).

This is not to say that there are not "in the head" dominance beliefs or attitudes. It is to say, however, that for certain evolutionary mechanisms involving selective mating to operate beliefs must be perceivable. Specifically it is in power-challenging social relationships that such information becomes available to perception in the form of affordances of relative dominance, much as to perceive the solubility of salt requires its interaction with a solvent (Turvey, 1992). Thus, although dominance and solubility differ in many ways, their perceivability as dispositions in each case rests on appropriate relational specification. Specifically, just as water might demonstrate the solubility of salt, so the appearance of a mating rival can call into play threatening acts by primates higher in the dominance hierarchy.

Relational Nature of Social Identity.   Social identity, like dominance, is not in the head, but in the relationships between individuals and groups. For example, a suitable circumstance for the manifestation of social identity is the availability of information that makes a person similar to certain people and dissimilar to others. In this context the reality of personal versus social identity parallels the context-contingent reality of electrons as either waves or particles depending on conditions of testing. That is, people can be in a personalized or depersonalized state, depending on circumstances. Moreover, which state they are in will be perceivable in terms of information specifying different social affordances; that is, one can do different activities with people in these different states much as different proxemic distances, body leans, or voice tones can afford sociopetal as opposed to sociofugal reactions, or a cup can more easily afford drinking than spilling the liquid, depending on its shape, distribution of weight, and other features.

Furthermore, we can view social identity in terms of the relative salience of affordances for personalization and depersonalization, a kind of relational spelling out of Brewer's (1991) opposing distinctiveness and assimilation motives. Within and between social configurations we can find both invariant and variant properties. For example, although differentiation–assimilation is a very basic dialectic, it is likely that social differentiation-assimilation probably begins at the level of the triad, the first social aggregation in which there can be a coalition, allies, enemies, and the rudiments of in-group–out-group. If this is correct, one has to have greater social complexity than a dyad for social identity to exist; that is, there has to be sufficient size for subgroup formation. Thus, the small face-to-face group is likely to be a possible basis for within versus between group social comparison even within a deme. As deme size increases, the tension between in-groups and out-groups within the deme likely increases, and group fissioning should be expected. In this context, cognitive processes serve to amplify and justify intergroup competition. Thus, although categorical thinking may have preceded stereotyping, it quickly became entrained to facilitate in-group–out-group differentiation.

Similarly, I-versus-we and we-versus-they can be viewed as both output and input for relationship dynamics. For example, at the within group level the tension is between minority and majority views, goals, beliefs, and attitudes. The problem is no different from that involved in forming a stable dyad: the coordination of individual goals to create a more complex social unity, given that complexity is necessary to solve adaptive problems. Note that at this level the key problem is likely the reciprocation of sharing and the chief cognitive problem became the ability to grasp transitivity in terms of a dominance hierarchy. Perhaps, at this level of social complexity, in-group–out-group differentiation becomes critical as a way of getting people to accept a power inferiority. That is, status–power asymmetries may trigger social identity in-group–out-group processes as a way of justifying the within group inequalities. For example, the group will protect you, so it is okay to take orders. The positive aspect of this is social identity that looks outward and does not allow people to dwell on what they lose within the group (Harris, 1995, p. 465). Thus social identity may emerge relationally, when a certain level of within group discrepancy exists between dominants and subordinates with regard to relative control and access to resources within the group.

Cognitive Consequences of the Relational View

One reason that the hierarchical selection approach appeals to us is its fit with existing research in psychology. Group identity, with its nested, hierarchical character (Turner, 1987) seems to be an extension of the biological hierarchy from macromolecules, genes, cells, and individuals to the next level, groups. In fact, Turner (1987) suggested that "collective processes are an adaptive achievement made possible by the capacity of the human self to vary [by psychologically taking on the character of the social whole" (p. 204).

We hypothesize that shifts in social identity are the crucial psychological process mediating sociocognition. For our current purposes, we distinguish between social, personal, and group identities. Personal identity is the locus of conscious awareness of goals, plans, and beliefs; group identity indicates membership in a specific group; and social identity refers to redefinitions of self that minimize distinctions between self and other(s) (Brewer, 1991). Dynamic shifts in social identity maintain the repeated assembly of core configurations; conversely, a core configuration can evoke a shift in social identity.

Turner (1987) emphasized that social identity is both cause and effect of the formation of a psychological group, a process rooted in the emergence of a shared social category based on intraclass similarities and between class differences (i.e., in-group similarities and out-group differences). In our view, social identity is both output and input to self organization. Specifically, it is the product of downward causation for groups of a certain size and higher. It then becomes an input at two levels: It is a control parameter that promotes certain kinds of social organizations involving increasing group cohesiveness, and identities are what have to be negotiated at annother level—social—between groups as in Aronson et al's (1978) jigsaw model of Sherif's (1966) superordinate goal paradigm.

The Role of Values in Self-Organizing Systems

Earlier we proposed that the present approach would lead to new perspectives on classic problems. At the global level one such problem in social psychology is understanding the processes and structures whereby society is psychologically represented in and mediated by individual minds. Turner (1987) observed that "individualistic minds are not individualistic in the ideological sense of being defined by some presocial psychological dynamic, but are socially structured—society is in the individual as much as individuals are in society" (p. 205). Indeed, recent attempts by Turner (1987) and others (Brewer, 1991) to revitalize the psychological reality of the group in terms of a self-categorization-social identity perspective, are a direct manifestation of the importance of this set of issues. In line with our self organization perspective however, it may be claimed that the issue of social identity is nested in a broader set of problems involving the role of values. For example, Sagiv & Schwartz (1995) recently linked value priorities to readiness for out-group social contact.

In this context we propose that the present perspective provides a still higher level of nesting by broadening the conception of values so as to move beyond the conventional view of values as subjective and arbitrary in the sense of "socially shared conceptions of what is good" (Sagiv & Schwartz, 1995, p. 437). Before attempting a definition, let us see how values operate both structurally and functionally. With regard to the classical view, our interpretation is that values are the source of the criteria for "what is good, better or best, among objects, actions, ways of life, and social and political institutions and structures" (Schwartz, 1990, p. 8). A concrete example is as follows: "An automobile stopped by a child standing in the road illustrates a value realizing relation. Such a relation is not arbitrary in the way stopping for a red light is, but not inviolable in the same way being stopped by a brick wall would be" (Hodges & Baron, 1992, p. 271).

We begin to look for objective sources of value by linking them to affordances. First, value, for Gibson (1979), is part of the definition of affordances: They are what the environment "offers an animal, what it provides or furnishes either for good or ill" (p. 127). Values in this sense are both antecedents and consequences of affordances, and they are embodied in affordances. Specifically, the values of a society are often manifested in the affordances of the artifacts of the society. For example, children in high chairs drinking from cups have an opportunity to learn about "the values that constrain and guide physical existence, and about their own identity as a physical and cultural being" (Hodges & Baron, 1992, p. 273). Constraints of scarcity, dependence, and responsibility are embodied in the design of the cup, and the height and restraints of the high chair.

Values also show up in coordination. Turvey (1990) noted that "Coordination necessarily involves bringing into proper relation multiple and different component parts . . . defined over multiple scales of space-time" (p. 238). A good example of how values are reflected in coordination, which ties directly into self-organization and repeated assembly, is a successful basketball team. First, in terms of Fig. 12.1, what is being organized at the local level is the transition from people functioning as individuals to people occupying roles in a social system—the team. This is the first step toward a good basketball team. Second, the roles need to be organized in a proper way. For example, on a good (e.g., cohesive) basketball team, such roles as "shooter," "picksetter," or "passer" are flexibly assigned depending on local conditions such as who is playing well that night, the other team's defense, or even the basketball. As the game as whole progresses, these roles are flexibly reassigned to optimize the effectiveness of the team as a whole (Hodges & Baron, 1992, p. 281). Note that this is also a description of a self-organizing system at work; the value underwriting this organization is the team winning aesthetically pleasing way. For basketball fans, the difference is one between "a good game" and "they didn't play very well but they won anyway." In Campbell's (1958) terms the team has high "entitativity"; it is realizing its "teamness." In effect there is a coalition of units that function cooperatively in a way that begins to mimic successful biological systems.

The cultural counterpart at the social psychological level becomes the tradition of a winning team that becomes a social identity. For example, winning becomes a standard analogous to the norm in the Sherif (1935) paradigm that modifies individual units through downward causation. Moreover, one can test the strength of this value—winning as a team—objectively by perturbing the system and seeing whether it still holds together. For example, does the system have the ability to reconfigure as a group (and with what latency) in the face of loss and keep on winning regardless of who is playing the roles? In *situ*, we can see how the system adjusts to the loss of a key player during a given game. Moreover, values operate in the sense that they define what success is. For example, the issue is not just winning, but winning so as to contribute to the continuity of the collective tradition, winning because the team works together to set up the person who is open, regardless of who that person is. In such coalitions subordinate and executive are not fixed but reversible: Control is in the system rather than an executive (Hodges & Baron, 1992, p. 281).

By value then we mean that which defines what is good (or bad) for that assembly of elements, given a consideration of both local and contextual conditions, including both the presence of other systems and the broader environments. For example, at the level of work team or family group, good can no longer be defined strictly at the lower level of the dyad just as in a good dyadic relationship the welfare of the component individuals is not sufficient. That is, in good dyad relationships people seek payoffs at the level of the couple or dyad. Similarly, bands as assemblies of families and dyads must succeed at the higher collective level of the band, not just at the family or dyad level.

Finally, modal tasks and proper functions for each core configuration reflect the value structure of each configuration. For example, in Table 12.1, the reason that sex and infant–adult interactions are modal tasks and microcoordination is the proper function is because tasks stabilize the system at the collective level in a way that the part is in the whole and the whole is in the part. For example, for adult partners, each component of the dyad perceives itself as belonging to a couple relationship in which the order parameter is commitment to the relationship as indexed by reciprocal time commitments. Similarly, at the group level the commitment must not be individual or dyadic, but at the group level. Stated more generally, there are nested commitments (social identities), with values defining the priorities. For example, with groupthink the collective commitment overrides all others, a kind of pathology of downward causation. An even more extreme collective pathology occurred in Russia at the nation–state level. The old Communist Party encouraged children to turn in disloyal parents. Specifically, this is improper or if you will, evil, because in realizing the collective level, the system should not disrespect or dissolve the lower levels of organization.

If the preceeding is a pathology of downward causation, then a failure to recognize the priority of the group level can also violate coordinated group functioning. For example, Campbell (1990a) suggested that conformity effects in Asch-type studies should be viewed as illustrating the morality of values involving "our ubiquitous dependence upon the reports of others for our knowledge of the world . . . rather than showing the character defect of suggestibility" (p. 39). More specifically at issue is the role of moral norms for social knowledge such as trust and honesty in social communication. Conformity, then, can be socially proper behavior if we recognize the nesting of individual minds in group values. For example, "It is our duty to report what we perceive honestly, so that others may use our observations in coming to valid beliefs" (Campbell, 1990a, p. 39). Thus social consensus based on trust, honesty and an attempt "to integrate one's own perceptions with the reports of others without deprecating them or ourselves" (Campbell, 1990a p. 39) is better than consensus as a test of group loyalty. It could then be argued at the level of group selection that groups operating as seekers of valid knowledge would be more likely to be selected for in a competition between groups (Sober, 1993). Indirect support for this interpretation can be found in evidence presented by Granovetter (1973) for the greater adaptive advantage of groups with weak within-group ties (over strong tie groups) in regard to the potentially greater ease of establishing good relationships between groups when global threats need to be faced. For example, could Bosnian Serbs and Croats work together in the face of an outbreak of a virulent disease? In effect does the Sherif (1966) solution to intergroup hostility break down when within-group ties are very strong? Such novel research questions arise from the present view of values as sources of cohesive coordination.

THE FUTURE OF EVOLUTIONARY PSYCHOLOGY

Darwinism regularly undergoes cycles of rejection and revival in the human sciences. In part the problem is that evolutionists have been presenting a feature list of natural items. Typically, the items on the list are drawn from everyday categories of experience and serve less as a guide to scientific research than as legitimization for various contentious (liberal and conservative) social arrangements (Caporael, 1994; Caporael & Brewer, 1991; Maynard Smith, 1987). Also, psychologists have lacked a conceptual vocabulary that allows us to avoid nature-nurture dualism and anthropomorphism. We believe a relational and dynamic evolutionary approach can overcome such obstacles.

One new claim in evolutionary psychology is that human cognition is domain specific. However, there is no specification for what constitutes a domain. Tooby and Cosmides (1992) indicated that there may be thousands of specialized domains with their own specialized mechanisms. One of the common objections to this (and modular approaches in general) is that there are no restrictions on developing a new domain or module (adaptation) to meet any need—the same objection that is raised against the instinct concept. In contrast, we have proposed that there are only four domains for the evolution of uniquely human attributes. Although we remain agnostic about the possible number of modules, we have proposed constraints. Human cognition has evolved to fit the proper functions and to perceive the affordances of those four domains. Once such cognitive processes have evolved, however, they may be extended to other (noncore) domains and recombined with other psychological processes. For example, although language is a proper function of macrodemes, it can be used in dyadic interaction, even though the proper functions of dyadic interaction do not require language. (Obviously, through evolutionary time, dyadic inter-action has been crucial to the development of prelinguistic behavior.)

An important advantage of our approach is that adaptive human behavior can be researched in the here and now; evolutionary psychologists need not be tied to speculations about life in the Pleistocene past. To go back to our chemical metaphor, there is no reason to believe that combustion occurred differently in the Pleistocene era from the way it occurs today. The range of combustible resources has been extended by human activity, but the relationships continue to obtain. Similarly, there are some fundamental aspects of behavior that continue to obtain. It is still true that only one person at a time holds a baby, even though the methods of feeding the baby have altered. Both phenomena, the first repeatedly assembled on an geologic scale and the second on a cultural scale, are amenable to research and lie within the scope of evolutionary psychology.

ACKNOWLEDGMENTS

We thank Jeff Simpson and Doug Kenrick for their helpful comments. This work was made possible by the National Science Foundation, Grant No. SBR-9321461 to the first author. Any opinions, findings, and conclusions or recom-mendations expressed in this material are those of the authors and do not necessarily reflect the views of the National Science Foundation.

REFERENCES

Alexander, R. D. (1979). *Darwinism and human affairs*. Seattle: University of Washington Press.
Aronson, E., Stephen, C., Sikes, J., Blaney, N., & Shapp, M. (1978). *The jigsaw classroom*. Newburg Park, CA: Sage.
Asch, S. (1952). *Social psychology*. New York: Prentice-Hall.
Barker, R. G. (1968). *Ecological psychology*. Stanford, CA: Stanford University Press.
Baron, R. M. (1991). A meditation on levels of structure. *Contemporary Psychology*, 36, 566–568.

Baron, R. M., Amazeen, P. G., & Beek, P. J. (1994). Local and global dynamcis of social relations. In R. R. Vallacher & A. Nowak (Eds.) *Dynamical systems in social psychology* (pp. 111–138). San Diego, CA: Academic Press.

Baron, R. M., & Misovich, S. J. (1993a). Dispositional knowing from an ecological perspective. *Personality and Social Psychology Bulletin, 19,* 541–552.

Baron, R. M., & Misovich, S. J. (1993b). An integration of Gibsonian and Vygotskian perspectives on changing attitudes in group contexts. *British Journal of Social Psychology, 32,* 53–70.

Brewer, M. (1991). The social self: On being the same and different at the same time. *Personality and Social Psychology Bulletin, 17,* 475–482.

Burgess, J. W. (1989). The social biology of human populations-spontaneous group formation conforms to evolutionary predictions of adaptive aggregation. *Ethology and Sociobiolog, 10,* 343–359.

Buss, D. M. (1995). Evolutionary psychology: A new paradigm for psychological science. *Psychological Inquiry, 6,* 1–30.

Buss, L. W. (1987). *The evolution of individuality.* Princeton, NJ: Princeton University Press.

Byrne, R. W., & Whiten, A. (1988). *Machiavellian intelligence.* Oxford: Clarendon Press.

Campbell, D. T. (1958). Common fate, similarity, and other indices of status aggregates of persons as social entities. *Behavioral Science, 2,* 14–25.

Campbell, D. T. (1990a). Asch's moral epistemology for socially shared knowledge. In I. Rock (Ed.), *The legacy of Solomon Asch: Essays in cognition and social psychology* (pp. 39–52). Hillsdale, NJ: Lawrence Erlbaum Associates.

Campbell, D. T. (1990b). Levels of organization, downward causation, and the selection-theory approach to evolutionary epistemology. In G. Greenberg & E. Tobach (Eds.), *Theories of the evolution of knowing* (pp. 1–17). Hillsdale, NJ: Lawrence Erlbaum Associates.

Caporael, L. R. (1994). Of myth and science: Origin stories and evolutionary scenarios. *Social Science Information, 33,* 9–23.

Caporael, L. R. (1995). Sociality: Coordinating bodies, minds and groups. *Psycoloquy,* 6(1), Available FTP: Hostname: princeton.edu Directory: pub/harnad/psycoloquy/1995.volume.6 File: psycoloquy.95.6.01.group-selection.1.caporael.

Caporael, L. R., & Brewer, M. B. (1991). The quest for human nature. *Journal of Social Issues, 47,* 1–9.

Caporael, L. R., & Heyes, C. M. (1996). Why anthropomorphise? Folk psychology and other stories. In R. W. Mitchell, N. S. Thompson, & H. L. Miles (Eds.), *Anthropomorphism, anecdotes, and animals* (pp. 59–74). Albany: State University of New York Press.

Caporael, L. R., Panichkul, E. G., & Harris, D. R. (1993). Tinkering with gender. *Research in Philosophy and Technology, 13,* 73–99.

Darwin, C. (1963). *The origin of species: by means of natural selection of the preservation of favoured races in the struggle for life.* New York: Washington Square Press. (Original work published 1859)

Darwin, C. (1972). *The variation of animals and plants under domestication.* New York: AMS Press. (Original work published 1896)

Dunbar, R. I. M. (1993). Coevolution of neocortical size, group size and language in humans. *Behavioral and Brain Sciences, 16,* 681–735.

Ellyson, S. L., & Dovidio, J. F. (1985). Power, dominance and nonverbal behavior: Basic conceptual issues. In S. L. Ellyson & J. F. Davidio (Eds.), *Power, dominance and nonverbal behavior* (pp. 1–28). New York: Springer-Verlag.

Endler, J. (1986). *Natural selection in the wild.* Princeton, NJ: Princeton University Press.

Forsyth, D. R. (1995). *Our social world.* Pacific Grove, CA: Brooks-Cole

Gibson, J. J. (1979). *The ecological approach to visual perception.* Boston, MA: Houghton-Mifflin.

Gould, S. J. (1980). Is a new and general theory of evolution emerging? *Paleobiology, 6,* 119–130.

Granovetter, M. S. (1973). The strength of weak ties. *American Journal of Sociology, 78,* 1360–1380.

Gray, R. (1992). Death of the gene: Developmental systems strikes back. In P. Griffiths (Ed.), *Trees of life* (pp. 165–209). Dordrecht: Kluwer Academic.

Griffiths, P. E., & Gray, R. D. (1994). Developmental systems and evolutionary explanation. *Journal of Philosophy, 91,* 277–304.

Hamilton, W. D. (1964). The genetical evolution of social behavior, I & II. *Journal of Theoretical Biology, 7,* 1–52.

Harris, J. R. (1995). Where is the child's environment? A group socialization theory of development. *Psychological Review, 102,* 458–489.

Heider, F. (1991). *The notebooks, Vol. 6: Units and coinciding Units.* Munich: Psychologie Verlags Union.

Ho, M.-W. (1991). The role of action in evolution: Evolution by process and the ecological approach to perception. *Cultural Dynamics, 4,* 336–354.

Hodges, B. H., & Baron, R. M. (1992). Values as constraints on affordances: Perceiving and acting properly. *Journal for the Theory of Social Behavior, 22,* 263–294.

Horgan, J. (1992). Trends in physics: Quantum philosophy. *Scientific American, 267,* 94–104.

Hull, D. L. (1988). *Science as a process.* Chicago: University of Chicago Press.

Hutchins, E. (1990). The technology of team navigation. In J. Galegher, R. E. Kraut, & C. Egido (Eds.), *Intellectual teamwork* (pp. 191–220). Hillsdale, NJ: Lawrence Erlbaum Associates.

Jarvenpa, R. (1993). Hunter-gatherer sociospatial organization and group size. *Behavioral and Brain Sciences, 16,* 712.

Jarvenpa, R., & Brumbach, H. (1988). Socio-spatial organization and decision-making processes: Observations from the Chipewyan. *American Anthropologist, 90,* 598–618.

Kenrick, D. T., McCreath, H. E., Govern, J., Kinj, R., & Bordin, J. (1990). Person-environment interactions: Everyday setting and common trait dynamics. *Journal of Personality and Social Psychology, 58,* 685–698.

Maynard Smith, J. (1987). Science and myth. In N. Eldredge (Ed.), *The Natural History reader in evolution* (pp. 222–229). New York: Columbia University Press.

McArthur, L. Z., & Baron, R. M. (1983). Toward an ecological theory of social perception. *Psychological Review, 90,* 215–238.

Millikan, R. G. (1984). *Language, thought, and other biological categories.* Cambridge, MA: MIT Press.

Newcomb, T. M. (1943). *Personality and social change.* New York: Dryden Press.

Newcomb, T. M. (1953). An approach to the study of communicative acts. *Psychological Review, 60,* 393–404.

Newcomb, T. M., Koenig, K. E., Flacks, R., & Warwick, D. P. (1967). *Persistence and change: Bennington College and its students after 25 years.* New York: Wiley.

Newtson, D. (1994). The perception and coupling of behavior waves. In R. R. Valacher & A. Nowak (Eds.), *Dynamical systems in social psychology* (pp.139–167). San Diego, CA: Academic Press.

Oyama, S. (1985). *The ontogeny of information.* New York: Cambridge University Press.

Oyama, S. (1991). Bodies and minds: Dualism in evolutionary theory. *Journal of Social Issues, 47,* 27–42.

Perper, T. (1985). *Sex signals: The biology of love.* Philadelphia: ISI Press.

Richards, R. J. (1987). *Darwin and the emergence of evolutionary theories of mind and behavior.* Chicago, IL: University of Chicago Press.

Sagiv, L., & Schwartz, S. H. (1995). Value priorities and readiness for out-group social contact. *Journal of Personality and Social Psychology, 69,* 437–448.

Schwartz, B. (1990). The creation and destruction of value. *American Psychologist, 45,* 7–15.

Sherif, M. (1935). A study of some social factors in perception [special issue]. In *Archives of Psychology, 187.*

Sherif, M. (1966). *In common predicament: Social psychology of intergroup conflict and cooperation.* Boston: Houghton-Mifflin.

Shotter, J. (1983). "Duality of Structure" and "Intentionality" in an ecological psychology. *Journal for the Theory of Social Behavior, 13,* 19–44.

Smith, K. C. (1992). The new problem of genetics: A response to Gifford. *Biology and Philosophy, 7,* 331–348.

Sober, E. (1993). *Philosophy of biology.* Boulder, CO: Westview Press.

Tooby, J., & Cosmides, L. (1992). The psychological foundations of culture. In J. H. Barkow, L. Cosmides, & J. Tooby (Eds.), *The adapted mind* (pp. 19–136). New York: Oxford University Press.

Tronick, E. Z., Winn, S., & Morelli, G. A. (1985). Multiple caretaking in the context of human evolution: Why don't the Efe know the Western prescription for child care? In M. Reite & T. Field (Eds.), *The psychobiology of attachment and separation* (pp. 293–322). New York: Academic Press.

Tuckman, B. W. (1965). Developmental sequences in small groups. *Psychological Bulliten, 63,* 384–399.

Turner, J. C. (1987). *Rediscovering the social group: A self-categorization theory.* Oxford: Basil Blackwell.

Turvey, M. T. (1990). Coordination. *American Psychologist, 45,* 938–953.

Turvey, M. T. (1992). Affordances and prospective control: An outline of the ontology. *Ecological Psychology, 4,* 173–187.

Vygotsky, L. S. (1962). *Thought and language.* E. Hanfmann & G. Vakar (Trans.). Cambridge, MA: MIT Press. (Original work published in 1934)

Warren, W. H. (1984). Perceiving affordances: Visual guidance of stair climbing. *Journal of Experimental Psychology: Human perception and performance, 10,* 683–703.

Wicker, A. W. (1979). *An introduction to ecological psychology.* Pacific Grove, CA: Brooks-Cole.

Wilson, D. S., & Sober, E. (1994). Reintroducing group selection to the human behavioral sciences. *Behavioral and Brain Sciences, 17,* 585–654.

Winterhalder, B., & Smith, E. A. (1992). Evolutionary ecology and the social sciences. In E. A. Smith & B. Winterhalder (Eds.), *Evolutionary ecology and human behavior* (pp. 3–23). New York: Aldine de Gruyter.

13

Incorporating Group Selection into the Adaptationist Program: A Case Study Involving Human Decision Making

David Sloan Wilson
Binghamton University

This chapter has both a general and a specific purpose. The general purpose is to show how to reason intuitively about natural selection as a multilevel process. The idea that higher-level units such as social groups can be well adapted, in the same sense that individuals are well adapted, has a long history in biology, the human sciences, and everyday thought. For example, the term "body politic" suggests that a human political organization is comparable to a single organism. In biology, group-level adaptations have been regarded as theoretically possible but sufficiently unlikely that they can be ignored for the majority of species in nature (Williams, 1966). Thus, when most evolutionary biologists reason intuitively about natural selection, they think that it is sufficient to ask, "What traits would maximize the fitness of individuals, relative to other individuals in the population?"

More recently, group selection has been reassessed by biologists and may be much more common and important than previously thought (review by Wilson & Sober, 1994; Sober & Wilson, in press). If so, then it is necessary to consider the effects of traits on the relative fitness of groups in addition to their effects on the relative fitness of individuals within groups. Previous discussions of this issue have been rather technical and have not had much impact on the way biologists think intuitively about adaptation and natural selection. It is therefore important to provide an intuitive framework for thinking about multilevel selection.

345

My specific purpose is to put the framework to use by examining the cognitive processes that allow humans to make adaptive decisions. Throughout their evolutionary history, people have been faced with the challenge of evaluating information from their environment to decide which of many ways to behave. The quality of a decision can have life-or-death consequences, not only for the relative fitness of individuals within a social group, but also for the fitness of entire social groups relative to other social groups. Thus, it is likely that the psychology of decision making has been strongly shaped by natural selection at both the individual and group levels. In addition, psychologists have been interested in decision making for many decades, which has resulted in literally hundreds of empirical studies that can be used as a data base for testing evolutionary hypotheses.

THINKING INTUITIVELY
ABOUT MULTILEVEL SELECTION

The individualistic view of adaptation and natural selection has been so influential over the last three decades that it is important to show how it can be fundamentally wrong. Consider the following passage from G. C. Williams (1966):

> Many biologists have implied, and a moderate number have explicitly maintained, that groups of interacting individuals may be adaptively organized in such a way that individual interests are compromised by a functional subordination to group interests.
>
> It is universally conceded by those who have seriously concerned themselves with this problem that such group-related adaptations must be attributed to the natural selection of alternative *groups* of individuals and that the natural selection of alternative alleles within populations will be opposed to this development. I am in entire agreement with the reasoning behind this conclusion. Only by a theory of between-group selection could we achieve a scientific explanation of group-related adaptations. However, I would question one of the premises on which the reasoning is based. Chapters 5 to 8 will be primarily a defence of the thesis that group-related adaptations do not, in fact, exist. (p. 92)

In this passage, Williams makes a theoretical claim that group-level adaptations require a process of natural selection at the group level. He also makes an empirical claim that group-level adaptations do not exist in nature because group selection is too weak. Finally, Williams showed how the evolution of any particular trait can be evaluated with respect to levels of selection. If the genes for the trait spread within a population, then they evolve by individual selection, and the trait is an individual-level adaptation. If the genes for the trait are selectively neutral or disadvantageous within populations, but cause entire populations to survive and reproduce better than other populations, then they evolve by group selection and the trait is a group-level adaptation.

Williams' theoretical claim has withstood the test of time. Group selection remains the only explanation for group-level adaptations, and all group selection models involve the evolution of genes by the differential survival and reproduction of groups. It might seem that Williams' empirical claim has also withstood the test of time, but a closer look reveals that the modern concept of "individual selection" has become very different from Williams' initial conception, as represented by the quoted passage. In particular, so-called theories of individual selection frequently assume that the evolving population is subdivided into a large number of groups but seldom separately examine the relative fitness of individuals within groups and the relative fitness of groups in the metapopulation. Instead, the fitness of each genotype (or gene) is averaged across all groups, which lumps both within- and between-group selection into a single measure of fitness (Wilson, 1989; Wilson & Sober, 1989, 1994; Sober & Wilson, in press). Whatever evolves in the model is said to evolve by "individual selection," and there is no effort to model group selection as an alternative hypothesis. After all, did Williams (1966) and others not show that group selection can be ignored?

Recent group selection models are sometimes referred to as the new group selection, but in many ways they are simply a return to Williams' original conception of group selection as outlined in the passage quoted earlier. When an evolving population is subdivided into a large number of groups, the "new group selectionist" examines the relative fitness of individuals within groups and the relative fitness of groups in the metapopulation. If the trait that evolves is neutral or selectively disadvantageous within groups but increases the relative fitness of groups, it is said to evolve by group selection and to be a group-level adaptation. When relative fitness is carefully partitioned into within- and between-group components, group selection emerges as an important evolutionary force, and many traits turn out to be group-level adaptations, even though the biologists who studied them imagined themselves to be individual selectionists.

Elsewhere, I and others documented the history of specific subjects in detail to show how the concept of individual selection has been stretched to include both within- and between-group selection as understood by Williams (for sex ratio, see Colwell, 1981; Sober & Wilson, in press; Wilson, 1983, for the eusocial insects, see Mitchell, 1993; Seeley, 1989, 1995; for disease virulence see Bull, 1994; Sober & Wilson, in press; for inclusive fitness theory, see Hamilton, 1975; Sober & Wilson, in press; Wilson & Sober, 1989, 1994; for game theory, see Sober & Wilson, in press; Wilson & Sober, 1994; for human social groups, see Sober & Wilson, in press; Wilson & Sober, 1994). For the purposes of this article, I will simply ask the reader to accept two basic points: (a) Group selection can no longer be dismissed as casually as it has been during the past three decades. Evolutionary biologists must seriously consider the possibility of adaptation at more than one level of the biological hierarchy. (b) The new group selection requires a reorganization of familiar theories in a way that shrinks the concept of individual-level adaptation and broadens the concept of group-level adaptation. For example, kin selection and evolutionary game theory are widely regarded as alternatives to group selection that explain altruism and cooperation in individualistic terms. However, both of these theories assume that the evolving population is subdivided into groups: kin groups in the case of inclusive-fitness theory and n-person groups in the case of game theory. When fitnesses are examined within and between these groups, it turns out that altruism and cooperation are always selectively disadvantageous within groups and evolve only by increasing the fitness of groups relative to other groups (Hamilton, 1975; Sober & Wilson, in progress; Williams and Williams, 1957; Wilson, 1983; Wilson and Sober 1989, 1994). These theories are therefore special cases of group selection rather than alternatives to group selection. It is difficult for some evolutionary biologists to make this transition, but there is no alternative if we wish to return to the concepts of within- and between-group selection as developed by Williams (1966).

The reorganization of familiar theories has led some biologists to advocate a form of pluralism in which each framework is equally legitimate (Dugatkin & Reeve, 1994; Holcomb, 1994). Which framework to employ then becomes a matter of taste, and those who prefer what Dugatkin & Reeve call *broad-based individual selection* do not need to change their intuition at all. It would be strange if this were true. After all, the rejection of group selection was treated as a momentous event in evolutionary biology, and for 30 years it has seemed terribly important to avoid thinking about groups as similar to individuals in their functional organization. If it turns out that group selection is important after all, should this not have some impact on the way we think about adaptation and natural selection?

The Adaptationist Program

The centerpiece of evolutionary theory is the concept of adaptation, or a fit between the properties of the organism and the properties of the environment, which evolves by natural selection. Adaptation is a central concept, not only because it is important, but also because it is so easy to employ. It is relatively simple to predict the traits that will maximize fitness in a given environment, at least compared to the task of unravelling the details of phylogeny, genetics, development, and physiology. That is why Darwin was able to achieve his fundamental insights despite his almost total ignorance of phylogeny or the mechanistic processes that make up organisms.

It is important to realize that adaptationist thinking does not deny the importance of other factors such as phylogenetic, genetic ,and developmental constraints. Even if a population is not well adapted to its environment, it is important to know what it would be like if it were, which allows deviations from the optimal phenotype to be interpreted (Orzack & Sober, 1994). Thus, even after we acknowledge that there is more to evolution than natural selection (and more to human nature than evolution), it is still useful to ask this simple question: "What would the population be like if it were adapted to maximize fitness in its environment?"

As previously mentioned, most biologists have been trained to ask this question at the individual level and to avoid asking it at the group level. If group selection is a legitimate possibility, however, we must explicitly ask the question at all relevant levels of the biological hierarchy. I now describe an appropriate procedure as a number of steps, although I do not wish to imply that the steps must be taken in order or that all of them are required for the study of every trait.

Step 1: Ask the question: What Would Groups be Like if Between-Group Selection Were the Only Evolutionary Force?

In this case, groups will be "superorganisms" functionally designed to maximize their survival and reproduction, relative to other groups. For example, if we are interested in predator defense, we might decide that the optimal group organization requires that at least one individual not be feeding and scanning the environment for predators at all times. If we are interested in decision making, we might decide that the optimal group organization will include one phase in which the group members separate to acquire information from different parts of their environment, a second phase in which they pool the information, a third phase in which they assess the information, and so on.

Step 2: Ask the question: What would groups be like if within-group selection were the only evolutionary force?

In this case, individuals will be functionally organized to maximize their survival and reproduction, relative to other individuals within the same group. For example, we might decide that it is adaptive for an individual to issue an alarm call when predators are absent, which distracts other members of the group and allows the individual to feed. If we are interested in decision making, we might decide that it is adaptive for individuals to withhold information or present false information that will increase their relative fitness.

This step of the procedure corresponds to what Williams (1966) understood as within-group selection. However, it is important to distinguish it from the broadened form of individual selection that has developed since Williams (1966). For example, consider a population that is subdivided into a large number of groups and a mutant behavior that increases the fitness of the actor by one unit and the fitness of everyone else in the actor's group by two units. If X is the baseline fitness, then the fitness of the mutant is $X + 1$, and the fitness of the average non-mutant is X because the vast majority of nonmutants exist in groups without the mutant. Many biologists would say that the behavior evolves by individual selection, but this conclusion is based on comparing the fitness of the mutant with the fitness of nonmutants averaged across groups. If we employ Williams' definition of within- and between-group selection we discover that the mutant has the lowest relative fitness within its own group ($X + 1$ vs. $X + 2$) but that groups with the mutant are more fit than groups without the mutant. This is a specific example of the general claim made earlier, that the modern version of individual selection includes both within- and between-group selection as defined by Williams. In any case, it is important to base step 2 of the procedure on a comparison of fitnesses within single groups.

Steps 1 and 2 bracket the possibilities of what can evolve by natural selection. We do not expect populations to lie at either extreme, but we do expect them to be somewhere between. The next step is to determine where a given population is likely to lie between the extremes.

Step 3: Examine the basic ingredients of natural selection at each level of the hierarchy.

The process of natural selection requires three basic ingredients: phenotypic variation among units, heritability, and differences in survival and reproduction that correlate with the phenotypic differences. To determine the balance between levels of selection, we need to examine these ingredients at each level.

Step 3a: Ask the question: What is the potential for phenotypic variation within and among groups?

Consider an asexual population that consists of two phenotypes, A and B, in frequency p and $(1 - p)$ respectively. Imagine that we subdivide the population into groups by allowing each individual to reproduce and form a clonal group of size N. The metapopulation now consists of a fraction p of groups that are pure A and a fraction $(1 - p)$ groups that are pure B. This kind of population structure is maximally conducive to between-group selection because there is no phenotypic (or genetic) variation within groups on which natural selection can act. Now imagine that we compose the groups by placing exactly p A-types and $(1 - p)$ B-types in each group. This kind of population structure is maximally conducive to within-group selection because there is no phenotypic (or genetic) variation between groups on which natural selection can act. Between these two extremes are a great range of population structures in which phenotypic variation is partitioned into a within-group component and a between-group component. For example, if we compose groups of size N by placing individuals randomly into each group, there will be a certain amount of phenotypic (and genetic) variation among groups that is given by the binomial distribution, whose variance is $p(1 - p)/N$. The partitioning of phenotypic variation within and among groups obviously has a strong influence on the balance between levels of selection.

Although kin selection theory is often described in terms of the proportion of shared genes that are identical by descent, the coefficient of relationship (r) is better understood as an index of variation within and among groups. This is not a controversial statement and has been accepted by virtually all theoretical biologists, including Hamilton (1975). For example, $r = 1$ represents the clonal population structure described above. Interactions among full siblings, or $r = .5$, represents a population structure in which groups of size N are composed by first picking a male and female at random from the metapopulation and then forming groups of size N from their gametes. The resulting variation among groups is far greater than if we had picked N individuals directly from the metapopulation, but less than the variation that results from asexual clonal reproduction ($r = 1$).

Kin selection theory sometimes makes it seem as if r is the only important factor determining the amount of variation within and among groups. In fact it is only one of many factors, which suggests that group selection can be a potent force even when the genetic relatedness among the interacting individuals is low. I briefly describe three other factors that can create large phenotypic differences among groups.

Complex Genotype–Phenotype Relationships. In a classic experiment, Wade (1976, 1977) created groups of flour beetles by picking $N = 16$ individuals at random from a large laboratory population and placing them in small vials with flour to reproduce. After 37 days he measured the total number of offspring produced by each group, which can be considered a group-level phenotypic trait. The variation among groups was enormous, ranging from 365 progeny for the most productive group to only 118 progeny for the least productive group. This amount of phenotypic variation could never have been predicted from the relatively large number of individuals that initiated each group. In other words, if we assume that group productivity is the sum of individual fecundities, randomly composed groups of $N = 16$ beetles could not possibly vary as much as observed in the experiment. Group-level variation must therefore reflect an interaction among individuals, rather than the additive sum of their properties. In a series of follow-up experiments, Wade (1979) and McCauley and Wade (1980) showed that group productivity depends on a complex interaction between a number of traits including development, cannibalism, and sensitivity to crowding. This example illustrates that the relationship between the genetic composition of a unit and the phenotype of that unit can be complex. Just as a small genetic change can have a large phenotypic effect at the individual level, small genetic differences between groups can lead to large phenotypic differences. This important point is missed by most population genetics models, including kin selection models, which tend to assume an additive relationship between genetic and phenotypic variation.

Assortative Interactions.   In kin selection models and many traditional group selection models, high phenotypic (and genetic) variation among groups is created by the process of reproduction. In other words, groups of size N are formed by initiating the groups with a smaller number of individuals, who then reproduce to form groups of size N. The variation among these groups is greater than among groups initiated directly with N individuals.

Another way to create high variation among groups is by assortative interactions. To pick an extreme example, imagine a very large population of individuals who differ in their cooperativeness. Suppose that every individual knows the cooperativeness of every other individual and can freely choose whom to associate with in groups of size N. Every individual will be rejected by more cooperative individuals, and each individual will reject less cooperative individuals, leading to a population structure in which there is no variation within groups and maximal variation among groups. This population structure is as favorable for group selection as clonal reproduction, even though group members are genealogically unrelated to each other. The assumptions of this example are obviously unrealistic, but more realistic assumptions can still lead to highly nonrandom groupings, even if they do not achieve the ideal of eliminating variation within groups (Wilson & Dugatkin, in press). Assortative interactions are likely to be an especially important mechanism for creating phenotypic variation among human groups (Frank, 1988).

Social Norms.   So far, I have assumed, along with most theoretical models, a strong form of genetic determinism in which individuals are altruistic (for example) because they have a gene for altruism. Evolutionary biologists are quick to admit that this is just a simplifying assumption and that a complex array of psychological mechanisms separate the phenotype from the genotype in many species. Nevertheless, the simplifying assumption is defended as useful because adaptive psychological mechanisms will prompt the organism to adopt the phenotype that would evolve in a given situation, given enough time and a simple genetic basis. We do not have a single gene for pulling our hand away from fires, but our adaptive psychological mechanisms cause us to behave as if we do.

This defense of genetic determinism as a useful simplifying assumption is warranted to a degree, but it is especially misleading when it comes to thinking about phenotypic variation within and among groups. In particular, human groups are often behaviorally homogenous in ways that could never be predicted from their genetic structure. Homogeneity is achieved, not by the assortative interactions of individuals with fixed phenotypes as outlined earlier, but by social norms that cause phenotypically plastic individuals to converge on a single behavior . The psychological mechanisms that structure social norms and phenotypic plasticity may well be biologically adaptive, but they also radically change the population structure, creating a pattern of phenotypic variation within and among groups that verges on the clonal, even when the groups consist of many thousands or even millions of genetically different individuals (see Boyd & Richerson, 1985, for a more thorough discussion of group selection and cultural evolution).

We may summarize step 3a by saying that phenotypic variation within and among groups can be influenced by many factors. The evolutionary literature sometimes gives the impression that genetic relatedness is the primary factor determining phenotypic variation among groups, leading to the expectation that group-level adaptations should be found only within groups of highly related individuals. This expectation is misleading because extreme phenotypic variation frequently exists among groups of genetically unrelated individuals. By focusing on phenotypic variation as one of the primary ingredients of natural selection, we begin to see new possibilities for group selection that were not forthcoming from kin selection theory.

Step 3b: Ask the Question: Is the Phenotypic Variation Heritable?

The simplest way to approach the issue of heritability is to imagine or actually perform a simple experiment. In a population of units that reproduces and varies phenotypically, divide the phenotypic distribution in half and allow each to reproduce as a separate population. If the phenotypic differences between the populations persist into the next generation, then they are least partially heritable.

The value of this experiment is that it does not attempt to identify the specific mechanisms that are responsible for heritability. The continuity between generations can be caused by genetic factors, cultural factors, or both. From the standpoint of natural selection, the specific mechanism does not matter as long as there is continuity between generations.

At the individual level, every offspring has exactly one (for asexual species) or two (for sexual species) biological parents. For sexual species, each parent contributes roughly the same amount of genetic material, although there are important exceptions (e.g., cytoplasmic genes, which are inherited only from the mother). Genetic inheritance becomes more complicated when we consider nonadditive genetic interactions (epistasis), which can cause offspring to become very different from their parents phenotypically, despite the fact that they share the same genes. When we consider cultural transmission, mechanisms of inheritance at the individual level become even more complex. A single offspring can have many parents who make unequal contributions, and so on. These mechanistic details are important, but they are also still poorly understood after more than a century of research. At the intuitive level we are forced to rely on the raw fact of heritability, much as Darwin did.

Heritability at the group level is even more complex than at the individual level. In some species, new groups are formed by a budding or fissioning process, such that each group has a single parent. In many other species, the members of any given group are derived from more than one group. Complex interactions among individuals in a group can have effects comparable to epistatic genetic interactions within individuals. The possibilities become so complex that we are forced to retreat to the more empirical question of whether phenotypic continuity exists, regardless of the mechanisms responsible for the continuity. For example, it would be almost impossible to know if the variation that Wade observed in his groups of flour beetles is heritable without actually forming new groups from old groups in a variety of ways. The empirical result is that group-level variation is heritable and can be selected, even though the group phenotype is caused by complex interactions among the group members and new groups are derived from multiple parental groups (Wade, 1977; Wade & McCauley 1980).

When we examine the phenotypic continuity of human groups, we find that it is often highly stable. In fact, many phenotypic differences between human groups persist despite extensive mixing of the individuals that make up the groups. This happens because individuals that move from one group to another are often expected to (and do) abandon the social norms of their old group and adopt the norms of their new group. Thus, phenotypic plasticity at the individual level, coupled with certain kinds of social norms, create a degree of heritability at the group level that could never be predicted from the size and mixing among groups. This is another example of how the simplifying assumption of genetic determinism, although useful for some purposes, can be misleading about the fundamental ingredients of phenotypic variation and heritability at the group level.

Step 3c: Ask the Question: What are the Fitness Consequences of Phenotypic Variation Within and Among Groups?

If heritable variation exists, then the differential survival and reproduction of units will cause evolutionary change, resulting in a fit between the properties of the unit and the environment that we call adaptation. The rate of evolutionary change and the degree to which natural selection at a given level prevails against opposing forces depends on the intensity of selection. For example, Wade's (1976, 1977) experiment included two group selection treatments. In the first, each group contributed to the next generation in proportion to its productivity. The second treatment was a form of truncation selection in which the least productive groups were discarded and only the most productive groups contributed to the next generation. Group selection occurred in both treatments but was more effective in the second because the fitness differences between groups were greater.

When a single trait (such as altruism) is selected against within groups but favored at the group level, the outcome will depend on the intensity of selection in addition to the amount of heritable variation at each level. For example, some environments are so harsh that it is difficult for a single individual to survive and reproduce on its own. The advantages of group-level functional organization can be so great in these environments that even behaviors with low relative fitness within groups can evolve. The same traits will not evolve in more benign environments because they do not produce the group-level advantages that compensate for the individual-level costs.

Fitness differences can be imposed, not only by the external environment, but also by social norms. Consider a behavior, x, that has a positive effect on the group but a substantial cost to the individual performing the behavior. An example might be sentry behavior, in which the sentry must refrain from eating (individual cost) in order to scan for predators (group benefit). Suppose that the balance between levels of selection is such that we would not expect this behavior to evolve by itself. Now consider a second behavior, y, that causes individuals to reward other individuals who perform x. For example, a female might be sexually attracted to males who perform sentry duty. Behavior y has a positive effect on the group to the extent that it causes others to perform x, and involves an individual cost, because rewarding x requires at least a small amount of time, effort, and allocation of resources that could have been used otherwise. Behavior y is therefore altruistic in the same sense as behavior x, but the balance between levels of selection is not necessarily the same. In particular, it is often possible for some individuals to provide very large rewards and punishments to others at trivial cost to themselves. Behaviors x and y may therefore evolve as a package when behavior x cannot evolve by itself (Sober & Wilson, in press). In general, the balance between levels of selection can be significantly altered when rewards and punishments are used to restructure the costs and benefits of other behaviors. These kinds of social controls are often interpreted as advantageous for the individuals who do the rewarding and punishing, but a careful analysis of fitness differences within and among groups shows that they are often group-level adaptations that evolve easily because the group-level benefits greatly outweigh the individual-level costs.

An Example of How the MultiLevel Framework Has Been Used Within Evolutionary Biology

The framework that I have sketched for thinking about multilevel selection is not new and, in fact, was employed by Williams (1966), who had to think carefully about group selection before rejecting it. To make the framework less abstract, I wll briefly show how Williams used it to address an important biological problem, before applying it myself to the subject of human decision making.

Williams' (1966) most convincing test of individual versus group selection involved the evolution of sex ratio. Fisher (1930/1950) had shown that natural selection within groups favors an even sex ratio because all offspring have exactly one father and one mother, so whichever sex is in the minority has the most offspring. Williams reasoned that an even sex ratio is not necessarily optimal from the standpoint of the group. If it is beneficial for the group to produce as many offspring as possible over several generations, it should have a highly female-biased sex ratio. Note that Williams was employing steps 1 and 2 of the procedure outlined earlier, by determining what would evolve by pure within- and between-group selection. Williams was unaware of any examples of female-biased sex ratios in nature, and therefore concluded that there was no evidence for group selection. He was so pleased with his empirical test that in the concluding section of his book he stated: "I would regard the problem of sex ratio as solved" (p. 272).

Williams' analysis of sex ratio shows that the multilevel framework can be useful even if it is not followed to completion. Williams completed only the first two steps of the procedure, but this was sufficient for him to make a prediction that could be tested empirically. In general, steps 1 and 2 of the framework are relatively easy because they involve thinking about adaptive design. Step 3 is more difficult because it involves thinking of the mechanistic details of the natural selection process. If adaptive design at the individual and group levels are sufficiently different from each other, then steps 1 and 2 can furnish testable predictions without one knowing much about step 3. Of course, step 3 is required for a deeper understanding.

One year after the publication of Williams (1966), Hamilton (1967) provided many examples of female-biased sex ratios and a theory to explain them, in which groups are colonized by a certain number of individuals whose offspring mate with each other before the daughters disperse to colonize new groups. Despite the fact that Hamilton's theory requires groups, it was interpreted as an example of individual selection for many years until Colwell (1981) showed that it is a mathematical version of Williams' verbally stated hypothesis, in which female-biased sex ratios have a low relative fitness within groups and evolve only because they increase the productivity of groups relative to other groups. Williams (1992, p. 49) has recently agreed with this interpretation. Hamilton's theory is an important advance over Williams' verbally stated hypothesis because it examines step 3 of the framework outlined earlier, allowing us to determine where any given population is likely to lie between the two extremes determined by steps 1 and 2. For example, if groups are initiated by single females, there is maximum variation among groups and the sex ratio is expected to be highly female biased. Increasing the number of individuals colonizing each group shifts the balance in favor of within-group selection, causing the sex ratio to converge on an even proportion of males and females. Subsequent theoretical models have elaborated on step 3 of the procedure, exploring the effects of local population regulation (Wilson & Colwell, 1981), multiple generations spent within groups (Aviles, 1993), and other factors. In general, the empirical data support the predictions of these more detailed models.

A careful reading of Williams shows that he has consistently employed the multilevel adaptationist program throughout his career, defining individual selection as gene frequency change within single groups. This has led him to interpret kin selection as a form of group selection (Williams & Williams, 1957), and to accept both female-biased sex ratios and the evolution of reduced virulence in diseases (Williams & Nesse, 1991, p. 8) as well documented empirical examples of group selection in nature. It is ironic that modern proponents of "individual selection" who think that they are operating in the tradition established by Williams have, in fact, departed from the framework that Williams actually defended. By interpreting both steps 1 and 2 as varieties of self interest, they have lost the ability to identify adaptation at the group level when it occurs.

HUMAN DECISION MAKING
AND MULTILEVEL SELECTION

Decision making is a process whereby a course of action is chosen from a number of alternatives. The ability to make decisions is centrally related to fitness and therefore should be one of the primary functions of the human brain. Decision making provides an ideal subject for the study of multilevel selection in humans for a number of reasons. First, if the quality of the alternatives can be ranked, then the adaptedness of a decision-making process can be accurately measured by its success at choosing the best course of action. Second, decision making is not as internal as other cognitive processes because it involves so much gathering and evaluation of information from the environment. The mechanistic details are therefore relatively accessible to study (examples are provided later). Third, decision making has been a favorite topic of psychologists for decades, resulting in literally hundreds of empirical studies that can be used as a database for testing evolutionary hypotheses. Fourth, even when these empirical studies involve problems that are not closely related to fitness (e.g., evaluating job applicants), the ability of a decision-making process to solve any problem probably correlates highly with its ability to solve problems related to fitness. Fifth, there is a tradition of thinking about groups as adaptive units in social psychology that dates back to the founding fathers of the discipline, who freely speculated about "the group mind" (Durkheim, 1915; LeBon, 1903; McDougall, 1920; Ross, 1908; Wundt, 1907). The group-level perspective in psychology was largely replaced by a more individualistic view, as in evolutionary biology, but it was not totally eclipsed. As a result, there is a small but vigorous modern literature on decision making as an adaptive group-level process in humans.

To employ the multilevel framework, we must first try to imagine what human decision making would look like as a product of pure group selection (step 1) and pure individual selection (step 2).

Groups as Adaptive Decision-Making Units (Step 1).   There are two senses in which human decision making can evolve to maximize the fitness of whole groups. First, individuals might function as independent decision makers whose goal is to benefit their group. This is the way that we usually think about altruism (Sober & Wilson, in press). Second, individuals might cease to function as independent decision makers and become part of a group-level cognitive structure in which the tasks of generating, evaluating, and choosing among alternatives are distributed among the members of the group. As with many other activities, decision making might benefit from the combined action, division of labor, and parallel processing made possible by groups of individuals interacting in a harmonious and coordinated fashion. At the extreme, the role of any individual in the decision-making process might become so limited that the group truly becomes the decision making unit, a group mind in every sense of the word.

The concept of a group mind is so strange, at least against the background of methodological individualism, that it may help to provide an example from nature. Social insect colonies such as honey bee hives are so highly organized at the group level that they can legitimately be called superorganisms (Seeley, 1989, 1996). To function adaptively, the hive must make decisions on almost a minute-by-minute basis about which flower patches to visit and which to ignore over an area of several square miles; whether to gather nectar, pollen, or water; the allocation of workers to foraging versus hive maintenance; and so on.

In an elegant series of experiments, Seeley and his colleagues have worked out in detail how some of these decisions are actually made (reviewed by Seeley, 1995). In one experiment, a colony with every bee individually marked was taken deep into the Adirondack woods where virtually no natural resources were available. The hive was then provided with artificial nectar sources whose quality could be experimentally manipulated. When the quality of a food patch was lowered below that of alternative patches, the hive responded within minutes by shifting workers away from the patch, yet individual bees visited only one patch and therefore had no frame of comparison. Instead, each individual contributed one link to a chain of events that allowed the comparison to be made at the hive level. Bees returning from the low-quality patch danced less and were less likely to revisit the patch themselves. With fewer bees returning from the poor resource, bees from better patches were able to unload their nectar faster, which they used as a cue to dance more. Newly recruited bees were therefore directed to the best patches. Adaptive foraging decisions were made by a decentralized process in which individuals acted more as neurons than as decision-making agents in their own right. Even the physical architecture of the hive such as the location and dimensions of the dance floor, honey comb, and brood chambers has been shown to play an important role in the cognitive architecture of adaptive decision making at the group level.

Although the social insects help us to grasp the concept of a group mind, we should not expect group-level cognition in humans to resemble the social insects in every detail. In particular, individual humans are such sophisticated cognitive units in their own right that they are unlikely to become neuronlike even when they have become integrated into a single adaptive unit. It therefore may help to supplement the social insect version of the group mind with the following metaphor. Imagine a room containing 30 microcomputers. Ask the question, "How much could the computational power of these computers be increased by connecting them into a network with a sophisticated communication system?" It should be obvious that an integrated network of 30 computers will be vastly superior to 30 isolated computers. Some of the advantages of the network will be mundane, such as increased memory capacity, whereas other advantages will be more synergistic, such as parallel processing, division of labor, sophisticated error-checking mechanisms and so on.

Imagining the human group mind as an integrated network of computers does justice to the sophistication of the individual as a cognitive unit but still allows us to appreciate the network as a cognitive unit in its own right that is vastly superior to any single component. It also makes it obvious that the communication system connecting the computers and individuals must be highly sophisticated for the network to function efficiently and is unlikely to be acquired by domain-general learning and cultural mechanisms. If human cognition is a product of group selection, we should expect individuals to be innately prepared (Tooby & Cosmides, 1992) to easily "hook up" with other individuals to form an integrated cognitive network.

Thinking of human groups as adaptive decision-making units leads to many specific predictions about social organization, some of which will be outlined later (see also Caporall & Baron, chap. 12, this volume). These predictions may turn out to be right or wrong, but at least they provide a conceptual anchor, allowing us to recognize what adaptive groups would look like if they have evolved.

Individuals as Adaptive Decision Making Units (Step 2).   If group selection can truly be ignored as a factor in human evolution, we should expect individuals to be highly adaptive as autonomous decision-making units, capable of performing the full range of activities from framing the problem, to generating alternatives, to evaluating the alternatives, to making the final decision. When individuals exist as members of groups, we should expect them to use others as sources of information, but only in ways that increase the individual's relative fitness within the group. We should not expect decisions that are adaptive for individual members to be necessarily adaptive for the group. Group-level benefits should occur only as the coincidental by-product of decisions that also maximize relative fitness within groups. After all, this is the entire thrust of Williams (1966). If "group-related adaptations do not, in fact, exist" (see passage quoted earlier), then regarding human groups as adaptive decision-making units is as wrongheaded as thinking that individuals evolve to benefit their species or that species evolve to benefit their ecosystem.

The same computer metaphor that allows us to imagine the human group mind as an integrated network also allows us to see why such networks can be undermined by within-group selection. Providing easy access to one's mind may be beneficial for the group, but it also makes one highly vulnerable to manipulation and exploitation by others within the same group. If the dangers of exploitation are sufficiently great, individuals should resist forming into adaptive decision-making units despite the advantages that all would share.

Determining the Balance Between Levels of Selection (Step 3). After we have imagined adaptive decision making at the group and individual levels, we must examine the factors that will determine where humans are likely to exist between the two extremes (step 3). At this point, it is important to appreciate that *Homo sapiens* is the most facultative species on earth and has existed in an enormous range of population structures throughout its evolutionary history. As a result, human decision making is unlikely to be located at a single point between the extremes. Rather, we might expect humans to have the capacity to function over the entire range, from autonomous decision-making units to components of a group-level mind, depending on the population structure in which they find themselves or build for themselves. We can also identify the kinds of groups in which adaptive group-level cognition will most likely be observed: groups with a high degree of genetic relatedness among the members, groups with a high degree of trust among the members, groups with social norms and other aspects of social organization that make exploitation within groups difficult, and groups in environmental situations that make the costs and benefits of group-level decision making large relative to the costs and benefits of exploitation within groups.

Once again, our computer metaphor provides an intuition about how human decision making might change along the continuum from extreme group level to extreme individual level functional organization. Imagine that you are an engineer who has built a computer network under the assumption that all of its users share a common goal. Then, to your dismay, you discover that some users are exploiting the network for their own gain, forcing you to implement features such as passwords and virus protection devices. The modified network is still useful, but not nearly as efficient as it was before, because the protection devices require resources and compromise the functional design of the network as a pure information processing device. Then, to your horror, you discover that those exploiting the network for their own gain are every bit as talented as you are and have discovered clever ways to bypass your defenses to plunder the other users. Ultimately you are forced to abandon the entire enterprise and sever all connections among the users. We must imagine human decision making as occupying this entire range of possibilities, from highly efficient networks that do not require protection, to compromised protected networks, to no networks at all.

Now that we have sketched steps 1 to 3 of the framework, we can evaluate the empirical literature on human decision making. To do this systematically, I employed the following procedure: First, I obtained the abstracts to 495 papers written between 1985 and 1994 by typing the keywords *group decision making* and *group problem solving* into the computerized literature service, Psych-Lit. The abstracts were downloaded and combined into a single file on my desktop computer. Second, all abstracts were read and summarized as short statements. Third, papers that address a specific topic (e.g., the effect of leader style) were found both by reviewing the statements and by using the "find" function on my word processor to locate appropriate key words (e.g., *leader*) in the file of full abstracts. Fourth, relevant papers were read in full along with other papers that they referenced. My ultimate goal is to test specific hypotheses that emerge from the multilevel framework with the formal methods of metaanalysis. In this article, however, space permits only an informal "guided tour" of research programs in psychology that are especially relevant to the evolutionary issues.

Individuals as Autonomous Decision Making Units

Individual humans do not always exist in groups, so they presumably have evolved to function as autonomous decision-making units in at least some situations. In a recent book entitled *The Adaptive Decision Maker*, Payne, Bettman, and Johnson (1993a; summarized in Payne, Bettman, & Johnson, 1993b) developed a research program that is highly conducive to the evolutionary perspective, even though the authors do not themselves relate their work to evolutionary theory. Payne and associates began by noting that there are many ways to evaluate and choose among alternative courses of action, which they called *alternative decision strategies*. For example, imagine that you are choosing among a number of habitats in which food, water, and safety from predators are all desirable attributes. One possibility is to evaluate each habitat for all three attributes, weighted appropriately, and then to choose the best habitat (the weighted additive strategy). Alternatively, you might first evaluate the habitats with respect to one attribute, evaluate the best of these for the second attribute, and so on (the elimination-by-aspects strategy). As a third possibility, you might simply choose the first habitat that exceeds an acceptable threshold for all three attributes (the satisficing strategy). Payne et al. identified six alternative decision strategies that tend to be used by human subjects.

Alternative decision strategies differ, not only in their ability to make good choices, but also in the time and cognitive effort that they require to work. The weighted additive strategy is the most accurate, but also the most costly, because all alternatives must be compared for all attributes that are weighted unequally. Because the adaptedness of a decision strategy depends on both its benefits (i.e., accuracy) and costs (i.e., time and cognitive effort), the most accurate strategy is not always the best. In fact, there is no single best strategy for all situations, and the adaptive decision maker must therefore have a collection of strategies that can be selectively employed in different decision environments.

Payne et al. measured the costs and benefits of the alternative decision strategies in impressive detail. They trained subjects to use all six strategies and to employ them in a computer-based information acquisition system called Mouselab. The subjects must use the mouse to acquire given pieces of information, allowing the computer to monitor the information they examined and the time they required to examine it (see Payne et al. 1993a for details). In this way, Payne et al. confirmed that the subjects were employing the prescribed strategy, the amount of time required to employ it, and the quality of their decisions.

To examine the cognitive effort required to employ a decision strategy, Payne et al. identified a number of elementary information processes (EIPs) that act as the building blocks for each strategy (Newell & Simon, 1972). Examples of EIPs include READ (read an alternative's value on an attribute into short-term memory), COMPARE (compare two alternatives on an attribute), and PRODUCT (weight one value by another). The cognitive effort required to employ a strategy in a given decision environment might depend on both the difficulty of each EIP (e.g., PRODUCT might be more difficult than COMPARE) and the number of times that each EIP is used. In addition, interaction effects might cause single EIPs to be more difficult in some strategies than in others. These possibilities were explored in a regression analysis that compared the number of each EIP used in a decision strategy as independent variables with response time and self-reports of effort by human subjects as dependent variables. The results suggest that the cognitive effort required to employ a decision strategy depends on the difficulty of the component EIPs and the number of times they are used, but not on interaction effects. In addition, subjects differed in the ease with which they could use single EIPs and therefore the effort required to employ a given decision strategy. This means that two individuals might adaptively employ different decision strategies in the same decision environment if their abilities to use single EIP's are sufficiently different.

The next step is to show that individuals can adaptively select the best decision strategy for a given decision environment. To determine what is the best decision strategy, Payne et al. modeled each of the six strategies as a production system (Newell & Simon, 1972) that combines the EIPs in a set of if-then rules that arrives at a decision. The production systems were then allowed to make decisions in Monte Carlo computer simulations and scored for both effort and accuracy. Finally, human subjects were confronted with the same decisions using MouseLab, and their choice of decision strategies was compared to the optimal strategies determined by the Monte Carlo simulations. In general, human subjects showed an impressive ability to select the appropriate decision strategy for a given decision environment.

Payne et al. (1993a, 1993b) provided an outstanding example of how the concept of adaptation can be used to inform the study of decision making as an individual-level process. However, they did not think about adaptation in the evolutionary sense and did not even use the word evolution in their book. This leaves a gaping hole in their research program. Where does all of this adaptive decision-making machinery come from? Payne et al. (1993b) addressed this question in a single sentence: "Individuals may have acquired different strategies through formal training or through experience" (p. 25). Evolutionary psychologists will immediately appreciate the inadequacy of this answer and how much the modular view of human cognition developed by Cosmides and Tooby (Cosmides & Tooby, 1992; Tooby & Cosmides, 1992) might add to the otherwise superb research of Payne et al. I return to the issue of innate mechanisms after discussing decision making as a group-level process.

Groups as Adaptive Decision-Making Units

If group selection has been an important force in human evolution, and if the process of decision making can benefit from coordinated action, then groups should make better decisions than individuals in at least some situations. As a newcomer to the psychological literature on decision making, I was struck by how many papers left the impression that groups are not effective decision making units compared to individuals. Yet, when I actually read the details of the studies, I discovered that groups frequently are much better than individuals at making decisions. It is important to explain this discrepancy and to establish the basic fact that groups can function as effective decision-making units before we proceed to the more detailed predictions of the multilevel framework.

Groupthink: Group Decisions as the Dysfunctional By-product of Internal Pressures. The concept of *groupthink* interprets group decision making as the largely dysfunctional by-product of internal forces (Janis, 1972, 1982):

> I use the term "groupthink" as a quick and easy way to refer to a mode of thinking
> that people engage in when they are deeply involved in a cohesive in-group, when

the members' strivings for unanimity override their motivation to realistically appraise alternative courses of action.Groupthink refers to a deterioration of mental efficiency, reality testing and moral judgement that results from in-group pressures (p. 9).

Janis made several predictions that are consistent with the within-group selection view of human decision making (step 2) and diametrically opposed to the between-group selection view (step 1). If groups can function as adaptive decision makers, then highly cohesive groups are more likely to realize this potential than less cohesive groups (step 1), yet Janis predicted that groupthink is especially characteristic of cohesive groups because of in-group pressures (step 2). Group decision strategies should be most strongly favored in environmental situations that threaten the group as an entire unit (step 1), yet these are the very situations that contribute to groupthink according to Janis. At least some kinds of strong and charismatic leadership should contribute to group function (step 1), yet Janis predicts that they will make groups dysfunctional by limiting the influence of other group members and creating an illusion of invulnerability (step 2).

The concept of groupthink has become enormously influential in the social sciences in addition to becoming a household word. Janis's work was cited over 700 times in the *Social Sciences Citation Index* between January 1989 and June 1992 (Aldag & Fuller, 1993). Many social scientists that I have talked to have accepted the conclusion that humans are poorly equipped to make decisions in groups. In evolutionary terms, this would imply that human psychology has not been strongly influenced by group selection and that step 2 provides a more accurate image of human decision making than step 1.

The concept of groupthink is based on retrospective analyses of famous policy failures, such as the Bay of Pigs invasion, the Vietnam War, and the space shuttle Challenger accident . This approach relies on hindsight and qualitative assessment and needs to be verified with more systematic studies that compare the processes associated with good versus bad decisions. A number of laboratory studies and quantitative analyses of historical events have been performed, whose results force a revision in the groupthink concept. I will describe one study in detail because it shows how the decision-making processes of real groups functioning in their environments can be analyzed quantitatively.

Tetlock, Peterson, McGuire, Chang, and Feld (1992) developed a method known as the Group-Dynamic Q-sort (GDQS) for converting verbal historical accounts of decision making into a systematic form that can be compared across groups. The GDQS consists of 100 statements about the properties of groups that are printed on a stack of cards. A person who has read a historical account of a decision-making event sorts the cards into 9 categories depending on how well each statement describes the event. The number of cards that can be placed in each category is held constant, creating a unimodal distribution. Thus, the first category includes the 5 cards that best describe the event; the second category consists of the next 8 most descriptive cards, and so on to the ninth category, which includes the 5 least descriptive cards. Forcing the cards into a single distribution eliminates variation in the judgmental style of the sorter, such as the tendency of some individuals to make middle-of-the-road distinctions and of others to jump to extremes.

To analyze any particular decision-making event, at least three historical sources describing the event were identified. Two independent observers read and performed a Q-sort for each source separately, allowing correlations between raters and between sources to be separately assessed. The average correlation between raters was .83, which is comparable to the reliability of widely used psychological tests. The average correlation among sources describing a single historical event was lower (averaging .52) but still respectable. Thus, Tetlock et al. succeeded in developing a method for converting unsystematic historical accounts of decision-making groups into a form that allows their internal structure to be analyzed quantitatively, which is an important achievement. For example, with little modification it could be used to analyze anthropological accounts of decision making in tribal societies (Boehm, in press).

No decision-making process is guaranteed to achieve the best solution, and there might not even be a good solution to many problems confronted by groups. For example, the fact that the Vietnam war turned out badly does not necessarily mean that Lyndon Johnson and his team of advisors functioned poorly as a decision-making unit. The most negative assessment of the groupthink concept would be that there are no structural differences between the groups that Janis classified as groupthink and nongroupthink, based on the outcome of historical events. Tetlock et al. show that this pessimistic assessment is unwarranted. Most of the policy disasters discussed by Janis do, in fact, reveal structural deficiencies in the groups as decision-making units. For example, Johnson was clearly an overbearing leader who greeted dissenting views from his advisors with the ominous statement, "I'm afraid he's losing his effectiveness" (Janis, 1982, p. 115), forcing many to leave his inner circle and those who remained to suppress their criticisms. By contrast, the Marshall plan (categorized by Janis as an example of nongroupthink) was formulated in a period of three weeks by a leader (George Kennan) who deliberately encouraged discussion and disagreement among his advisors (McCauley, 1989, provided capsule summaries of these events in addition to a perceptive assessment of groupthink).

In addition to the policy disasters caused by dysfunctional groups, Tetlock et al. also discovered some exceptions to the rule. For example, the Mayaguez rescue mission during the Ford administration and the Iran rescue mission during the Carter administration were evidently formulated by groups that functioned well as decision-making units, but nevertheless failed because of bad luck. Of course, an imperfect correlation between process and outcome is exactly what we should expect for something as inherently stochastic as decision making.

Although Tetlock et al. show that there is something to the groupthink concept, they also show that Janis' account needs to be substantially revised. In particular, there is no evidence that group cohesion or the importance of the situation make groups less functional as decision-making units. On the contrary, cohesion and salience tend to enhance the performance of groups in decision making tasks. This conclusion has been reached by virtually every careful analysis of the groupthink concept, including qualitative assessments, quantitative analyses of historical events, and controlled studies that attempt to duplicate the conditions of groupthink in the laboratory. The most recent review (Aldag & Fuller, 1993) states it thus:

> On the basis of our review, it seems clear that there is little support for the full groupthink model. . . . Furthermore, the central variable of cohesiveness has not been found to play a consistent role. Flowers (1977) went so far as to state that "a revision of Janis's theory may be justified, one which would eliminate cohesiveness as a critical variable" (p. 895). This suggestion is diametrically opposed to Janis's (1982) view that high cohesiveness and an accompanying concurrence-seeking tendency that interferes with critical thinking are "the central features of groupthink." (p. 9)

We can summarize the groupthink literature from the standpoint of the multilevel selection framework as follows: Decision making at any level is a complex process that requires a number of components interacting in the right way to function properly. It would be astonishing if human groups always functioned as efficient decision-making units, regardless of their social organization. Furthermore, certain behaviors that obviously would increase relative fitness within a single group, such as leaders attempting to assert their social dominance, are especially likely to disrupt the ability of the group to function as an efficient decision-making unit. This illustrates the basic conflict between levels of selection, in which a group of individuals behaving adaptively does not make an adaptive group (Williams, 1966). If groups always behaved dysfunctionally as decision-making units, or were most likely to behave dysfunctionally in the very situations in which it is important to perform well (e.g., high cohesion, high salience), this would be powerful evidence that human mentality is largely the product of within-group selection. However, the groupthink literature does not warrant this conclusion. In fact, it provides compelling evidence for two of the basic ingredients of natural selection at the group level: phenotypic variation among groups in decision-making ability (step 3a) with important consequences for fitness (step 3c).

Benchmarks for Comparing Groups and Individuals.   Psychologists tend to judge the decision-making ability of groups against a number of benchmarks, such as the ability of the average member acting alone (mean-of-n) or the ability of the best member acting alone (best-of-n). Often the purpose of a study is to demonstrate synergistic processes that cause groups to perform even better than the best individual, in which case the appropriate benchmark for comparison is best-of-n. Synergistic processes are highly interesting and are discussed below, but it is important to realize that they are not required for a decision-making process to count as adaptive at the group level. In fact, the distinction between within- versus between-group selection is different from either the mean-of-n or best-of-n benchmarks commonly employed by psychologists.

To see this, consider a group of individuals who vary in their knowledge about a particular subject that is required to make a good decision. If individuals are acting to maximize their relative fitness within groups (step 2), then experts should resist sharing their decision and the information on which it is based, causing the group to lie far below its best-of-n potential. We might even expect individuals to engage in sabotage by spreading false information, causing others to make even worse decisions than they would by themselves. This might appear dysfunctional to a psychologist (because the group lies below its mean-of-n potential) but an evolutionist would regard it as adaptive at the individual level. If individuals are acting to maximize the fitness of their group (step 1), we might expect them to actively meet to compare their private decisions and determine which is best if they do not all initially agree. The sharing of knowledge and joint evaluation of alternatives would eliminate fitness differences within groups and cause the group to approach its best-of-n potential. However, the group would not actually reach the best-of-n benchmark unless the most knowledge-able member can be determined with certainty. Furthermore, the group may well be unable to arrive at the best decision unless at least one member has reached it privately. This result might be a disappointment for a psychologist looking for synergistic effects (because the group fails to exceed its best-of-n potential), but it is gratifying for an evolutionist interested in adaptation at the group level.

When viewed from this perspective, the experimental evidence is over-whelming that groups usually exceed the decision-making ability of the average member (mean-of-n benchmark); they often approach and sometime exceed the decision-making ability of the best member (best-of-n benchmark). These results are especially remarkable because the vast majority of experiments were performed on extremely short-lived groups formed for the purposes of the experiment, whose members were complete strangers prior to group formation. If these minimal groups can outperform individuals, then the decision-making potential of real groups whose members are intimately familiar with each other might be very great indeed.

Two examples will illustrate how groups can surpass individuals in their decision-making ability, yet be judged as inferior to individuals by psychologists who employ a different set of benchmarks. Michaelsen, Watson, and Black (1989; see also Watson, Michaelsen, & Sharp, 1991) taught college courses in which students formed into learning groups that lasted the entire semester. The majority of class time was spent on group problem-solving tasks, including six objective and at least two essay exams that accounted for more than 50% of the course grade. Groups also met frequently outside of class to study and complete projects. This is one of the few research programs in which groups are presented with contextually relevant tasks and group dynamics have time to develop.

Exams were administered first to individuals and immediately afterward to groups (see Hill, 1982, for a discussion of this and other experimental designs in group decision research). In other words, after group members handed in their answer sheets, they were given an additional answer sheet to fill out as a group. For a total of 222 groups from 25 courses taught over a 5-year period, the mean individual test score was 74.2; the mean score of the best individual in each group was 82.6; and the mean group score was 89.9. A total of 215 groups (97%) outperformed their best member; four groups (2%) tied their best member; and three groups (1%) scored lower than their best member.

This study seems to provide overwhelming evidence for a group decision-making process that exceeds the best-of-n benchmark, but it has been criticized by Tindale and Larson (1992a, 1992b). The gist of their argument is that the best-of-n benchmark is exceeded only for the total test score and not for individual questions. If groups can do no better than select among the individual answers, and if all group members are wrong some of the time, then the best-of-n benchmark can only be approached at the level of single questions but can be exceeded at the level of the entire test. For Tindale and Larson, groups must be able to answer a question right *when every member answered it wrong* to demonstrate what they call assembly bonus effect.

Michaelsen, Watson, Schwartzkopf, and Black (1992) replied that their learning groups do display an assembly bonus effect, even as stringently defined by Tindale and Larson. From the multilevel perspective, however, even the most critical interpretation is good enough because it shows that groups can evaluate which of their members is most likely to be correct on a case-by-case basis. This error-checking capacity is available only to individuals who interact in cooperative groups.

My second example is the literature on brainstorming (reviewed by Mullen, Johnson, & Salas, 1991; Stoebe & Diehl, 1994), in which groups of individuals who suggest ideas to each other in a relaxed and uncritical atmosphere are supposed to be both more productive and more creative than individuals who attempt to generate ideas on their own. More than 50 psychological studies have compared the performance of brainstorming groups with the performance of so-called *nominal* groups, whose members generate ideas by themselves and do not interact with each other at all. The experimenter removes the redundant ideas generated by members of the nominal groups and compares their productivity and creativity with the brainstorming groups. These studies show un-equivocally that brainstorming groups are less productive and are not more creative than nominal groups. The productivity loss of brainstorming groups is caused by a variety of factors involving both the motivation and coordination of group members. Ideas must be presented one at a time in the brainstorming groups, whereas individuals in the nominal groups can write their ideas simultaneously. Members of brainstorming groups can feel uncomfortable suggesting outlandish ideas despite their instructions to feel relaxed in front of each other. Standard social norms of politeness can prevent the most creative and productive individuals from taking over the brainstorming session (Stroebe & Diehl, 1994).

These studies definitely cast doubt on the magic spark of productivity and creativity that is sometimes claimed for brainstorming groups. From the multilevel perspective, however, brainstorming groups and nominal groups must be considered as two kinds of cooperative social organization. In one case the ideas are generated in isolation and then pooled for evaluation, whereas in the other case they are generated and pooled simultaneously. In both cases they are pooled, which means that individuals benefit from the ideas generated by other individuals. From the multilevel perspective, we need to compare the productivity and creativity of a single isolated individual, or a group whose members are attempting to maximize their relative fitness, with the productivity and creativity of a group whose members freely share their ideas. For example, consider an experiment in which many individuals are asked to generate ideas on a topic in isolation. It is possible that each individual can easily conceive of all the possibilities and that pooling ideas merely adds redundancy. If so, then the number of different ideas generated by nominal groups of size n should quickly plateau as n increases. Alternatively, if each individual can conceive of only a small fraction of the possibilities, then the number of different ideas generated by nominal groups will not plateau until large values of n are achieved. Figure 13.1 shows the results of one typical study (Bouchard & Hare, 1970; see Mullen, et al., 1991 for a review) which shows no hint of a plateau over the range of $n = 1 - 9$, for either nominal or brainstorming groups. If the number of ideas generated at the beginning of the decision-making process contributes to the quality of the final decision, there is little doubt that groups will surpass individuals; large groups will surpass small groups; and cooperative groups whose members freely share their ideas will surpass noncooperative groups whose members do not.

Multiple Group Decision Strategies. Just as individuals are faced with a range of decision environments that require an arsenal of decision strategies (Payne et al, 1993a, 1993b), so groups must have an arsenal of decision strategies to function as adaptive decision-making units (Steiner, 1972, 1976).

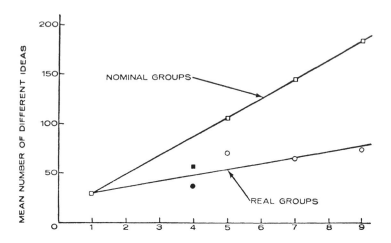

FIG. 13.1. The number of nonredundant ideas generated by brainstorming and nominal groups as a function of group size. From "Size, Performance, and Potential in Brainstorming Groups," by T. J. Bouchard and M. Haire, 1970, *Journal of Applied Psychology, 54*, p. 53. Copyright 1970 by American Psychological Association. Reprinted with permission.

The psychological literature is replete with examples of individuals who made stupid decisions because they employed the wrong heuristic, especially when the experiment caused them to frame the problem in the wrong way. We can expect the same kind of error at the group level. One interesting example involves a research program called *hidden profiles*, in which the information required to make the best decision is distributed among members of the group. Because individuals cannot get the right answer except by guessing, members of the group must interact synergistically to function adaptively. The first hidden profile studies suggested that groups do not perform better than individuals at this task. Members of the group seemed to focus on the information that they shared, which prevented them from integrating their unshared information. More recently, however, Stasser and Stewart (1992) distinguished between two decision-making strategies, only one of which is appropriate to the hidden profile task (see also Laughlin, 1980; Laughlin & Ellis, 1986). If the task is likely to have a demonstrably correct answer (an intellective task), then the appropriate strategy is to search until the answer is obtained. If the task does not have a demonstrably correct answer (a judgmental task), then the appropriate strategy is to reach a consensus. After all, if everyone is equally likely to be correct or incorrect, then the way to minimize error is to pool the estimates. Stasser and Stewart reasoned that groups perform poorly in the hidden profile task because they believe it is a judgmental problem rather than an intellective problem. They therefore changed the instructions to make it clear that the problem could be solved and obtained the results shown in Figure 13.2. Three factors were varied in a 2 x 2 x 2 design yielding eight treatments: (a) individual versus group, (b) "all shared" (all individuals receive all the clues) versus "hidden profile" (clues distributed among group members), and (c) subjects instructed that it is a judgemental task versus subjects instructed that it is an intellective task. Groups outperformed individuals in all conditions except hidden profile/judge. Groups who believed that they were solving an intellective task and had to merge their clues (groups/hidden profile/solve) performed as well as individuals who believed that they were solving an intellective task and were provided all the clues (individual/all shared/solve).

Groups as Decision Making Units in the Real World. Although the majority of psychological studies involve minimal groups that exist only for the purpose of the experiment, a few psychologists have made a special effort to study groups that actually function in the real world. Perhaps the most fascinating example is a recent book entitled *Cognition in the Wild* (Hutchins, 1995; see also Hutchins, 1991a, 1991b), which focused on the navigation methods of ancient and modern sailors. Unlike the decision-making tasks we have considered so far, that emphasize creativity and the solution of novel problems,

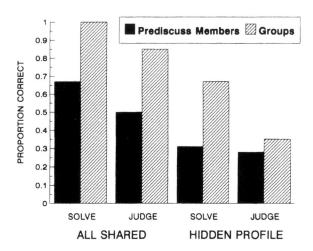

FIG. 13.2. The performance of individuals ("prediscuss members") and groups on a task in which the information was either available to all individuals ("all shared") or distributed among group members ("hidden profile") and the subjects believed that the problem was an judgmental task ("judge") or an intellective task ("solve"). From, "Discovery of Hidden Profiles by Decision-Making Groups: Solving a Problem Versus Making a Judgment," by G. Stasser and D. Stewart, 1992, *Journal of Personality and Social Psychology*, 63, p. 430. Copyright 1992 by American Psychological Association. Reprinted with permission.

estimating one's position on the earth's surface is a thoroughly routine proce-dure that must be performed again and again. Nevertheless, it involves a large number of decisions that are subject to error, with important consequences for fitness. When a boat is within sight of land, its position on a chart can be estimated by finding the compass directions of three landmarks, drawing lines from the landmarks on the chart at the appropriate angles, and noting where they intersect. The intersection of two lines is theoretically sufficient but does not provide an estimate of error. The intersection of three lines creates a triangle whose size indicates the likelihood of error.

On modern ships that are comfortably away from the shore, the entire task of estimating position is performed by a single person. When the ship enters a harbor, however, this person is replaced by a six-person team whose social organization has evolved over a period of centuries. Hutchins describes in detail how the team functions as an integrated cognitive unit, swiftly gathering and transforming information in a way that allows spatial position to be noted at a glance at three-minute intervals. The functional organization of the team includes a number of subtle but important design features. The members are spatially positioned and communicate in such a way that mistakes made by one person can be caught by at least one other person. There is redundancy of function so that the team can perform smoothly if one of its members is called away. The sequence in which the tasks are learned allows the team to function when its experienced members leave permanently and are replaced by new initiates. Even the architecture of the chart room contributes to the adaptive decision-making process, much as the architecture of the hive contributes to group-level cognition in bees. Hutchins argued that this is only one of many examples of cognition as an adaptive distributed group-level process in humans.

Our computer metaphor, which compares an integrated network of microcom-puters with the same number of isolated microcomputers, makes the advantages of group-level cognition appear obvious. Nevertheless, a superficial reading of the psychological literature gives the strong impression that groups seldom function better than individuals as decision-making units. Because this impression is mis-leading, I devoted considerable space to asking the most basic possible question: "Can groups make better decisions than individuals?" The answer to this question is "yes," allowing us to proceed to a more sophisticated set of questions. The multilevel framework allows us to make many specific predictions about the design features of adaptive decision-making groups (step 1) and the factors that allow group-level organization to prevail over individual-level organization (step 3). Unfortunately, space permits the exploration of only a few of these factors.

Social Norms That Regulate the Group Decision Process

Decision making at any level is a complex process that requires multiple components working in a coordinated fashion to function well. This means that decision making at the group level requires a social organization whose structure can be predicted from the nature of the decision environment. Hutchins (1995) provided a good example for the case of navigation.

In many groups, the social organization required for adaptive decision making should be surrounded by a set of social norms that guide individuals through the process and make it difficult for them to subvert the process for their own gain. The details will depend on the decision environment. Adaptive norms governing groups attempting to solve a novel problem may be very different from adaptive norms governing groups attempting to solve a routine problem such as navigation. Nevertheless, a few generalizations can be made that furnish testable predictions.

Phase-Dependent Social Norms. Most decision-making processes require a sequence of phases, such as idea generation, idea evaluation, idea selection, and idea implementation (Aldag & Fuller, 1993). Because each phase requires a different set of activities, no single set of social norms can be optimal for the entire sequence. For example, pressures to conform may allow the group to act decisively (idea selection and implementation) but may inhibit the group from fully exploring the alternatives (idea generation). It follows that a really adaptive set of social norms must change dynamically as the decision-making process unfolds. Members of the group must be encouraged to be nonconformists during the idea generation stage, conformists during the idea selection stage, and so on.

Is it reasonable to expect human social norms to be as sophisticated and dynamically changing as they must be to function adaptively from the decision-making standpoint? This question is seldom explicitly asked, but at least some biologists and social scientists have implicitly assumed that the answer is "no." For example, Janis (1972, 1982) assumed that group cohesion inevitably leads to pressures to conform, which compromises the early stages of the decision-making process. Among biologists, Boyd and Richerson (1985) modeled conformity as an inflexible social norm that has an important effect on cultural evolution. Nevertheless, a number of studies in the psychological literature suggest that the social norms surrounding the decision-making process are phase dependent, encouraging both diversity and conformity at the appropriate times. Perhaps the most interesting example involves a naturalistic study of Israeli scout groups deciding between two sites for a work camp, one of which was clearly better than the other (Kruglanski & Webster, 1991). Individuals had previously filled out a sociometric scale, rating other members of their group for liking, appreciation, and respect. These three measures correlated highly with each other and were averaged to yield a single index of social status. For each group, a member whose score was at the median of the distribution was approached to become a confederate of the experiment and was instructed to advocate the better (conformist) or worse (deviant) site, either early or late in the decision making process. After the decision was made, members were informed that their previous sociometric ratings had been lost and were asked to again fill out the same scale, enabling the experimenters to measure the social status of each member before and after the decision-making event. The only change in social status was a decrease that occurred when the confederate expressed a deviant opinion late in the decision-making process. This study provides clear evidence that social norms can be phase dependent, as they must to function adaptively at the group level.

Hierarchical Social Organization and Leader Control.   In an early experiment, Guetzkow and Simon (1955) had members of five-person problem solving groups communicate with each other in a variety of ways. They found that the most efficient communication network was hierarchical, in which four members communicated directly with a central fifth member. More richly connected groups, in which all members could communicate with each other, were less efficient and actually reverted to the hierarchical structure over time by breaking some of the connections (see Collins & Raven, 1969 for a review of this research tradition). This study illustrates an important principle: Well designed information structures must often (although not always) be hierarchically organized (Simon, 1981). Purely from the standpoint of group fitness (step 1), we should often expect to find some individuals that we would call "leaders," others that we would call "followers," and so on.

Of course, social hierarchies can also be explained purely at the individual level, as the outcome of reproductive striving within groups (step 2). If individuals differ in their fighting ability, then the strongest will simply subjugate the weakest and take most of the resources. If the weakest are also the youngest, they might optimally bide their time as subordinate individuals until they grow large enough to challenge the dominants and so on. Thus, the multilevel framework provides two very different pictures of hierarchical social organization. Which picture best describes human groups and how they interact with each other when they both apply depends on the balance between levels of selection (step 3).

A systematic review of the psychological literature on leadership is beyond the scope of this chapter (see Bass, 1990; Hogan, Curphy, & Hogan, 1994), but at least two general conclusions seem warranted. First, when social hierarchies are required from the standpoint of group-level functional design, the conflict between levels of selection is likely to be severe because leaders, by definition, are in a position of power that allows them to exploit other members of their own group. For this reason, the social norms that control the behavior of leaders must be especially strong if the group is to function as an adaptive unit. There seems to be abundant evidence that leaders in human social groups are often controlled at least as much as they are in control (Boehm, 1982, 1993). They are selected on the basis of their ability to lead the group and are required to prove their leadership abilities by deed in addition to word. They are often required to take the same risks as other group members and are subject to exceptionally high moral standards. Often their power is confined to a single decision-making domain beyond which they have no special authority. None of these features of human leadership are expected on the basis of within-group selection and are easily interpreted as group-level adaptations. Of course, there is also abundant evidence that leaders frequently escape social control to exploit members of their own group. The point is that both steps 1 and 2 are required to explain the hierarchical nature of human groups.

Second, when adaptive group-level decision making requires a social hierarchy, the leader usually does not function as an autonomous decision maker who acts as the "brains" for the group. Decision making remains a group-level process in which the leader functions as a component. A study by Anderson and Balzer (1991) provides a good example, in which leaders of decision-making groups were instructed to announce their opinion either early or late in the decision-making process. The group generated more ideas when the leader's opinion was delayed and the final decision was often based on ideas that were not generated by the leader. In fact, a general rule for effective leadership from the group standpoint might be *act as an organizer and moderator of group-level processes and refrain from exercising too much personal control* (Hogan et al., 1994).

Innate Psychology, Learning, and Culture

The psychological literature on decision-making is remarkable for its lack of curiosity about where decision-making abilities come from, at either the group or individual level. The entire field seems to implicitly assume what Tooby and Cosmides (1992) referred to as the standard social science model (SSSM) in which all skills are learned from scratch by domain-general cognitive mechanisms. As an alternative to the SSSM, Tooby and Cosmides developed a view of human psychology in which the major activities that influenced fitness in ancestral environments, such as mate choice and social exchange, are governed by innate specialized cognitive modules that have evolved for millions of years. The human mind presumably consists of dozens of Darwinian algorithms invoked by environmental stimuli, much as a jukebox consists of many records played by pressing the appropriate buttons. Tooby and Cosmides realized that the mind seen as a juke box is as much a caricature as the mind seen as a single general-purpose computer, so the truth presumably lies somewhere between.

The study of human-decision making affords an excellent opportunity to explore the middle ground between these two extreme views of human psychology. The ability to make decisions is as central to biological fitness as mate choice, social exchange, and language acquisition, so there is every reason to expect it to be governed by innate specialized psychological mechanisms. It is impossible for an evolutionary psychologist to read Payne et al.'s (1993a, 1993b) account of multiple decision strategies without thinking of Tooby and Cosmides'

juke box metaphor. However, just because the six decision strategies studied by
Payne et al. function as modules does not imply that they are innate modules.
One of the most fundamental conclusions that emerges from the decision-making literature is that adaptive decision strategies are as diverse as decision
environments (Steiner 1972, 1976), making it highly unlikely that an innate
Darwinian algorithm exists for every decision strategy. This does not mean that
we must abandon the jukebox metaphor entirely, but we do need to supplement
it with more open-ended mechanisms of learning and cultural transmission that
fine tune human decision-making ability to the particular decision environment.

Adaptive decision making at the group level must similarly reflect a mixture
of innate psychology, open-ended learning, and cultural transmission (which
themselves can be guided by innate psychological mechanisms). When I began
my inquiry on this subject, numerous colleagues suggested that a review of the
psychological literature was a waste of time because the study of minimal groups
that exist only for the purpose of the experiment says nothing about the function
or structure of real groups. It is true that real groups deserve far more attention
from psychologists than they have received, but I am also impressed by the
information that has been forthcoming from minimal groups. It seems that
humans are innately prepared to function adaptively in groups under appropriate conditions and that learning and cultural transmission merely improve on
a scaffolding that is already present. A small literature exists on group-level
decision making in children (French & Stright, 1991; Kinoshita, 1989; Messer,
Joiner, Loveridge & Light, 1993), which may allow the appearance and development of this scaffolding to be examined in more detail.

GENERAL IMPLICATIONS
OF THE MULTILEVEL PERSPECTIVE

I have attempted to outline a general framework for thinking about multilevel
evolution and to apply the framework to the specific subject of human decision
making. I conclude by discussing a few of the most important implications of
the multilevel perspective for the disciplines of psychology, anthropology, and
evolutionary biology.

Implications for Psychology

It should be obvious that I regard the psychological literature as a gold mine of
empirical information for testing evolutionary hypotheses about human behavior. However, extracting this information is a bit like trying to assemble a large
jigsaw puzzle using the pieces from many small jigsaw puzzles. Dozens of research
programs address specific issues (e.g., what accounts for production-blocking in

brainstorming experiments?), but in a larger sense they are conceptually adrift. A glance at some of the most general models of group decision making in the psychological literature reveals the nature of the problem. Aldag and Fuller (1993) separated the decision-making process into six compartments, each of which can be influenced by a large number of variables. For example, the compartment labeled *group structure* can be influenced by cohesiveness, homogeneity, insulation, leader impartiality, leader power, history of the group, probability of future interaction, stage of group development, and group type. The entire model includes 53 variables, which induces a sort of combinatorial paralysis. With 1,378 dyadic combinations and 23,426 triadic combinations of variables, where do we begin? Hirokawa and Johnson's (1989) general theory of group decision making also celebrates complexity without providing a guide for managing complexity.

The multilevel evolutionary framework is conceptually general and also appreciates the importance of multiple factors, but it leads to highly specific predictions that are not forthcoming from the other models. Steps 1 and 2 allow us to use the concept of functional design to predict exactly what adaptive decision making would look like if it had evolved at the group and individual levels respectively. Step 3 allows us to identify the major variables that will facilitate or disrupt adaptive decision making at the group level. Tooby and Cosmides' modular view of human psychology encourages us to look for cognitive foundations of decision-making abilities in a way that seems to have been almost entirely ignored by psychologists, if my sample of over 500 papers is representative. These elements of the multilevel framework act as guides for managing complexity by identifying specific regions of the parameter space that deserve attention. For example, relationships characterized by a high degree of trust should lie closest to the ideal of an integrated computational network (step 1) unburdened by the protective devices required in less trusting relationships. Individuals in trusting relationships should therefore be superior at decision making and other cognitive tasks. To my knowledge, this computational theory of intimate relationships has not been developed by psychologists interested in either decision making or intimate relationships (reviewed by Brehm 1992), with the notable exception of Wegner and his colleagues (Wegner, 1986; Wegner, Erber & Raymond, 1991; Wegner, Giuliano & Hertel, 1985) who explicitly adopt a group-level approach to human cognition. Thus, the multilevel framework easily locates an important region of the parameter space that has been previously missed by most psychologists.

Another major implication of the multilevel framework concerns the importance of studying human mentality as a group-level process. The concept of a group mind was popular among the first social psychologists, but was largely replaced by a more individualistic perspective, as in evolutionary biology. Those who continue to think about group-level cognition clearly regard themselves as endangered species within their own disciplines (Campbell, 1994; Hutchins,

1995; Wegner et al., 1985). The multilevel framework makes it obvious that the study of adaptive group-level processes needs to occupy a more central position in psychology. Thinking of groups as organisms is more than just a metaphor. It is literally correct, to the extent that the traits being studied are the product of group selection. To say that adaptive group-level processes are merely the outcome of interactions among self-interested individuals is exactly as wrongheaded as saying that individual physiology is just the outcome of self-interested organs. The multilevel framework gives equal and equivalent status to individuals and groups as potentially adaptive units, which is a powerful justification for moving the study of group-level mentality to center stage. In fact, in our modern world of books and computers, it is difficult for us to imagine how deeply communal human thought must have been throughout our evolutionary history, when all knowledge had to be stored and passed between human heads (Ong, 1982).

The multilevel framework has implications for applied psychology in addition to that of basic psychology. Cognition at the individual level requires communication among neurons. Cognition at the group level requires communication among individuals. Individuals must physically encounter each other to communicate. Encounter rates are heavily influenced by the basic patterns of daily life and by the spatial architecture of buildings and villages. Women who wash their clothes together by a river will have more opportunity to communicate than women who wash their clothes by themselves in their basements. I have witnessed numerous incidents in which an academic group such as a laboratory or a department moves from one building to another and experiences a radical change in its intellectual quality of life, usually for the worse. The new building causes individuals to bump into each other less often, and scheduled encounters (e.g., meetings) are not as effective as continuous and spontaneous encounters for certain kinds of communication. Thinking of human mentality as a group-level process makes it obvious that mundane and pervasive aspects of human life, such as the placement of offices and the scheduling of activities, should be arranged to facilitate certain kinds of communication among members of a group. If architects and social planners regarded themselves as group-level brain surgeons, perhaps they would think in a way that results in fewer group-level lobotomies.

Implications for Anthropology

Anthropological studies have a special status in evolutionary psychology because they allow us to see how human behavior functions in environments closer to the ancestral condition than our own modern world. It is therefore important to know how well tribal societies (and especially hunter–gatherer societies) function as adaptive decision-making units (Knauft, 1991). At one extreme, we might discover that tribal groups are best described by step 2 of our procedure and that group-level functional organization is a recent historical event. At the other extreme, we might discover tribal social organizations that are intricately adapted at the group level (step 1), including mechanisms that prevent individual-level selection within groups in the same way that the rules of meiosis prevent gene-level selection within individuals (step 3).

Boehm (1996) recently surveyed the ethnographic literature on the way tribal societies make decisions, especially in crisis situations. His main purpose was to show that decisions are often made by a rational process, as opposed to superstition or culturally determined traits inherited by simple transmission rules. For our purpose, his examples are remarkable for showing how decision making is functionally organized at the group level. Meggitt's (1977) description of Mae Enga raiding and warfare in highland New Guinea reads like a textbook example of nongroupthink. The major Big Man solicits responses from those present without offering his own opinion. All men who have passed through the bachelor's association are eligible to participate and, "being among their own clansmen, can speak with complete freedom." Anyone possessing pertinent information has "a moral obligation to contribute so that the group may reach the best possible decision in the circumstances." This procedure is followed because "only in this way, it is believed, can each clansman truly ascertain the thoughts of his fellows and the evidence behind them." The Big Men have their own opinion, but "none of them, especially in the early sessions, reveals much of his hand or tries patently to push for the acceptance of his suggestions." After hours of argument, a consensus usually emerges and the Big Man "incisively summarizes the main arguments, indicates which have been rejected, and finally announces the decision reached by the clan." Although freedom of expression is encouraged during the early stages of the decision-making process, the social norms eventually shift to favor conformity because "even as both parties are making clear their positions, everyone knows that, because the clan's survival may be at stake, once combat begins the doves will almost certainly take their places fighting alongside the hawks" (p. 76). All three examples of decision making reviewed by Boehm seem to lie remarkably close to the group-adapted end of the continuum (step 1). In another review article, Boehm (1993) showed that leaders of tribal groups are often highly controlled by their constituents, as expected for societies that are functionally organized at the group level.

Early anthropologists such as Evans-Prichard (1940/1979), and Levi-Strauss (1949/1969) speculated freely about social organizations and cultural practices that allow tribal groups to function as corporate units. For example, Evans-Prichard (1940/1979) described a segmented social organization for the Neur that allowed them to become functionally organized at a nested hierarchy of levels. Normally the smallest units in the hierarchy acted in a corporate fashion to accomplish the tasks of daily life but even these were large and genetically heterogeneous by biological standards. The smallest units fought among themselves but easily united into higher-level corporate units to fight other higher-level units. The only hint of social stratification was a person known as the leopard-skin chief, whose authority was limited to the single domain of conflict resolution. According to Evans-Prichard (1940/1979), "his function is political, for relations between political groups are regulated through him. His activities are chiefly concerned with settlement of blood-feuds, for a feud cannot be settled without his intervention, and his political significance lies in this fact. Chiefs sometimes prevent fights between communities by running between the two lines of combatants and hoeing up the earth here and there. The older men then try to restrain the youths and obtain a settlement of the dispute by discussion . . . On the whole, we may say that Nuer chiefs are sacred persons, but that their sacredness gives them no authority outside specific social situations . . . I have often heard remarks such as this: 'We took hold of them and gave them leopard skins and made them our chiefs to do the talking at sacrifices for homicide.'" (p. 173).

This way of thinking about tribal society is still common among anthropologists, but not among those who have been most influenced by evolutionary theory. Yet, it is highly compatible with step 1 of the multilevel framework. Thus, viewing evolution as a multilevel process makes it even more relevant to anthropology than it has been so far.

Implications for Evolutionary Biology

Evolutionary biologists who regard themselves as individual selectionists overwhelmingly view nature (including human nature) through the lens of an individual organism attempting to maximize its inclusive fitness. This does not correspond to step 2 of the multilevel framework because the concept of inclusive fitness does not correspond to natural selection within single groups. As Williams and Hamilton were among the first to appreciate (Hamilton, 1975; Williams & Williams, 1957), inclusive fitness theory assumes a multigroup population structure, with natural selection occurring both within and among groups. Thus, even though modern individual selectionists think that they are operating in the tradition of Williams and Hamilton, their concept of individual selection has been stretched to include elements of steps 1, 2, and 3 of the multilevel framework.

An article entitled "Group Composition: An Evolutionary Perspective" (Harvey & Greene, 1981) provides a good example because it was published in a multidisciplinary volume on group cohesion (Kellerman, 1981) and therefore shows how evolutionary biologists have contributed to a subject that is relevant to group-level decision making. The paper begins with this statement: "Evolution through natural selection has produced patterns of group living among animals that maximize the inclusive fitness of individuals" (p. 149). The authors correctly define group selection as "natural selection operating on groups as opposed to individuals" (p. 151), but then immediately dismiss it as "theoretically feasible, but only under extreme conditions which are unlikely to be found in nature" (p. 151). Dominance hierarchies and other social structures are interpreted as follows:

> If we accept that animals modify their behavior in the presence of others for adaptive reasons . . . we might ask whether emergent social structures exist which cannot be reduced to an explanation at the individual level. The simple answer, from the adaptationist standpoint, must be *no*. For example, many social structures exist which have been described as emergent characteristics of group living, such as dominance hierarchies. But such arguments, when they provide any explanation of the existence of dominance hierarchies, rely on group selection. In animal societies, however, such social structures can profitably be analyzed in terms of individual reproductive strategies. Williams (1966) has pointed out that dominance hierarchies are no more than the "compromise made by each individual in its competition for food and mates, and other resources. . . ." (Harvey & Greene 1981, p. 163)

In this passage, Harvey and Greene seem to be explicitly denying the possibility that social hierarchies might function as decision making structures that benefit the whole group. The group-level speculations of early thinkers are described as follows:

> Evolutionary analyses of social form have frequently used the group as the unit of analysis, with a direct analogy to organic evolution in the formulation of society as a "superorganism". . . . For example, Bagehot (1872) argued that groups had a clear advantage over individuals in the struggle for survival, and that group cohesion was essential for group success. Gumplowicz (1885) continued this reasoning with strong group selectionist arguments—that social evolution proceeds through differential *group* survival. (Harvey & Greene, 1981, p. 164)

These passages sound for all the world as if Harvey and Green are employing the multilevel framework outlined earlier and are claiming that step 1 of the procedure should be avoided at all costs because group selection is such a weak evolutionary force. However, a more careful reading reveals that they have simply forgotten that Williams defined individual selection as natural selection within single groups. Their own conception of "individual evolutionary interests" includes a myriad of traits that do not increase relative fitness within groups and do require the differential fitness of groups to evolve, exactly as outlined by the early thinkers that Harvey and Green reject. In this way, group-level adaptations are barred from the front door but welcomed through the back door as a variety of self-interest.

There are two major problems with this conceptual framework. First, it is logically indefensible. If the tradition established by Williams is going to be followed, it should be followed well. Merely returning to Williams' own defini-

382 WILSON

tions of individual and group selection causes the modern concept of individual selection to collapse. Second, it is heuristically counterproductive. Group-level adaptations are difficult to see when they are first explicitly denied and then accepted in a disguised form. To see them clearly, we must return to Williams' original framework and ask the question that evolutionary biologists have avoided for so long: What would groups look like if group selection *were* the only force acting on them?

ACKNOWEDGMENTS

I thank C. Boehm, M. Brewer, D. Campbell, A.B. Clark, B. Knauft, B. Smuts, E. Sober, K. Sterelny, and Binghamton University's Evolution, Ecology, and Behavior group for helpful discussions

REFERENCES

Aldag, R. J., & Fuller, S. R. (1993). Beyond fiasco: A reappraisal of the groupthink phenomenon and a new model of group decision processes. *Psychological Bulletin, 113*, 533–552.

Anderson, L. E., & Balzer, W. K. (1991). The effects of timing of leaders' opinions on problem-solving groups: a field experiment. *Group and Organization Studies, 16*, 86–101.

Aviles, L. (1993). Interdemic selection and the sex ratio: A social spider perspective. *American Naturalist, 142*, 320–345.

Bagehot, W. (1872). *Physics and politics.* New York: Appleton and Co.

Bass, B. M. (1990). *Bass and Stogdill's handbook of leadership* (3rd ed.). New York: Free Press.

Boehm, C. (1982). The evolutionary development of morality as an effect of dominance behavior and conflict interference. *Journal of Social and Biological Structures, 5*, 413–422.

Boehm, C. (1993). Egalitarian behavior and reverse dominance hierarchy. *Current Anthropology, 34*, 227–254.

Boehm, C. (in press). Cultural selection mechanics in human groups. *Current Anthropology.*

Bouchard, T. J., & Hare, M. (1970). Size, performance, and potential in brainstorming groups. *Journal of applied psychology, 54*, 51–55.

Boyd, R., & Richerson, P. J. (1985). *Culture and the evolutionary process* . Chicago: University of Chicago Press.

Brehm, S. S. (1992). *Intimate relationships* (2nd ed.). New York: McGraw-Hill.

Bull, J. J. (1994). Perspective: Virulence. *Evolution, 48*, 1423–1437.

Campbell, D. T. (1994). How individual and face-to-face-group selection undermine firm selection in organizational evolution. In J. A. C. Baum & J. V. Singh (Eds.), *Evolutionary dynamics of organizations* (pp. 23–38). New York: Oxford University Press.

Collins, B. E., & Raven, B. H. (1969). Group structure: attraction, coalitions, communication and power. In G. Lindzey, & E. Aronson (Eds.), *The handbook of social psychology* (pp. 102–204). Reading, MA: Addison-Wesley.

Colwell, R. K. (1981). Group selection is implicated in the evolution of female-baised sex ratios. *Nature, 290,* 401–404.

Cosmides, L., & Tooby, J. (1992). Cognitive adaptations for social exchange. In J. H. Barkow, L. Cosmides, & J. Tooby (Eds.), *The adapted mind: Evolutionary psychology and the generation of culture* (pp. 163–228). Oxford: Oxford University Press.

Dugatkin, L. A., & Reeve, H. K. (1994). Behavioral ecology and levels of selection: dissolving the group selection controversy. *Advances in the study of behavior, 23,* 101–133.

Durkheim, E. (1915). *Elementary forms of the religious life.* New York: McMillan.

Evans-Prichard, E. E. (1979). *The Nuer.* Oxford: Oxford University Press. (Original work published 1940)

Fisher, R. A. (1950). *The genetical theory of natural selection.* New York: Dover. (Original work published 1930)

Frank, R. H. (1988). *Passions within reason.* New York: Norton.

French, D. C., & Stright, A. L. (1991). Emergent leadership in children's small groups. *Small Group Research, 22,* 187–199.

Guetzkow, H., & Simon, H. A. (1955). Communication patterns in task-oriented groups. *Management Science, 1,* 233–250.

Gumplowicz, L. (1885). *Outline of sociology.* Philadelphia: American Academy of Political and Social Sciences.

Hamilton, W. D. (1967). Extraordinary sex ratios. *Science, 156,* 477–488.

Hamilton, W. D. (1975). Innate social aptitudes in man, an approach from evolutionary genetics. In R. Fox (Ed.), *Biosocial anthropology* (pp. 133–153). London: Malaby Press.

Harvey, P. H., & Greene, P. J. (1981). Group composition: An evolutionary perspective. In H. Kellerman (Ed.), *Group cohesion: Theoretical and clinical perspectives* (pp. 148–169). New York: Grune & Stratton.

Hill, G. W. (1982). Group versus individual performance: Are N+1 heads better than one? *Psychological Bulletin, 91,* 517–539.

Hirokawa, R. Y., & Johnston, D. D. (1989). Toward a general theory of group decision making: Development of an integrated model. *Small group behavior, 20,* 500–523.

Hogan, R., Curphy, G. J., & Hogan, J. (1994). What we know about leadership: Effectiveness and personality. *American Psychologist, 49,* 493–504.

Holcomb, H. R. I. (1994). Empirically equivalent theories. *Behavioral and Brain Sciences, 17,* 625–626.

Hutchins, E. (1991a). The social organization of distributed cognition. In L. Resnick, J. Levine, & S. Teasley (Eds.), *Perspectives in socially shared cognition* (pp. xx–xx). Washington, DC: American Psychological Association Press.

Hutchins, E. (1991b). The technology of team navigation. In J. Galegher, R. E. Kraut, & C. Egido (Eds.), *Intellectual teamwork* (pp. 191–220). Hillsdale, NJ: Lawrence Erlbaum Associates.

Hutchins, E. (1995). *Cognition in the wild.* Cambridge, MA: MIT Press.

Janis, I. L. (1972). *Victims of groupthink.* Boston: Houghton Mifflin.

Janis, I. L. (1982). *Groupthink* (2nd ed.). Boston: Houghton Mifflin.

Kellerman, H. (1981). *Group cohesion: Theoretical and clinical perspectives*. New York: Grune & Stratton,

Kinoshita, Y. (1989). Developmental changes in understanding the limitations of majority decisions. *British journal of developmental psychology, 7*, 97–112.

Knauft, B. M. (1991). Violence and sociality in human evolution. *Current Anthropology, 32*, 391–428.

Kruglanski, A. W., & Webster, D. M. (1991). Group member's reactions to opinion deviates and conformists at varying degrees of proximity to decision deadline and of environmental noise. *Journal of Personality and social psychology, 61*, 212–225.

Laughlin, P. R. (1980). Social combination processes of cooperative problem-solving groups on verbal intellective tasks. In M. Fishbein (Ed.), *Progress in social psychology* (pp. 127–155). Hillsdale, NJ: Lawrence Erlbaum Associates.

Laughlin, P. R., & Ellis, A. L. (1986). Demonstrability and social combination processes on mathematical intellective tasks. *Journal of experimental social psychology, 22*, 177–189.

LeBon, G. (1903). *The Crowd*. London: Allen & Unwin.

Levi-Strauss, C. (1969). *The elementary structures of kinship*. Boston: Beacon. (Original work published 1949)

McCauley, C. (1989). The nature of social influence in groupthink: Compliance and internalization. *Journal of personality and social psychology, 57*, 250–260.

McCauley, D. E., & Wade, M. J. (1980). Group selection: The genotypic and demographic basis for the phenotypic differentiation of small populations of *Tribolium castaneum*. *Evolution, 34*, 813–821.

McDougall, W. (1920). *The group mind*. New York: MacMillan.

Meggitt, M. (1977). *Blood is their argument*. Palo Alto: Mayfield.

Messer, D. J., Joiner, R., Loveridge, N., & Light, P. (1993). Influences on the effectiveness of peer interaction: Children's level of cognitive development and the relative ability of partners. Special Issue: Peer interaction and knowledge acquisition. *Social Development, 2*(3), 279–294.

Michaelsen, L. K., Watson, W. E., & Black, R. H. (1989). A realistic test of individual versus group consensus decision making. *Journal of applied psychology, 74*, 834–839.

Michaelsen, L. K., Watson, W. E., Schwartzkopf, A., & Black, R. H. (1992). Group decision making: How you frame the question determines what you find. *Journal of Applied Psychology, 77*, 106–108.

Mullen, B., Johnson, C., & Salas, E. (1991). Productivity loss in brainstorming groups: A meta-analytic integration. *Basic and Applied Social Psychology, 12*, 3–24.

Newell, A., & Simon, H. A. (1972). *Human problem solving*. Englewood Cliffs, NJ: Prentice-Hall.

Ong, W. (1982). *Orality and literacy: The technologizing of the word*. London: Methuen.

Orzack, S., & Sober, E. (1994). Optimality models and the test of adaptationism. *American Naturalist, 143*, 361–380.

Payne, J. W., Bettman, J. R., & Johnson, E. J. (1993a). *The adaptive decision maker*. Cambridge, England: Cambridge University Press.

Payne, J. W., Bettman, J. R., & Johnson, E. J. (1993b). The use of multiple strategies in judgment and choice. In N. J. J. Castellan (Ed.), *Individual and group decision making* (pp. 19–39). Hillsdale, NJ: Lawrence Erlbaum Associates.

Ross, E. A. (1908). *Social Psychology*. New York: MacMillan.

Seeley, T. (1989). The honey bee colony as a superorganism. *American Scientist, 77,* 546–553.

Seeley, T. (1995). *The wisdom of the hive* . Cambridge, MA: Harvard University Press.

Simon, H. A. (1981). *The sciences of the artificial* (2nd ed.). Cambridge, MA: MIT Press.

Sober, E., & Wilson, D. S. (in press). *Unto others: The evolution of altruism.* Cambridge, MA: Harvard University Press.

Stasser, G., & Stewart, D. (1992). Discovery of hidden profiles by decision-making groups: solving a problem versus making a judgment. *Journal of Personality and Social Psychology, 63,* 426–434.

Steiner, I. D. (1972). *Group process and productivity.* New York: Academic Press.

Steiner, I. D. (1976). Task-performing groups. In J. W. Thibaut, J. T. Spence, & R. C. Carson (Eds.), *Contemporary topics in social psychology* (pp. 393–422). Morristown, NJ: General Learning Press.

Stroebe, W., & Diehl, M. (1994). Why groups are less effective than their members: On productivity losses in idea-generating groups. *European Review of Social Psychology, 5,* 271–303.

Tetlock, P. E., Peterson, R. S., McGuire, C., Chang, S., & Feld, P. (1992). Assessing political group dynamics: a test of the groupthink model. *Journal of Personality and Social Psychology, 63,* 403–425.

Tindale, R. S., & Larson, J. R. J. (1992a). Assembly bonus effect or typical group performance? A comment on Michaelsen, Watson and Black (1989). *Journal of applied psychology, 77,* 102–105.

Tindale, R. S., & Larson, J. R. J. (1992b). It's not how you frame the question, it's how you interpret the results. *Journal of applied psychology, 77,* 109–110.

Tooby, J., & Cosmides, L. (1992). The psychological foundations of culture. In J. H. Barkow, L. Cosmides, & J. Tooby (Eds.), *The adapted mind: Evolutionary psychology and the generation of culture* (pp. 19–136). Oxford: Oxford University Press.

Wade, M. J. (1976). Group selection among laboratory populations of Tribolium. *Proceedings of the National Academy of Sciences, 73,* 4604–4607.

Wade, M. J. (1977). An experimental study of group selection. *Evolution, 31,* 134–153.

Wade, M. J. (1979). The primary characteristics of Tribolium populations group selected for increased and decreased population size. *Evolution, 33*(2), 749–764.

Wade, M. J., & McCauley, D. E. (1980). Group selection: the phenotypic and genotypic differentiation of small populations. *Evolution, 34,* 799–812.

Watson, W., Michaelsen, L. K., & Sharp, W. (1991). Member competence, group interaction, and group decision making: a longitudinal study. *Journal of applied psychology, 76,* 803–809.

Wegner, D. M. (1986). Transactive memory: a contemporary analysis of the group mind. In B. Mullen, & G. R. Goethals (Eds.), *Theories of group behavior* (pp. 185–208). New York: Springer-Verlag.

Wegner, D. M., Erber, R., & Raymond, P. (1991). Transactive memory in close relationships. *Journal of Personality and Social Psychology, 1991,* 923–929.

Wegner, D. M., Giuliano, T., & Hertel, P. T. (1985). Cognitive interdependence in close relationships. In W. J. Ickes (Ed.), *Compatible and incompatible relationships* (pp. 253–276). New York: Springer-Verlag.

Williams, G. C. (1966). *Adaptation and Natural Selection: a critique of some current evolutionary thought.* Princeton, NJ: Princeton University Press.

Williams, G. C., & Nesse, R. M. (1991). The dawn of darwinian medicine. *Quarterly Review of Biology, 66,* 1–22.

Williams, G. C., & Williams, D. C. (1957). Natural selection of individually harmful social adaptations among sibs with special reference to social insects. *Evolution, 11,* 32–39.

Wilson, D. S. (1983). The group selection controversy: History and current status. *Annual Review of Ecology and Systematics, 14,* 159–187.

Wilson, D. S. (1989). Levels of selection: An alternative to individualism in biology and the social sciences. *Social Networks, 11,* 257–272.

Wilson, D. S., & Colwell, R. K. (1981). Evolution of sex ratio in structured demes. *Evolution, 35*(5), 882–897.

Wilson, D. S., & Dugatkin, L. A. (in press). Group selection and assortative interactions. *American Naturalist.*

Wilson, D. S., & Sober, E. (1989). Reviving the superorganism. *Journal of Theoretical Biology, 136,* 337–356.

Wilson, D. S., & Sober, E. (1994). Reintroducing group selection to the human behavioral sciences. *Behavioral and Brain Sciences, 17,* 585–654.

Wundt, W. (1907). *Outlines of psychology* (C. H. Judd, Trans.). Leipzig: W. Engelmann. (Original work published in 1905)

VII

Capstone

14

The Emergence of Evolutionary
Social Psychology

David M. Buss
University of Texas at Austin

Although Darwin's theory of evolution by selection has been around since 1859, it is a plain fact that scientists for more than a century afterward could not figure out how to use the theory to study the most complex organic creation yet discovered—the human mind. Future historians of science will have fun figuring out why it took so long. Perhaps it was the reign of radical behaviorism, which dominated the social sciences for a good part of the 20th century. Perhaps it was the persistent refusal to dethrone our species from an exalted plane and view us through the same lens used for other animals. Perhaps it was religious resistance. Perhaps scientists recoiled in horror at imagined pernicious political fallout from viewing humans through the lens of evolutionary theory. Yet perhaps it was the sheer complexity of the subject matter: The human mind, encased in a 1,400 cubic centimeter brain, may be the most intricate and formidable product of the evolutionary process on earth.

Evolutionary psychology represents a new synthetic discipline, a true blend of evolutionary biology and modern psychology that provides powerful models for understanding the complexities of the human mind. Evolutionary psychology borrows profound insights from psychology and combines them with insights from evolutionary theory. One of the key insights is that it is at the psychological level that many adaptations are most incisively described (Buss, 1995; Symons, 1979; Tooby & Cosmides, 1992). This is not a "cannibalizing" of psychology by biology, but a true synthesis of the best insights from each discipline.

A decade ago, it would have been inconceivable to assemble the fine collection of chapters represented in this volume. Would-be editors then would have faced a number of hurdles. They would have been unable to locate enough psychologists to render 12 independent contributions. Among the few they might have been able muster, most would have been speculators rather than empirical scientists, and they would have inhabited the fringes of academia rather than coming from strong empirical programs housed in major colleges and universities.

In this chapter, I note several promising trends within evolutionary social psychology that are signaled by the preceding chapters and offer some reflections for the future of the field.

PROMISING TRENDS
IN EVOLUTIONARY PSYCHOLOGY

The Emergence of Testable Hypotheses

There is nothing wrong with speculation; it is an important aspect of science. Science progresses by varying combinations of theory, observation, intuition, hunch, formalization, dreams, and data. A speculation can be extraordinarily valuable and lead to an avalanche of new discoveries. As scientific fields mature, however, speculations get more rigorous, resulting in formal hypotheses articulated with sufficient precision to render them testable. The chapters in this volume attest to the maturation of evolutionary social psychology.

Each chapter in this book offers propositions that are eminently testable. Krebs and Denton, (chapter 2) for example, develop a fascinating model of social illusions and self-deception, and propose a predictable set of contexts in which each should occur. Daly and Wilson (chapter 10) offer a fascinating set of testable predictions about the psychology of kinship, a sorely neglected domain in mainstream psychology. David Sloan Wilson (chapter 13) proposes a set of predictions about human decision making from a group selection model. Shackelford (chapter 14) develops a testable model of betrayal that offers specific predictions about which events will constitute betrayals, depending on the specific relationship context.

Evolutionary approaches in the past have been criticized—justly, in my opinion—for having a high ratio of speculation to testable propositions. Judging from the contributions to this volume, this criticism is no longer valid.

The Emergence of Competing Hypotheses

Another sign of the maturing of a scientific discipline is the emergence of competing hypotheses about important and cutting-edge phenomena. In the field of astronomy, for example, there are competing cosmological theories of the origins of our universe. Within biology proper, there are competing theories about the origins of sexual reproduction from asexually reproducing ancestors.

Now within evolutionary psychology, there are competing theories in several important domains.

Graziano and his colleagues (chapter 6), for example, challenge Kenrick's evolutionary hypothesis about the importance of dominance, and offer instead an evolutionary model of interpersonal attraction based on the interaction between dominance and agreeableness. Cunningham and his colleagues (chapter 5) present a model of attractiveness that includes configural combinations of neonatal features, sexual maturity, expressivity, and grooming. Gangestad and Thornhill (chapter 7), in contrast, argue for the importance of fluctuating asymmetry as a signal of genetic quality and perhaps freedom from environmental insults during development.

Within the domain of human mating, Miller and Fishkin (chapter 8) challenge Buss and Schmitt's (1993) sexual strategies theory by arguing that humans have evolved primarily for monogamous mating. Buss and Schmitt, in contrast, propose that humans have evolved both long-term and short-term mating strategies, and pursue different strategies (sometimes both long-term and short-term, or mixed mating strategy, simultaneously) depending on contexts and individual difference variables such as mate value. The key point is that the two models can be pitted against one another in empirical competition. As in mature sciences, the empirical data will be the final arbiter.

As more scientists become drawn into the field of evolutionary psychology, we can expect further increases in the prevalence of competing hypotheses. As a result, we can expect a commensurate increase in the acquisition of knowledge about the evolutionary psychology of the human mind.

The Anchoring in Empirical Evidence

A related trend is the anchoring of evolutionary arguments in bodies of empirical data. This telling trend signals progress, because new hypotheses can no longer remain untethered. They must square with the empirical evidence, or at a minimum, provide an account of why the existing evidence is flawed.

The contributions in the domains of mating and attraction are the best exemplars of this trend. These domains, previously peripheral areas within social psychology, have drawn more research attention from evolutionary social psychologists than any other area. The "obsession" with attraction and mating is not coincidental. The process of natural selection operates by differential reproductive success, and hence events that surround mating are theoretically of pivotal importance. If selection has not fashioned psychological mechanisms in the domain most closely linked with reproduction, it would have been unlikely to fashion mechanisms in domains seemingly more distant from reproduction.

Over the past decade, evolutionary scientists have documented a number of phenomena that must be accounted for by any theory of attraction and mating (see Symons, 1995, for an excellent review of this literature). We know, for example, that cues to youth and health are key components of attractiveness. We know that "average" faces are more attractive that faces that depart from the statistical average (Langlois & Roggman, 1990). We know that kindness and agreeableness are central to the selection of a long-term mate (Buss et al., 1990; Graziano, chapter 6, this volume) We know that both men and women elevate the importance they attach to physical appearance in short-term, as contrasted with long-term, mating contexts (Buss & Schmitt, 1993; Kenrick Sadalla, Groth, & Trost, 1990; Gangestad & Thornhill, 1994). We know that men place a higher premium on youth and physical attractiveness than do women in both mating contexts, whereas women place greater value on a potential mate's resources (Buss, 1989). Evolutionary theories of mating and attraction must be anchored in this well-developed body of empirical evidence.

Aside from occasional exceptions, evolutionary psychological theories are increasingly grounded in empirical evidence. Those that remain untethered from the existing empirical data violate a fundamental function of good theories—to account for existing evidence. As more and more scientists start to use the conceptual tools of evolutionary psychology, these bodies of evidence will grow.

FUTURE DIRECTIONS FOR EVOLUTIONARY SOCIAL PSYCHOLOGY

In this section, I suggest several important directions for the future of this field, most of which are signaled by the chapters in this volume.

Relationships as the Core of Social Psychology

In the history of social psychology, enduring social relationships have been relegated to the periphery of the field. There may be several reasons for this. First, relationships do not lend themselves to studies that use undergraduates, who constitute the bulk of the participants in the studies published in the mainstream social psychology journals. The participants in undergraduate subject pools are mostly strangers to each other. Second, relationships do not lend themselves easily to experimental manipulation and control. One cannot "assign lovers" randomly, or experimentally manipulate a marital infidelity.

The emergence in mainstream social psychology of the study of social cognition and person perception might have heralded the study of relationships. But it did not. The "social" aspect of social cognition often turned out to be trait terms presented in milliseconds to undergraduates on computer screens.

In contrast to the mainstream of the field, the vast majority of chapters in this volume focus on enduring social relationships. Shackelford (chapter 4) provides a c mpelling argument and summarizes empirical data suggesting that what constitutes a "betrayal" differs depending on the type of relationship.

Having sex with someone outside of the relationship, for example, is a betrayal of a long-term mateship, but not a betrayal of a long-term friendship (unless the friend has sex with one's long-term mate!). This suggests the existence of domain-specific psychological mechanisms to deal with the different adaptive problems posed by different sorts of relationships.

Zeifman and Hazan (chapter 9) provide a crisp and cogent model of the evolutionary psychology of adult attachment. Based on sound evolutionary reasoning, they argue that failures at adult attachment would have jeopardized the survival of children of a bonded couple. There is empirical evidence to support their contention, beyond the evidence they mention. Hill and Hurtado (1995), for example, found that mortality rates among children are 10% higher without an investing father. Investing fathers may not only aid the survival of their children, but also can teach them skills, aid their negotiation of social hierarchies, and give them a mating advantage (Buss, 1994).

The field of "attachment," as articulated by its founder John Bowlby (1969), was firmly anchored in evolutionary principles. Bowlby focused primarily on mother–infant attachment, and proposed a clear function: proximity maintenance to increase the survival of children. Subsequent work on attachment, however, lost its evolutionary anchor. The excellent work of Zeifman and Hazan reestablishes that link within the more sophisticated framework of modern evolutionary psychology.

Another jewel, chapter 10 in this volume by Daly, Salmon, and Wilson, focuses on a sorely neglected domain of close relationships: the evolutionary psychology of kinship. From an evolutionary perspective, with the possible exception of mating, there is no other domain more likely to be the target of evolutionary selection pressure. Hamilton's (1964) theory of inclusive fitness paved the way for a bounty of hypotheses about kinship. Daly, Salmon, and Wilson provide a fascinating agenda for reaping this bounty.

They suggest that humans have evolved a specific kinship psychology, consisting of specialized psychological mechanisms for solving adaptive problems linked with different forms of kinship relationships. Being a mother poses different adaptive problems than being a father, and both sets are distinct from adaptive problems posed by being a daughter, a brother, or a grandparent. Mothers, for example, are 100% certain that they are the genetic parents of their children. Because fertilization occurs within the mother, however, paternity is less certain. Men can channel investments to extrapair matings in ways that are more directly reproductively advantageous compared with extrapair matings by women (i.e., in a direct increase in offspring production). These straightforward reproductive differences suggest that the psychology of being a mother will differ from that of being a father, and perhaps may account for the large sex differences observed in all cultures in the amount of direct parenting, a trend that has persisted despite the social pressure for mothers to work and fathers to parent.

Mates may come and go, but kinship ties often last a lifetime. The next decade should witness a flowering of research in the area of kinship, and great empirical fruits will go to those who harvest this ripe field.

The notion of different categories of relationships, each with a relationship-specific psychology, gains additional credence from the work of Haslam (chapter 11), based on the pioneering work of Fiske (1991). Four fundamental types of relationships are proposed, each with its own "grammar": communal sharing (e.g., coalitions), authority ranking (e.g., social hierarchies), equality matching (e.g., friendships), and market pricing (e.g., getting paid for performing a service).

It will be interesting for future research to determine whether these four types of relationships map onto the enduring relationships that follow from evolution-minded conceptions: mateship, kinships, coalitions, dyadic reciprocal alliances, and hierarchical relationships (Buss, 1986). Because just one of these—kinships—really subsumes several qualitatively different types of relationships (e.g., father–son, sibling), an evolutionary psychological perspective would suggest that there are more than four distinct types of evolved relationship-specific psychology.

Collectively, the chapters in this volume herald a major shift in social psychology—restoring the study of enduring relationships to its proper place at the center of the field.

Groups

Humans evolved in the context of small groups. Although estimates vary, group size probably averaged between 50 and 200 (Caporael & Baron, chapter 12). Judging from the prevalence of tribal warfare, these groups were likely to be in conflict with neighboring groups, at least some of the time (Chagnon, 1983). Within these larger groups, smaller coalitions invariably form, and these smaller coalitions often compete with each other for access to reproductively relevant resources. Thus, it is reasonable to suggest that humans have evolved specific psychological mechanisms to deal with the unique problems posed by the formation and consolidation of coalitions, groups of people who cooperate with each other, but sometimes compete with rival coalitions of cooperators (Alexander, 1979, 1987; Tooby & Cosmides, 1988).

Krebs and Denton (chapter 2) present impressive evidence for precisely such psychological adaptations. The tendency to make sharp distinctions between in-group and out-group, for example, may be critical to the formation of strong coalitions. In-group members are treated more benevolently, being credited with their successes and excused for their failures. Out-group members are treated less charitably, their successes attributed to luck, their misfortunes attributed to enduring personal failings. So pervasive are these negatively biased out-group attributions that Pettigrew (1979) dubbed this phenomenon the *ultimate attribution error*. Krebs and Denton, in a chapter crackling with creative sparks, marshal powerful arguments for the evolutionary basis of the many fascinating cognitive biases discovered by social psychologists.

Two other chapters focus on the evolutionary psychology of groups. Caporael and Baron (chapter 12) argue that the group is the "mind's natural environment," echoing the arguments of Alexander (1979, 1987). But they go further. They argue against the neo-Darwinian revolution brought about by Hamilton's (1964) theory of inclusive fitness, and against the "gene-centered" perspective that follows from it. They propose that four, and only four, "core configurations" have provided all uniquely human adaptations: dyads, teams, demes, and macrodemes. Furthermore, they argue for a hierarchical perspective in which complex dynamic systems produce self-organizing emergent properties.

Although interesting in the abstract, Caporael and Baron do not provide a description of the specific adaptive problems linked with each of the "core configurations." It will be exciting to see whether their approach will generate any specific empirical predictions and an attendant program of research.

David Sloan Wilson (chapter 13), a long-time champion of the importance of group selection, provides a fascinating overview of the logic of group selection, and illustrates it with a summary of research on group decision making. Unlike Caporael and Baron, Wilson does not oppose the neo-Darwinian paradigm that has emerged since Hamilton's theory of inclusive fitness. Rather, he argues that group selection may be a more important force in some specific domains than is commonly believed.

The standard view since Williams' (1966) classic treatise on the topic is that group selection is possible, but that the conditions for it to occur are stringent and rarely met. Group selection requires "shared fate" of the members of the group, much like the genes within one's body usually have a shared fate, succeeding or failing together. Interbreeding between groups, however, tends to break down the shared fate, and so reproductive isolation is often required for group selection to take place. Moreover, the selection pressure at the group level has to be recurrent over a large enough number of generations for selection to forge an adaptation at the level of the group.

David Sloan Wilson's arguments are refreshing and cogently presented. Even if he turns out to be wrong in the end, I expect that his perspective will lead to the discovery of new social psychological adaptations not discovered by the more traditional gene-centered perspectives, thereby deepening our understanding of the evolutionary psychology of groups.

Understanding Individual Differences

Boundaries within psychology have created a separation between the study of social psychological processes and the study of individual differences, with the latter usually relegated to personality psychology. This division is more historical than logical. First, individual differences occur in social phenomena, even when they are not directly investigated. In the classic Milgram (1974) study on obedience to authority, for example, two thirds of the participants conformed and delivered the highest level of shock to the "learners" when ordered to do so by the white-coated experimenter. But one third refused to deliver that shock. Thus, important individual differences emerge in social phenomena, even when they are not directly investigated.

Second, an examination of the major taxonomies of individual differences show that many are primarily social or have profound consequences for social interaction. Individual differences in dominance–submissiveness—a trait represented in virtually every personality taxonomy—is defined by one's hierarchical stance toward others in one's social environment. Even individual differences not so defined, such as differences in emotional stability, have consequences of interactions with others, including the tactics one uses to manipulate others (Buss, 1992) and the ways in which people irritate and anger their spouses (Buss, 1991). Thus, individual differences and social psychology need to be integrated into evolutionary accounts.

There exist a variety of theoretical options for integrating individual differences and evolutionary social psychology. Zeifman and Hazan (chapter 9) describe one option. In their analysis of attachment, they describe the three attachment styles in functional context. Specifically, they argue that the "securely attached" individuals learn that caregivers are dependable, so their trust of others is an adaptive response to a reliable social environment. The "avoidant" individuals, in contrast, experience repeated rejections from caregivers, so they develop an attachment style in which they avoid relying on others, which is also an adaptive response, according to Zeifman and Hazan. The "ambivalent" individuals, they argue, learn that caregivers are unpredictable, which leads them to either crank up their demands or cling to others as social tactics for eliciting resources.

In essence, Zeifman and Hazan are proposing a species-typical psychological mechanism, attachment, which has three evolved settings. The settings are established early in life as a function of interactions with primary caregivers, and serve as models for the social environment in adulthood. This is a plausible account of individual differences, although one can imagine evolved systems with greater flexibility—(e.g., those that shift the model of the social environment in adulthood to accommodate changes in the social relationships experienced).

These do not exhaust the options for integrating individual differences with evolutionary social psychology. Other options include enduring heritable differences in interaction styles due to frequency-dependent selection and facultative shifts in strategy as a function of current environment occupied (see Buss, 1991). It is also possible that some individual differences are not adaptively patterned at all, but represent either "noise" (adaptively irrelevant variations) or "incidental by-products" of other individual differences that are adaptively patterned. The future of evolutionary psychology will achieve greater concep-tual power by an inclusion of the major ways in which individuals differ.

Self-Delusion and Social Influence

The study of social influence—long central to the field of social psychol-ogy—needs to be evolutionized, and several chapters in this book illustrate ways in which this can be done. From an evolutionary perspective, social signals have been reinterpreted as "social manipulation" rather than "information commu-nication" (Dawkins & Krebs, 1978; Krebs & Dawkins, 1984). On this account, signals are designed to influence the psychological mechanisms of others in particular ways for the benefit of the signaler. We understand this function intuitively when it comes to advertisements, which are clearly designed to persuade rather than to inform. Dawkins and Krebs argue that this view applies to all social signals.

"Brother, can you spare a dime?" Daly, Salmen, and Wilson (chapter 10) outline a few of the fascinating ways in which kin terms are invoked outside of kin contexts for the goal of social influence. By invoking the concept of "brother," the panhandler has learned that triggering kin schema leads to better outcomes than failing to do so. Similar functions may be involved when sorority members call each other "sisters."

Krebs and Denton (chapter 2) provide some fascinating examples of the use of social categorization and even self-attribution for the goals of social influence. Projecting hostile intentions on our enemies, for example, may simultaneously rally others against them and provide a social justification for exploiting them. When the halo of love within which we view our spouses turns to the soured scourge of divorce, we denigrate the same person we previously praised. Pre-sumably, this impression management functions to minimize the reputation damage we might experience as a result of divorce and hence maintain or enhance our perceived "mate value."

Kinship may interact with these attempts to staunch the hemorrhaging of reputation. Why do mothers of serial killers insist on the innocence of their sons? Why do parents blame teachers for their children's poor performance? Self-delusion and social manipulations are intimately intertwined, one serving the function of the latter.

The "illusion of optimism," Krebs and Denton argue, may generate motiva-tional power, helping to drive us to goals we could not achieve if we had a more realistic appraisal of outcomes. Furthermore, the commonly observed "self-handicapping" may function to lower social expectations of us, thereby increas-ing the credit we get when a success is attained.

Social biases, out-group attributions, and self-delusions are all ripe for evolutionary analysis, and we can anticipate great advances in this area of evolutionary social psychology in the coming decade.

Task Analysis, Computational Models, and Function

As the field of evolutionary social psychology matures, we can expect an increased sophistication in the analysis of evolved psychological mechanisms. The chapters in this volume are advances on previous analyses, and the field needs to continue to move in this direction.

One method for increasing the sophistication about function is to develop a "computational theory" within each of the focal domains (Cosmides & Tooby, 1994; Marr, 1982). A computational theory specifies in detail the function of an information processing devise: the specifics of the adaptive problem, why there is an evolved mechanism for solving that problem, and the strategy by which the computations are carried out. Intuitive notions of "adaptive" or "beneficial" cannot substitute for formal analyses of function.

Computational theories require a precise task analysis of the nature of the information processing problem. Consider the adaptive problem of pursuing a short-term mating strategy. Among the informational processing problems that must be solved, at a minimum, are these: (a) identifying from among the range of potential mates those that possess desirable characteristics, such as cues to fertility; (b) assessing which potential mates from this subset are potentially sexually accessible, or amenable to a short-term mating; (c) assessing one's capacity to attract that person, which includes an assessment of one's own mate value and the relative value of the potential mate; and (d) assessing the potential costs of the short-term mating, including the presence of a potentially violent spouse and reputation damage.

As this example suggests, each adaptive problem often can be analyzed in detail for the specific subproblems that must be solved. This sort of task analysis, specifying with greater precision the information processing problems that must be solved, leads to a more incisive description of the nature and function of evolved psychological mechanisms.

Included in these task analyses will be descriptions of precisely which sorts of information are available for solving the adaptive problem. This dovetails nicely with Springer and Berry's (chapter 3) arguments about the role of evolution in an ecological model of social perception. They argue for a Gibbsonian view, with a central role for social "affordances," the invariant (or recurring) properties of the social environment that provide information about opportunities for action. Clearly, evolved mechanisms must capitalize on these invariants and embody procedures for acting on them.

Many of the social psychological phenomena described in this volume will ultimately be viewed as "first passes" in an ultimate description of our evolved mechanisms. Computational theory and task analysis represent promising avenues for deepening our descriptions of the nature and functions of these mechanisms.

Modularity and Domain Specificity

At the core of evolutionary psychology is a firm position on the modularity or domain specificity of evolved psychological mechanisms. This position is implicit in many chapters in this volume, but I would argue that it needs to be more explicit for future progress.

Because there is still controversy about the issue of domain specificity, it is worth examining several of the arguments for it (see Cosmides & Tooby, 1994; Sperber, 1994 for more extensive discussions). First, what qualifies as an "adaptive solution" differs from domain to domain. Consider the problem of food selection, choosing objects with cues to the presence of sugar, fat, and protein, but avoiding objects with high concentrations of toxins. Solutions to this problem are fundamentally different from successful solutions to the problem of mate selection, such as detecting and choosing persons showing cues to high fertility and willingness to commit. It is extraordinarily unlikely that a domain-general mechanism, lacking specialized procedures for the respective domains of food and mates, could solve both problems. Given the large number and different nature of the adaptive problems humans must solve to survive and reproduce, it is extraordinarily unlikely that a small number of general mechanisms will provide all of the qualitatively different successful solutions.

A second, and related, argument pertains to the search through "design space" for successful solutions. Given the vastness of possible solution space, an organism needs mechanisms to guide it to the minuscule pockets that actually provide successful solutions. Domain-general mechanisms, by failing to provide guides through design space, get paralyzed by combinatorial explosion (Cosmides & Tooby, 1994).

Adaptive problems require solutions in real time. A venomous snake, a jealous husband, a hungry child, an impending raid by a neighboring village—solutions to these problems cannot await the careful search and evaluation of the billions of possible options. The combinatorial explosion endemic to domain-general mechanisms renders them useless for solving these real-time problems.

A third argument against domain-general, and hence content-independent, mechanisms pertains to whether content of the information needed for solving the adaptive problem can be gleaned ontogenetically. It has been well documented in the domains of vision and language that our evolved mechanisms "assume" things about the nature of the world that cannot be extracted in a single lifetime. In Roger Shepard's (1987) terms, informational invariants are "internalized" over evolutionary time. Our evolved mechanisms thus supply information needed to solve the adaptive problem.

Consider the problem of paternity uncertainty, a problem faced by males in species where fertilization occurs internally within the female. This problem is exacerbated in our species in contexts where men invest heavily in offspring.

From an investing man's point of view, a cuckoldry could result in the loss of all the effort he put into courting and attracting his mate, the loss of all of the woman's parental effort, which now gets channeled to another man's children, and the loss of his own parental effort, which also gets channeled to another man's children.

Male sexual jealousy has been proposed as a specific evolved solution to this problem, albeit an imperfect solution. Could a domain-general mechanism, a content-free capacity to draw inferences, solve this problem and just "figure it out"? Consider the sorts of information it would need to do so: A person would have to observe a reasonable sample of cases in which other men having sex with other men's wives produced children who were not genetically related to the husband and compare them with another sample of cases in which sexually faithful wives bore children who were genetically related to their husbands.

Is it reasonable to suppose that this information is available to each man ontogentically? Is it reasonable to suppose that men have access to this sample space of cases, given the fact that infidelity is cloaked in greater secrecy than nearly any other human behavior? Is it reasonable to suppose the people can witness babies being born, accurately appraise their paternity, and then correlate that information back to unseen events that might have happened nine months ago? It strains credulity to imagine that men have access to the information needed to solve this adaptive problem ontogenetically.

Furthermore, even if they did, and men could figure out rationally that other men having sex with their wives would jeopardize their paternity, the domain-general account would be led to the following prediction: Men whose wives were taking birth control, or had their tubes tied, should not get the least bit jealous if other men have sex with their wives, since their paternity and hence investments are not jeopardized. The relevant study has not been done, but the prediction is clear.

Evolved domain-specific mechanisms, coming richly packed with information about statistical regularities of the social world, supply information that cannot be gleaned ontogenetically. Domain-general mechanisms lacking content-specialized procedures lack the information needed to solve specific adaptive problems.

CONCLUSIONS

Among all the branches of psychology, there may be no branch more ripe for evolutionizing than social psychology. It is unlikely that our huge brains—the 900 cubic-centimeter advantage we have over chimpanzees—have evolved to help us pick better berries or avoid snakes. These survival problems are all solved by chimps with a much smaller brain.

It is far more likely that humans evolved such large brains as a consequence of the complexities of social living and social competition that includes forming coalitions, executing a rich repertoire of short-term and long-term mating strategies, negotiating the intricacies of complex kin networks and social hierarchies, forming long-term reciprocal alliances, and socializing children for years or decades (Alexander, 1987; Byrne & Whiten, 1988; Humphry, 1976). Social psychology in general, and the study of relationships in particular, belong at the center of evolutionary psychology because so many of the uniquely human adaptive problems are inherently social ones.

The chapters in this volume attest to the value of integrating social psychology with evolutionary psychology. They are limited to be sure. Undoubtedly they will be surpassed in the future by those delving deeper into the specific tasks required for solving the social adaptive problems, being more explicit about the concept of function, applying more rigorous criteria for specifying evolved psychological mechanisms, and incorporating individual differences into the models.

This collection represents an intellectually exciting stocktaking and agenda-setting for the next decade of evolutionary social psychology. We are poised on the brink of a scientific revolution—a revolution that realizes, after more than a century, Darwin's vision of a new foundation for psychology.

REFERENCES

Alexander, R. D. (1979). *Darwinism and human affairs.* Seattle: University of Washington Press.

Alexander, R. D. (1987). *The biology of moral systems.* New York: Aldine de Gruyter.

Bowlby, J. (1969). *Attachment.* New York: Basic Books.

Buss, D. M. (1986). Can social science be anchored in evolutionary biology? *Revue Europeen des Sciences Sociales, 24,* 41–50.

Buss, D. M. (1989). Sex differences in human mate preferences: Evolutionary hypotheses tested in 37 cultures. *Behavioral and Brain Sciences,12,* 1–49.

Buss, D. M. (1991). Conflict in married couples: Personality and the evocation of anger and upset. *Journal of Personality, 59,* 663–688.

Buss, D. M. (1992). Manipulation in close relationships: Five factors of personality in interactional context. *Journal of Personality,*

Buss, D. M. (1994). *The evolution of desire: Strategies of human mating.* New York: Basic Books.

Buss, D. M. (1995). Evolutionary psychology: A new paradigm for psychological science. *Psychological Inquiry, 6,* 1–30.

Buss, D. M., Abbott, M., Angleitnor, A., Asherian, A., Biaggio, A., & Blanco-Villasenor, A. (1990). International preferences in selecting mates: A study of 37 societies. *Journal of Cross-Cultural Psychology, 21,* 5–47.

Buss, D. M., & Schmitt, D. P. (1993). Sexual strategies theory: An evolutionary perspective on human mating. *Psychological Review, 100,* 204–232.

Byrne, R., & Whiten, A. (Eds.). (1988). *Machiavellian intelligence.* Oxford: Clarendon.

Chagnon, N. (1983). *Yanomamo: The fierce people.* New York: Holt, Rinehart, & Winston.

Cosmides, L., & Tooby, J. (1994). Beyond intuition and instinct blindness: Toward an evolutionary rigorous cognitive science. *Cognition, 50,* 41–77.

Dawkins, R., & Krebs, J. R. (1978). Animal signals: Information or manipulation. In J. R. Krebs & N. Davies (Eds.), *Behavioural ecology* (pp. 282–309). Oxford: Blackwell.

Fiske, A. P. (1991). *Structures of social life: The four elementary forms of social relationship.* New York: Free Press.

Gangestad, S. W., & Thornhill, R. (1994). Human facial attractiveness and sexual selection: The roles of averageness and symmetry. *Ethology and Sociobiology, 15,* 73–85.

Hamilton, W. D. (1964). The evolution of social behavior. *Journal of Theoretical Biology, 7,* 1–52.

Hill, K., & Hurtado, M. (1995). *Demographic/life history of Ache foragers.* New York: Aldine de Gruyter.

Humphrey, N. K. (1976). The social function of intellect. In P. P. Bateson & R. A. Hinde (Eds.), *Growing points in ethology* (pp. 303–317). Cambridge: Cambridge University Press.

Kenrick, D. T., Sadalla, E. K., Groth, G., & Trost, M. R. (1990). Evolution, traits, and the stages of human courtship: Qualifying the parental investment model. *Journal of Personality, 58,* 97–116.

Krebs, J. R., & Dawkins, R. (1984). Animals signals: Mind-reading and manipulation. In J. R. Krebs & N. B. Davies (Eds.), *Behavioural ecology* (2nd ed., pp. 380–402). Oxford: Blackwell.

Langlois, J., & Roggman, L. S. (1990). Attractive faces are only average. *Psychological Science, 1,* 115–121.

Marr, D. (1982). *Vision: A computational investigation into the human representation and processing of visual information.* San Francisco: Freeman.

Milgram, S. (1974). *Obedience to Authority.* New York: Harper & Row.

Pettigrew, T. F. (1979). The ultimate attribution error: Extending Allport's cognitive analysis of prejudice. *Personality and Social Psychology Bulletin, 5,* 461–476.

Shepard, R. N. (1987). Toward A universal law of generalization for psychological science. *Science, 237,* 1317–1323.

Sperber, D. (1994). The modularity of thought and the epidemiology of representations. In S. Gelman & L. Hirshfeld (Eds.), *Mapping the mind: Domain specificity in cognition and culture.* New York: Cambridge University Press.

Symons, D. (1979). *The evolution of human sexuality.* New York: Oxford University Press.

Symons, D. (1995). Beauty is in the adaptations of the beholder: The evolution of human female sexual attractiveness. In P. R. Abramson & S. D. Pinkerton (Eds.), *Sexual nature, sexual culture.* Chicago: University of Chicago Press.

Tooby, J., & Cosmides, L. (1988). *The evolution of war and its cognitive foundations.* Institute for Evolutionary Studies, Technical Report 88–1.

Tooby, J., & Cosmides, L. (1992). Psychological foundations of culture. In J. Barkow, L. Cosmides, & J. Tooby (Eds.), *The adapted mind: Evolutionary psychology and the generation of culture.* New York: Oxford University Press.

Williams, G. C. (1966). *Adaptation and natural selection.* Princeton, NJ: Princeton University Press.

Author Index

Subject Index

Printed and bound by CPI Group (UK) Ltd, Croydon, CR0 4YY

17/10/2024

01775682-0006